T0418824

Unconventional Oilseeds and Oil Sources

Unconventional Oilseeds and Oil Sources

Abdalbasit Adam Mariod
Mohamed Elwathig Saeed Mirghani
Ismail Hussein

ACADEMIC PRESS

An imprint of Elsevier

Academic Press is an imprint of Elsevier
125 London Wall, London EC2Y 5AS, United Kingdom
525 B Street, Suite 1800, San Diego, CA 92101-4495, United States
50 Hampshire Street, 5th Floor, Cambridge, MA 02139, United States
The Boulevard, Langford Lane, Kidlington, Oxford OX5 1GB, United Kingdom

Notices
Knowledge and best practice in this field are constantly changing. As new research and experience
broaden our understanding, changes in research methods, professional practices, or medical treatment may become necessary.

Practitioners and researchers must always rely on their own experience and knowledge in evaluating
and using any information, methods, compounds, or experiments described herein. In using such
information or methods they should be mindful of their own safety and the safety of others, including parties for whom they have a professional responsibility.

To the fullest extent of the law, neither the Publisher nor the authors, contributors, or editors, assume
any liability for any injury and/or damage to persons or property as a matter of products liability,
negligence or otherwise, or from any use or operation of any methods, products, instructions, or
ideas contained in the material herein.

Library of Congress Cataloging-in-Publication Data
A catalog record for this book is available from the Library of Congress

British Library Cataloguing-in-Publication Data
A catalogue record for this book is available from the British Library

ISBN: 978-0-12-809435-8

Publisher: Nikki Levy
Acquisition Editor: Nina D. Bandeira
Editorial Project Manager: Mariana L. Kuhl
Production Project Manager: Susan Li
Designer: Alan Studholme

Typeset by Thomson Digital

Contents

Part C
Unconventional Oils from Insects

42. *Tessaratoma papillosa* Longan Stink Bug

43. *Copris nevinsoni* Dung Beetle

44. *Schistocerca gregaria* (Desert Locust) and *Locusta migratoria* (Migratory Locust)

45. *Oecophylla smaragdina* Fabricius Weaver Ant

46. *Aspongopus viduatus* Watermelon Bug As Source of Edible Protein and Oil

Part A

Principles of Oil Extraction, Processing, and Oil Composition: Unconventional Oils From Annuals, Herbs, and Vegetable

Chapter 1

Chrozophora brocchiana (Argessi)

BOTANICAL DESCRIPTION

Species of *Chrozophora* genus (family: Euphorbiaceae) are annual plants, whose leaves, stems, and fruits have been utilized in food and conventional pharmaceutical. *Chrozophora brocchiana* (Vis.) Schweinf, is a shrubby monoecious herb up to 60–150 cm tall. The leaves of this plant are alternate and straightforward with three veined at the base, a scantily bushy upper surface and a smooth shaggy lower surface is. The fruit is 3-seeded, 1-cm long, 3-lobed case, (Fig. 1.1) thickly secured with white or violet-tinged, and glossy stalked scales. The seeds are ovoid, smooth, yellowish tan, secured by a slender, pale, gleaming aril (Ahmed, Mariod, Hussein, & Kamal-eldin, 2014).

GEOGRAPHICAL DISTRIBUTION

The plant is found in Sudan (where it is known as Argessi and other African countries like Mauritania, Algeria, and Egypt. In eastern Sudan, the seeds are boiled and utilized for food and the seed oil is concentrated utilizing conventional mills to get clear yellow edible oil. *C. brocchiana*, demonstrated high substance of protein and oil with high percentages of fatty acids (Hussein, Mirghani, & CheMan, 2006; Mirghani, Hussein, Dagne, & Bekele, 1996).

MEDICINAL USES

C. brocchiana has several interesting medicinal uses; the plant ash is applied to sores, and the crushed leaves are rubbed on the affected sites to treat stitch in the side. The aerial parts are taken in decoction to strengthen lactating mothers and their children, and to treat fever and dysentery, while powdered dried leaves in water are taken to treat diarrhea. Root sap in water is used as ear drops to treat otitis (Schmelzer, 2007). Analysis of the chemical content shows no particular reason for a beneficial action as a wound dressing; however, there is unusually high silica content.

Unconventional Oilseeds and Oil Sources. http://dx.doi.org/10.1016/B978-0-12-809435-8.00001-9

FIGURE 1.1 *Chrozophora brocchiana* plant, flowers, and seeds.

Hawas (2007) isolated and identified the analog of brevifolin carboxylic acid and its methyl ester from the aqueous ethanolic extract of the leaves of *C. brocchiana*; in addition he identified eight known compounds as gallic acid, methyl and ethyl gallate, ellagic acid, monomethoxy and methylenedioxy ellagic acid, apigenin 7-O-glucoside, and luteolin 7-O-glucoside. This author also determined the structures by ESI-MS and NMR spectroscopy. He preformed the assignment of NMR signals by means of 1H–H COSY, HMQC, and HMBC experiments. The isolated compounds showed high antioxidant activity against DPPH free radicals.

In Mali and Niger, the plant ash is applied to sores of humans and camels. In Niger the Hausa people rub crushed leaves on the affected sites to treat stitch in the side. The aerial parts are taken in decoction to strengthen lactating mothers and their children, and to treat fever and dysentery. Root sap in water is used as ear drops to treat otitis. In central Sudan sweet nondrying oil is pressed from the seeds (Schmelzer, 2007).

Successful prediction of antibacterial activity from plant material is largely dependent on the type of solvent used in the extraction procedure (Parekh, Jadeja, & Chanda, 2005). The antimicrobial activity of methanol, ethyl acetate and hexane extracts of leaves and stem, and oil methanol extract of *C. brocchiana* plant were studied using agar well diffusion method by measuring the diameter of the growth inhibition zone (Ahmed et al., 2014). The methanol extract of leaves showed a positive significant antibacterial and antifungal activity against all the test organisms with inhibition zone between 1.5 and 2.5 cm at 0.1–0.01

concentration, and showed high significance at all concentrations against *Staphylococcus aureus*. The ethyl acetate extract of leaves showed a positive significance against *Bacillus cereus* with inhibition zone 1.5 cm at concentration 0.1 mg and *S. aureus* with inhibition zone between 3.0 and 2.3 at all concentrations and fungal resistance. The hexane extract of leaves showed a positive significance against *Bacillus subtitle* with inhibition zone of 1.1–2.0 cm at the concentration of 0.1–0.01 mg of extract and *S. aureus* with inhibition zone between 1.5 and 2.5 cm in the same concentration and fungal resistance. All the three extracts (leaves, stem, and oil) of *C. brocchiana* showed dose dependent activity. The leaves methanolic extract was highly active against all bacterial and fungal strains at high dose, while the ethyl acetate extract was active against *S. aureus* strain. The stem extract showed medium antimicrobial activity when compared with other extracts. The leaves, stem, and oil methanolic extracts of *C. brocchiana* should further be studied for their phytochemical constituents in order to elucidate the active principle within the extract, which can turn out to be a novel antimicrobial agent of the future (Ahmed et al., 2014).

SEED CHEMICAL COMPOSITION

Physical Characteristics of *Chrozophora brocchiana* Seeds

C. brocchiana seeds resemble those of sorghum in shape and size but are somewhat darker in color. The average weight of 100 seeds was 3.27 g, which was similar to that reported in the literature by Hussein et al. (2006). The weight of 100 kernels was 2.07 g and the weight of 100 hulls was 1.2 g. The ratio of the hull to the kernel was 43:57 and the percentage of kernel in the whole seed was 63.3% while it was 36.7% for the hull. Hussein et al. (2006) reported that the average weight of 1000 seeds stored for more than 18 months was 34.5 g with moisture content of 3%–5%. *C. brocchiana* seeds contain about 26% protein and 37%–40% oil containing 11 triacylglycerols with low melting point. The examination of fatty acid composition of the seed oil of *C. brocchiana* showed linoleic acid as the major fatty acid followed by oleic, stearic, and palmitic acids, and the following triacylglycerol composition: LLL (15.8%), OLL (14.9%), PLL (7.5%), OOL (7.7%), PLO (16.9%), OOO (2.8%), POO (11.7%), POP (3.9%), SOO (4%), SOP (6%), and SOS (2.3%) (Hussein et al., 2006; Mirghani et al., 1996).

The Physicochemical Properties of the Seed Oil

The seed contains 4.9% as moisture, 18.20 protein, 21.7 crude fiber, and 1.05% as ash. On the other hand, the percentages of fat and carbohydrates were 42.9 ± 0.14 and 11.28 ± 0.70%, respectively. The results showed that *C. brocchiana* seed is a good source of oil (42.9%) and protein (18.2%). The oil extracted from the seeds had 2.7% of free fatty acids. The free fatty acids were high when compared to the results of Hussein et al. (2006) who reported that freshly extracted oil from *C. brocchiana* initially had a low free fatty acid value that increased upon long storage. The unsaponifiable matter of the oil extracted from *C. brocchiana* seeds was 0.22% and the refractive index was 1.4720 at 25°C (Ahmed, 2015).

Fatty Acid and Tocopherols Composition of *Chrozophora brocchiana* Oil

Ahmed (2015) reported that the percentage of myristic (C14:0), palmitic (C16:0), palmitoleic (C16:1), stearic (C18:0), oleic (C18:1), and linoleic (C18:2) acids in *C. brocchiana* seed oil were 0.1, 8.2, 0.2, 16.6, 24.9, and 49.3%, respectively. Their results in fatty acid composition are in good agreement with Hussein et al. (2006) and Mirghani et al. (1996). Other plants of the same family contain higher amounts of unsaturated fatty acids, for example, *Caryodendron orinocense* contains 75.13% linoleic acid (Mde & de Padilla, 1994), while *Chrozophora plicata* contains 59.3% linoleic acid (EL Bassam, 2013). The amounts of tocopherols (vitamin E) in *C. brocchiana* seed oil compared with three commercial oils mainly used in human diet, viz. sunflower, sesame, and groundnut oils (CODEX 1999). The total vitamin E contents were 87.1 mg/100 g; this amount is higher than those reported by Codex (1999) for sunflower oil (44.0 mg/100 g), sesame (33.0 mg/100 g), and groundnut (17.0 mg/100 g) oils. The main tocopherol of the *C. brocchiana* oil was gamma-tocopherol, representing 84.6% of the total tocopherols, followed by delta-tocopherol, which represented up to 15.4% of total tocopherols. The amounts of gamma- and delta-tocopherols were 73.6 and 13.5 mg/100 g, respectively (Ahmed, 2015). Tocopherols induce a protective effect against oxidative stress linked to metabolic syndrome and are also essential for normal neurological function (Dias, 2012). Ahmed (2015) subjected *C. brocchiana* oil to successive heating of 70°C for 3 days and the oxidative stability was monitored using peroxide value and FTIR analysis. They found that peroxide values of *C. brocchiana* oil increased gradually from 1.1 ± 0.14 mEq. O_2/kg oil at the start time to 23.2 ± 0.14 after 3 days of storage at 70°C. The PVs in the oil were increased significantly ($P < 0.05$) after 72 h of storage to 23.2 ± 0.14 mEq. O_2/kg oil, as incubation at 70°C accelerates the oxidation in the oils (Mariod et al., 2012).

Fig. 1.2 shows the FTIR spectra of the fresh oils extracted from germinated (A) and ungerminated (B) *C. brocchiana* seeds. The prevailing groups in these oils are the same as in other edible oils. There were 14 visible peaks in ungerminated *C. brocchiana* seed oil at frequencies of 3473, 3008, 2923, 2854, 2677, 1745, 1649, 1463, 1377, 1236, 1163, 1099, 914, and 723 cm^{-1}. Henna and Tan (2009) reported that absorption peaks at 3600–2800 and 1800–700 cm^{-1} are the dominant bands in vegetable oils. When the results of germinated seed oils were compared with those of oils from ungerminated seeds, there were several changes in the peak intensities (absorbances). For example, (1) the band at frequency 723 cm^{-1}, associated to bending of $—(CH_2)_n—$, HC$=$CH$—(cis)$

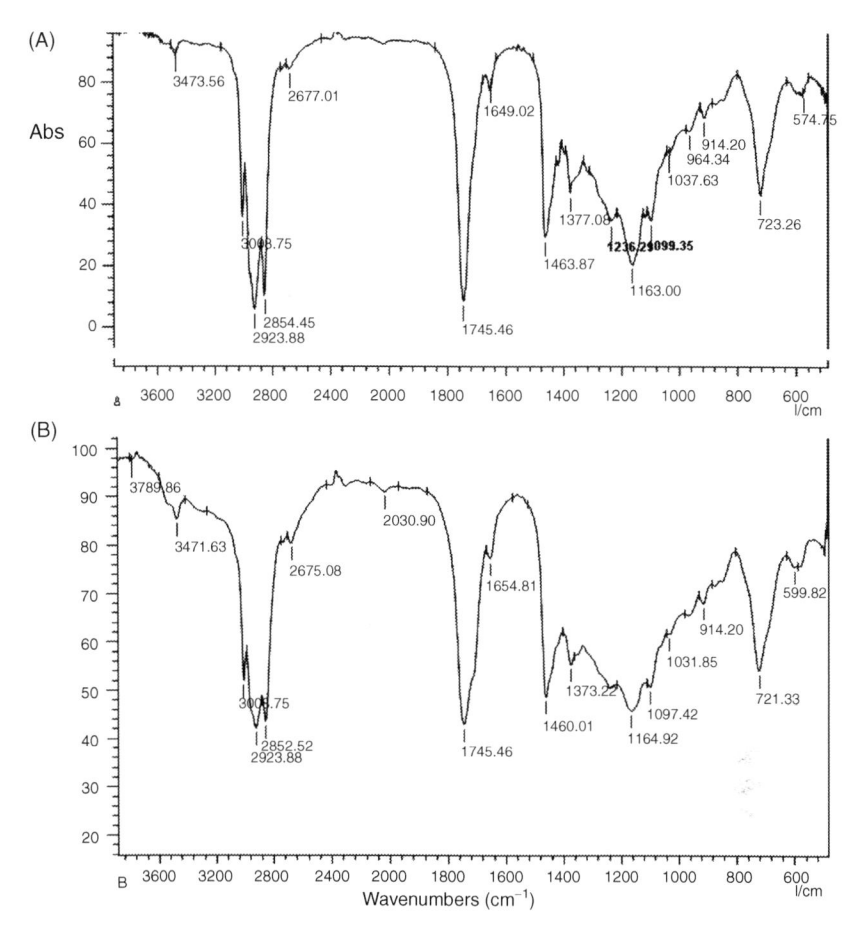

FIGURE 1.2   Fourier transform infrared spectra of oil extracted from germinated (A) and unger-minated (B) *Chrozophora brocchiana* seeds stored at 70°C for 0 h.

(Henna & Tan, 2009), was changed to 721 cm^{-1} and its intensity was increased to 56, (2) the band near 1163 cm^{-1} experienced an increased in wave number and intensity and gave a broad band as affected by germination, while (3) the band near 1236 cm^{-1} disappeared as a result of germination (Fig. 1.2).

MINERAL COMPOSITION OF *Chrozophora brocchiana*

Potassium was the predominant element in the seeds followed by magnesium and sodium then calcium, manganese, and iron. The seeds of *C. brocchiana* had a high content of potassium (32.01 mg/100 g), magnesium (9.77 ± 0.014), and sodium (1.589 ± 0.001). The content of calcium, manganese, iron, zinc, and copper were 0.340, 0.304, 0.548, 0.828, and 0.145 mg/100 g (Ahmed, 2015).

REFERENCES

Ahmed, M.A. (2015). Nutritional value and characteristics of oil, protein and bioactive components of *Chrozophora brocchiana* seed. *PhD Thesis*. Khartoum North, Sudan, College of Agricultural Studies, Sudan University of Science & Technology.

Ahmed, M. A., Mariod, A. A., Hussein, I. H., & Kamal-eldin, A. (2014). Review: biochemical composition and medicinal uses of *Chrozophora* genus. *International Journal of Pharmacy Review Research*, *4*(4), 227–232.

Codex (1999). Codex Standard for named vegetable oils CODEX-STAN 210-1999.

Dias, J. S. (2012). Major classes of phytonutriceuticals in vegetables and health benefits: a review. *Journal of Nutritional Therapeutics*, *1*, 31.

EL Bassam, N. (2013). *Energy plant species: their use and impact on environment and development*. UK: Routledge Taylor & Francis Group Ltd.

Hawas, U. W. (2007). Antioxidant activity of brocchlin carboxylic acid and its methyl ester from *Chrozophora brocchiana*. *Natural Products Research*, *21*(7), 632–640.

Henna, Lu F. S., & Tan, P. P. (2009). A comparative study of storage stability in virgin coconut oil and extra virgin olive oil upon thermal treatment. *International Food Research Journal*, *16*, 343.

Hussein, I. H., Mirghani, M. E. S., & CheMan, Y. B. (2006). Physicochemical characteristics of Argessi (*Chrozophora brocchiana*), Kenaf (*Hibiscus cannabinus*) and Loofah (*Luffa cylindrica*) seed oils. *Sudan Journal of Agricultural Research*, *6*, 53–60.

Mariod, A. A., Ahmed, S. Y., Abdelwahab, S. I., Cheng, S. F., Eltom, A. M., Yagoub, S. O., & Gouk, S. W. (2012). Effects of roasting and boiling on the chemical composition, amino acids and oil stability of safflower seeds. *International Journal of Food Science and Technology*, *47*, 1737–1743.

Mde, J. A., & de Padilla, F. C. (1994). Physico-chemical characteristics of the Barinas nut (*Caryodendron orinocense* Karst. Euphorbiaceae) crude oil. *Archivos Latinoamericanos de Nutrición*, *44*(3), 172.

Mirghani, M. E. S., Hussein, I. H., Dagne, E., & Bekele, T. (1996). A comparative study of seed oils of *Chrozophora brocchiana* and *Guizotia abyssinica*. *Bulletin of the Chemical Society of Ethiopia*, *10*(2), 161–164.

Parekh, J., Jadeja, D., & Chanda, S. (2005). Efficacy of aqueous and methanol extracts of some medicinal plants for potential antibacterial activity. *Turkish Journal of Biology*, *29*, 203–210.

Schmelzer, G.H. (2007). *Chrozophora brocchiana* (Vis.) Schweinf. [Internet] Record from PROTA4U. In Schmelzer, G. H. & Gurib-Fakim, A. (Eds,), *PROTA (Plant Resources of Tropical Africa / Ressources végétales de l'Afrique tropicale)*, Wageningen, Netherlands. Available from: http://www.prota4u.org/search.asp.

Chapter 2

Argemone mexicana (Argemone Seed)

BOTANICAL DESCRIPTION

Argemone mexicana is an annual herb, growing up to 150 cm with a slightly branched tap root (Fig. 2.1). Its stem is branched and usually extremely prickly. It exudes a yellow juice when cut. It has showy yellow flowers. Leaves are thistle-like and alternate, without leaf stalks (petioles), toothed (serrate), and the margins are spiny. The gray-white veins stand out against the bluish-green upper leaf surface. The stem is oblong in cross-section. The flowers are at the tips of the branches (are terminal) and solitary, yellow and of 2.5–5 cm diameter. Fruit is a prickly oblong or egg-shaped (ovoid) capsule. Seeds are very numerous, nearly spherical, covered in a fine network of veins, brownish black, and about 1 mm in diameter (Nacoulma, 1996; Bosch, 2007).

ORIGIN AND DISTRIBUTION

A. mexicana is native to Mexico and the West Indies, but has become pantropical after accidental introduction or introduction as an ornamental. It is naturalized in most African countries, from Cape Verde east to Somalia, and south to South Africa (Bosch, 2007). Distribution: Africa, Northeast Tropical Africa, Ethiopia, Socotra Asia-Temperate, Arabian Peninsula, North Yemen, Oman, Saudi Arabia, South Yemen Europe, Middle Europe, Austria, Germany, Switzerland, Southeastern Europe, Bulgaria, Southwestern Europe, France, Portugal, Spain, Southern America, Brazil, Piaui, Northern South America, Venezuela, southern South America, Paraguay, Western South America, Bolivia, Colombia, Ecuador (Bosch, 2007).

MEDICINAL USES

The alkaloid 6-acetonyldihydrochelerythrine has recently been isolated from whole plant extracts of *A. mexicana* and was found to have significant anti-HIV activity. The alkaloids berberine, protopine, protopine hydrochloride, sanguinarine, and dihydrosanguinarine have been isolated from the seeds. Protopine is

Unconventional Oilseeds and Oil Sources. http://dx.doi.org/10.1016/B978-0-12-809435-8.00002-0

FIGURE 2.1 *Argemone mexicana* **plant flower and seeds**. *(Data taken from http://luirig. altervista.org/pics/index5.php?recn=44214&page=1)*

considered a narcotic and it reduces morphine-withdrawal effects significantly. Protopine and sanguinarine showed molluscicidal properties against *Lymnaea acuminata* and *Biomphalaria glabrata*. Berberine has improved effects on the circulation in small doses and also has hallucinogenic properties. An overdose, however, produces death by paralysis of the central nervous system. Other pharmacological effects of berberine include spasmolytic, antibacterial, and to some degree antifungal and antiprotozoal activities. Most berberine is formed in the flowers. The alkaloid fraction from the roots showed antiinflammatory activity in rabbits and rats. Leaf extracts showed in vitro antiplasmodial activity (Bosch, 2007; Neuwinger, 2000; Avello, 2009).

Argemone oil was claimed to be used as antifungal agent. The seed also contains very small amounts of starch and free sugars. The seed oil also has a significant nematicidal effect on larvae of the genus *Meloidogyne*. An aqueous mixture of the oil (0.2%) applied to the soil of okra significantly reduced nematode infection and nematode concentrations in roots and soil, thereby increasing okra growth. When sprayed on the leaves, the effect was even more striking, showing the systemic effect of the spray (Neuwinger, 2000). Dried plant extracts significantly reduced nematode damage on seedlings of tomato and eggplant. Tomatoes treated with a leaf extract showed significantly less fruit rot caused by

Aspergillus niger. A flower extract induced a high level of resistance to tomato virus × in *Chenopodium album* L. Extracts also showed antibacterial activity in vitro against *Bacillus subtilis*, *Escherichia coli*, and *Streptococcus faecalis* (Neuwinger, 2000). Aqueous extract of aerial parts of *A. mexicana* at a dose of 200 and 400 mg/kg body weight was reported to have hypoglycemic efficacy in alloxan induced diabetic rats; significant reduction in blood glucose levels, plasma urea, creatinine, triacylglyceride, cholesterol values, and recovery in body weight compared to diabetic control rats and the standard drug treated rats (Nayak, Kar, & Maharana, 2011). The hydroalcoholic extract of aerial parts of *A. mexicana* also reduces fasting blood glucose levels in streptozotocin-induced hyperglycemic Wistar albino rats at a dose of 200 and 400 mg/kg body weight (Rout, Kar, & Mandal, 2011).

ARGEMONE SEED PROXIMATE ANALYSIS

The proximate analysis of argemone seed showed oil content of 35% and protein content of 24%. However, their use for edible purposes is doubtful as both products may be contaminated with seed toxic ingredients, such as alkaloids. The oil may, however, be used for nonedible purposes, such as a solvent for pesticide preparations, biodiesel, soap manufacture, etc.

The seeds of *A. mexicana* contain 35%–40% of orange-yellow oil which consists mainly of linoleic acid (54%–61%) and oleic acid (21%–33%) (Bosch, 2007). The saponification value of *A. mexicana* seed oil was 200 and it is higher than the values obtained from groundnut oil (188–195) and Niger oil (186–195). Ashwath (2010), who studied argemone seed oil as a new source for biodiesel production, reported that the seeds contain 24% oil, which is reddish in color and is dominated by linoleic and oleic acids. The oil has very high acid value (141). Despite its high acidity, the oil can be converted to biodiesel with a conversion efficiency of 97%.

FATTY ACID COMPOSITION OF *Argemone mexicana* OIL

The fatty acids of *A. mexicana* oil was determined as palmitic acid (14.1%–15.3%), stearic acid (3.7%–9.8%), oleic acid (34.3%–45.8%), linoleic acid (29.0%–44.2%), myristic acid (0%–0.1%), linolenic acid (0%–0.3%), palmitoleic acid (0%–1.3%), arachidic acid (0%–0.3%), and behenic acid (0%–0.2%). The fatty acids present in *A. mexicana* oils do not vary significantly from the other vegetable oil sources already reported; variations in their proportions and their unsaturation degree are noted (Rao, Zubaidha, Kondhare, Reddy, & Deshmuk, 2012). The fatty acid composition of argemone seed oil from Bangladesh showed that the saturated and unsaturated fatty acids were 14.2%–14.5% and 84.2%–84.8%, respectively depending on the areas in which the plant grows. The major fatty acids were oleic acid (23%), linoleic acid (58%), palmitic acid (7%), and ricinoleic acid (10%). Unsaturated fatty acids were mainly linoleic (58%) and oleic (23%) acids, which altogether accounted for 81% of the total fatty acids (Ahmed et al., 2011).

Ahmed et al. (2011) fractionated the total lipids of argemone oil into three major lipid groups, neutral lipids, glycolipid, and phospholipids by silicic acid column chromatography. They reported that, the neutral lipids were varied from 92.1% to 92.3%, glycolipids from 5.5% to 5.8%, and phospholipids from 1.5% to 1.7% of the total oil of the lipid applied. They also fractionated the oil into mono-, di-, and triglyceride by silicic acid column chromatography. They reported that, the triglycerides were varied from 90.1% to 90.3%, diglycerides from 2.3% to 2.8%, and monoglycerides from 1.5% to1.8%.

REFERENCES

Ahmed, G. M., Rahman, M. S., Zaman, M. R., Hossain, M. A., Uddin, M. M., & Yeasmin, S. (2011). Studies on the physico-chemical properties of siyal kanta (*Argemone mexicana* Linn) seed oil. *Bangladesh Journal of Science Industrial Research, 46*(4), 561–564.

Ashwath, N. (2010). *Evaluating biodiesel potential of Australian native and naturalised plant species* (p. 84). Kingston, Australia: Rural Industries Research and Development Corporation.

Avello, S. P. C. (2009). *Investigation of antiplasmodial compounds from various plant extracts.* *Thèse de doctorat.* Univ. Genève. no. Sc.

Bosch, C. H. (2007). Argemone mexicana L. [Internet] record from PROTA4U. In G. H. Schmelzer & A. Gurib-Fakim (Eds.), PROTA (plant resources of tropical Africa/Ressources végétales de l'Afrique tropicale). Wageningen, Netherlands. Available from: http://www.prota4u.org/search.asp.

Nacoulma, O. (1996). P lantes médicinales et pratiques médicales traditionnelles au Burkina Faso Cas du plateau central. *TOME II. Thèse d'Etat* (332 p.). Univ Ouaga.

Nayak, P., Kar, D. M., & Maharana, L. (2011). Antidiabetic activity of aerial parts of *Argemone mexicana* L. in alloxan induced hyperglycaemic rats. *Pharmacologyonline, 1*, 889–903.

Neuwinger, H. D. (2000). *African traditional medicine: a dictionary of plant use and applications.* Stuttgart, Germany: Medpharm Scientific, 589 p.

Rao, R. Y., Zubaidha, P. K., Kondhare, D. D., Reddy, N. J., & Deshmuk, S. S. (2012). Biodiesel production from *Argemone mexicana* seed oil using crystalline manganese carbonate. *Polish Journal of Chemical Technology, 14*(1), 65–70.

Rout, S. P., Kar, D. M., & Mandal, P. K. (2011). Hypoglycaemic activity of aerial parts of *Argemone mexicana* L. in experimental rat models. *International Journal of Pharmacy and Pharmaceutical Sciences, 3*, 533–540.

Chapter 3

Cassia obtusifolia (Senna or Sicklepod Seed)

PLANT DISTRIBUTION

Cassia obtusifolia family Leguminosae (Fabaceae) is generally distributed in Africa and the Americas. In Sudan it is found mostly on the clay plains of the central rain lands and in the southern regions. *Senna obtusifolia* is native to tropical South America but has become widespread throughout the tropics and subtropics. However, the extent of its original distribution in the neotropics is unknown. In the USA the range of *S. obtusifolia* is similar to that of 150 years ago although infestations are still increasing (Teem, Hoveland, & Buchanan, 1980). This model additionally gives an anticipated world conveyance outline districts, such as much of southern Europe, Southeast Africa, Madagascar, and North New Zealand as having a high potential to support populations of the species. There is also some evidence that a cold tolerant biotype may be evolving and this would further increase the potential range of the weed (Mackey, Miller, & Palmer, 1997).

BOTANICAL DESCRIPTION

The plant has a compound, pinnate leaf, composed of three pairs of leaflets. Flowers are yellow to orange and the pod is slender, slightly curved, and may reach 20 cm in length. The plant has a slender stem and may grow up to 2.0 m height (Mariod & Matthäus, 2008). *S. obtusifolia* is an erect, bushy, annual, or short-lived perennial herb growing to heights of 1.5–2.5 m (Parsons & Cuthbertson, 1992; Holm, Doll, Holm, Pancho, & Herberger, 1997). The stems are obtusely angled to cylindrical, smooth, often highly branched. The robust taproot is about 1-m long, with several descending laterals. Unlike many legumes, the roots of *S. obtusifolia* and *Senna tora* do not support nodules of nitrogen-fixing bacteria (https://en.wikipedia.org) (Fig. 3.1).

Two stipules, about 15-mm long, are present where the alternate leaves join the stem. The first true leaves are pinnate with two pairs of leaflets. Leaves in mature plants are even pinnately compound, 8–12 cm long, with three pairs of leaflets. The leaves of *S. tora* are rank-smelling when crushed (Holm et al., 1997)

Unconventional Oilseeds and Oil Sources. http://dx.doi.org/10.1016/B978-0-12-809435-8.00003-2
13

FIGURE 3.1 *Cassia obtusifolia* **plant and seeds.**

but *C. obtusifolia* is less pungent. Leaflets are obovate to oblong-obovate with asymmetrical bases, increasing in size from the base to the apex of the leaf, up to 6-cm long and 4-cm wide. The tips of the leaflets are bluntly oval to round, with a very small point at the tip of the main vein. A small, rod-like gland is situated on the rhachis between the lower pair of leaflets but, in *S. tora*, a second gland is present between the middle pair of leaflets. A second gland is sometimes present in *S. obtusifolia* but only on lower leaves (Brenan, 1967).

Flowers are solitary or in pairs, in leaf axils, on pedicels 1–3 cm long. The calyx has five free, unequal sepals, keeled on the back. The corolla has five free, spreading, yellow petals, obovate to obovate-oblong, narrowed at the base, and rounded at the tip, except for the standard (uppermost petal) which has two lobes. There are 10 stamens of which 7 are fertile and 3 are staminodes. The ovary has numerous ovules. In *S. obtusifolia*, the stigma is oblique with an acute rim; in *S. tora*, it is straight with two rolled back lips (ACTA, 1986a,b). The fruit is a brownish-green, slender, curved, compressed pod, 10–25 cm long and 2–6 mm wide, containing 25–30 seeds. Pods are slightly indented between the seeds. There are two major variants of *S. obtusifolia* in the Americas, differing primarily in pod type. Plants from the Antilles and the USA have pods 3.5–6 mm in diameter, as do African specimens and those from India, Indo Malaya, and

China (Irwin & Barneby, 1982). In South America and the Philippines, the pod is narrower (2–3.5 mm) in diameter and strongly curved. Seeds are rhomboidal, 4–5 mm long, shiny, and yellowish brown to dark red. In *S. obtusifolia*, the areole (marking on the seed coat) is very narrow (0.3–0.5 mm wide); in *S. tora*, it is large (1.5–2 mm wide) (Brenan, 1967). The plant is commonly known in Sudan as Kawal (Dirar, 1993). The fruit is a slender pod up to 18-cm long, 5-mm wide, four-angled in cross section, and usually curved downward. The pods are green and turn brown as the seeds mature, the seeds are 4–6 mm (Hall, Vernon, & Jason, 2006). *C. obtusifolia* and its seeds, common contaminants of agricultural commodities, are toxic to cattle and poultry. Toxicity has been attributed to anthraquinones which are major constituents of *C. obtusifolia* (Kenneth & Lucas, 1991). *C. obtusifolia* leaves, seeds, and root are used medicinally, primarily in Asia. It is believed to possess a laxative effect, as well as to be beneficial for the eyes. As a folk remedy, the seeds are often roasted, then boiled in water to produce a tea. Roasted and ground, the seeds have also been used as a substitute for coffee (http://www.wikipedia.com).

DIFFERENT USES OF *Cassia obtusifolia*

Food and Feed Uses

C. obtusifolia is reported to have a number of foods and feed uses; fermented leaves are used as a meat substitute (Kawal) in Sudan (Dirar, 1984) and as a mineral and vitamin supplement by certain tribes in Kenya and Senegal (Becker, 1986). Leaves of *C. obtusifolia* are high in protein (14.4%) and are highly palatable to poultry, but excessive consumption can be detrimental. Modest quantities of *C. obtusifolia* seeds can be used as dietary supplements in livestock but too much can have adverse effects (Holm et al., 1997). Seeds can be used as a substitute for coffee and as a mordant in dyeing (Staples, Imada, & Herbst, 2003; DeFilipps, Maina, & Crepin, 2004). Nuha, Isam, and Elfadil (2010) investigated the changes in chemical composition, antinutritional factors and minerals content, and extractability of two samples of sicklepod leaves of different origin. They fermented and dried the samples into powder and then cooked in boiling water for a certain period of time (5 min).

Fermented leaves of *C. obtusifolia* "Kawal" were mixed in the food of broiler chicks at concentrations of 25, 50, and 100 g/kg and then fed to chicks from 1 day to 8 weeks of age. Growth rate was depressed in relation to the concentration of Kawal. Lesions of an inflammatory-degenerative type were seen in the proventriculus, intestine, liver, heart, lungs, and kidneys, their severity increasing with the amount of Kawal eaten. These were accompanied by similar increases in the activities of lactate dehydrogenase, alkaline phosphatase, glutamate oxaloacetate, and glutamate pyruvate transaminases and in the concentrations of bilirubin, potassium, phosphorus, total lipids, and carotenes in the blood and dose-related decreases in total protein, albumin, cholesterol, globulin, sodium, calcium, and alkaline phosphatase in the blood. Birds fed on Kawal tended to become anemic but white blood cell count increased. It is concluded that Kawal even at an inclusion rate of 25 g/kg is unacceptable as a protein supplement (Suliman, Shommein, & Shaddad, 1987).

Medicinal Uses

C. obtusifolia is reported to have a number of medicinal uses; in traditional Chinese medicine it has the efficiency to remove "heat," improve eyesight, and relax bowels (Pharmacopoeia of the People's Republic of China, 2005). Modern pharmacological research has revealed that the seed has multiple functions, such as liver protection, bacteriostasis, catharsis, diuresis, neuroprotection, antitumor, and antioxidation, and it is highly valued for the treatment of hypercholesterolemia, hypertension, and hyperlipidemia (Ju et al., 2010; Xu, Tang, Zhou, Zhou, & Wang, 2015). The seeds have long been used in traditional eastern medicine and more recently the ethanolic fraction of the seeds has been shown to attenuate memory impairments in mice. The ethanolic fraction of the seeds was found to be neuroprotective against the mitochondrial toxin 3-NP (1 mM), while having no significant effect on cell death induced by incubation with naturally secreted oligomers of Abeta (8.2 pg/mL) (Drever et al., 2008).

An experimental diet was designed by Neils, Augustine, Mojaba, and Dazala (2013) who fed birds with *C. obtusifolia* seed meal for a period of 8 weeks. They noticed that the groups fed with 2.5 and 5.0% of the seed meal had normal blood biochemical parameters while group fed 7.5% did give normal range of biochemical values except for cholesterol level (172.86 mg). The increase showed a level of significance when compared to the normal range (52–148 mg). The change in the concentration of cholesterol that occurred with inclusion of *C. obtusifolia* seed meal was made at the higher percentage of 7.5%. This means that the level of cholesterol in chickens would be increased and becomes a risk factor to those who have preference for broiler meat (Neils et al., 2013).

SEED AND LEAVES COMPOSITION

Leaves and stems contain sennosides, D-mannitol, myricyl alcohol, and β-sitosterol. Leaves also contain emodin, a flavanol glycoside, triacontan-1-ol, stigma sterol, β-sitosterol-β-D-glucoside, friedelin, palmitic, stearic, succinic and D-tartaric acids, uridine, myo-inositol, D-ononitol, kaempferol, quercetin, juglanin, astragalin, quercitrin, and isoquercitrin. Pods have been reported to contain sennosides. Seeds contain anthraquinones and anthraquinone glycosides, chrysophanic acid, rhein, emodin, gluco-obtusifolin, cascaroside, rubrofusarin, chrysophanol, torosachrysone, questin, naphthalenic lactones, isotoralactone, toralactone, and cassialactone. Methanolic extract of the seeds yielded pure chrysophanol, chryso-obtusin, aurantio-obtusin, obtiosin,

2-glucosyl obtusifolin, cassiaside and rubro-fusarin-gentiobioside. Seeds contain new naphtha-a-pyrone–toralactone and an oxytocic principle. Roots contain anthraquinones and β-sitosterol (Ghani, 2003).

The plant's seeds are a source of cassia gum, a food additive usually used as a thickener. Vadivel and Janardhanan (2002) studied the chemical composition of *C. obtusifolia* seeds, collected from Western Ghats, South India; they found that the crude protein ranged from 18.56% to 22.93%, crude lipid was between 5.35% and 7.40%, crude fiber ranged from 6.83% to 9.45%, ash content ranged from 5.14% to 5.83%, and carbohydrate varied from 57.00% to 60.69%. Globulins constituted the bulk of the seed protein as in most legumes. Mineral profiles, namely, sodium, potassium, calcium, magnesium, phosphorus, iron, copper, zinc, and manganese ranged from 42.92 to 84.83, 758.05 to 1555.79, 559.92 to 791.72, 456.36 to 709.47, 629.13 to 947.79, 8.42 to 12.35, 0.93 to 2.06, 10.60 to 30.04, and 2.12 to 4.12 mg/100 g seeds flour, respectively. Seed proteins exhibited relatively high levels of nonessential and essential amino acids, with the exception of threonine. The in vitro protein digestibility of the legume ranged from 74.66% to 81.44%. Antinutritional substances, such as total free phenolics ranged from 0.34% to 0.66%; tannins were between 0.47% and 0.60%; trypsin inhibitor activity varied from 11.4 to 13.5 TIU/mg protein; and chymotrypsin inhibitor activity ranged from 10.8 to 12.3 CIU/mg protein. Seed meal of *S. obtusifolia* collected from Mubi area of Adamawa State, Nigeria were found to contain protein (21.89%), energy (3171.61 kcal/kg), crude fiber (12.45%), ether extract (9.50%), Nitrogen free extract (44.55%), dry matter (96.80%), and ash (4.16%) (Augustine, Abdulrahman, Masudi, & Ngiki, 2014).

The leaves of *C. obtusifolia* contained comparatively and significantly ($P < 0.05$) higher crude protein (27.84%), total ash (13.0%), calcium (3.50%), and potassium (0.72%) than the seeds. The ether extract (6.10%), organic matter (95.26%), crude fiber (11.52%), neutral detergent fiber (40.32%), acid detergent fiber (17.28%), hemicellulose (23.04%), and phosphorus (0.36%) levels were comparatively and significantly ($P < 0.05$) higher in the seeds than in the leaves (Agbo, 2004).

Nuha et al. (2010) studied the effect of fermentation and cooking on the leaves of *C. obtusifolia*. These authors noticed that the protein (21.5%), oil (3.9%), fiber (18.7%), and carbohydrates (36.4%) contents were fluctuated during processing of *C. obtusifolia* leaves. However, the protein and ash contents increased after cooking of the leaves. The total energy was decreased. The antinutritional factors (tannin, phytate, and total polyphenols) of the leaves were significantly ($P \leq 0.05$) decreased after fermentation and cooking. Tannin content significantly ($P \leq 0.05$) increased after fermentation but decreased after cooking. Ca (averaged 2933.96 mg/100 g) and K (1924.97 mg/100 g) were the major mineral constituents of raw leaves. Total major minerals were significantly ($P \leq 0.05$) increased after processing. However, the trace minerals content was fluctuated. HCl extractability of minerals for both samples was very low and fluctuated (Nuha et al., 2010).

PHYSICOCHEMICAL PROPERTIES, FATTY ACIDS, AND TOCOPHEROLS

Sudanese *C. obtusifolia* seed was investigated by Mariod and Matthäus (2008); the oil content was 7.0% which was very low in comparison with any promising new oil seed. The free fatty acids of the obtained oil was found as 2.2%, with refractive index (30°C) as 1.4740 and specific gravity (60°C) as 0.9215 ± 0.04 while the unsaponifiables of the obtained oil was found as 0.81% which is considered very low. The fatty acid composition of *C. obtusifolia* was investigated and 14 fatty acids were identified, among which linoleic acid (18:2Δ6) contributed 38.2% to the total fatty acids followed by oleic acid (18:1Δ9) at 24.4%, palmitic acid (16:0) at 20.0%, and stearic acid (18:0) at 9.6%. The remaining fatty acids contributed only few percentages to the total fatty acids present (Mariod & Matthäus, 2008).

REFERENCES

ACTA. (1986a). *Cassia obtusifolia* L. In *Adventices tropicales [tropical weeds]*. Paris, France: ACTA Publications.

ACTA. (1986b). *Cassia tora* L. In *Adventices tropicales [tropical weeds]*. Paris, France: ACTA Publications.

Agbo, J. Y. (2004). Proximate nutrient composition of sickle-pod (*Cassia obtusifolia*) leaves and seeds. *Plant Products Research Journal, 8*(1), 13–17.

Augustine, C., Abdulrahman, B. S., Masudi, B., & Ngiki, Y. U. (2014). Comparative evaluation of the proximate composition and anti-nutritive components of tropical sickle pod (*Senna obtusifolia*) and coffee senna (*Senna occidentalis*) seed meals indigenous to Mubi area of Adamawa State Nigeria. *International Journal of Management and Social Sciences Research, 3*(2), 2319–4421.

Becker, B. (1986). Wild plants for human nutrition in the Sahelian Zone. *Journal of Arid Environments, 11*(1), 61–64.

Brenan, J. P. M. (1967). Leguminosae subfamily Caesalpinioideae. In E. Milne-Redhead, & R. M. Polhill (Eds.), *Flora of Tropical East Africa*. London, UK: Crown Agents for Oversea Governments and Administrations.

DeFilipps, R. A., Maina, S. L., & Crepin, J. (2004). *Medicinal plants of the Guianas (Guyana, Surinam, French Guiana)*. Available from: http://www.mnh.si.edu/biodiversity/bdg/medicinal

Dirar, H. A. (1984). Kawal, meat substitute from fermented *Cassia obtusifolia* leaves. *Economic Botany, 38*(3), 342–349.

Dirar, H. A. (1993). *The indigenous fermented food of the Sudan: A study of African food and nutrition*. Wallingford, UK: CAB International, pp. 467–468.

Drever, B. D., Anderson, W. G., Riede, G., Kim, D. H., Ryu, J. H., Choi, D. Y., & Platt, B. (2008). The seed extract of *Cassia obtusifolia* offers neuroprotection to mouse hippocampal cultures. *Journal of Pharmacological Sciences, 107*(4), 380–392.

Ghani, A. (2003). Medicinal plants of Bangladesh with chemical constituents and uses. *Asiatic Society Bangladesh, 2*, 402.

Hall, D. W., Vernon V. V., Jason., A. F. (2006). Sickle pod, *Cassia obtusifolia* L. in weeds in Florida, SP 37, Florida Cooperative Extension Service, Institute of Food and Agricultural Sciences, University of Florida. Publication.

Holm, L., Doll, J., Holm, E., Pancho, J., & Herberger, J. (1997). *World weeds. Natural histories and distribution*. New York, NY: John Wiley and Sons, Inc.

Irwin, H. S., & Barneby, R. C. (1982). The American Cassiinae. *Memoirs of the New York Botanical Garden, 25*, 1–918.

Ju, M. S., Kim, H. G., Choi, J. G., Ryu, J. H., Hur, J. Y., Kim, Y. J., & Oh, M. S. (2010). Cassiae semen a seed of *Cassia obtusifolia*, has neuroprotective effects in Parkinson's disease models. *Food and Chemical Toxicology, 48*, 2037–2044.

Kenneth, A. V., & Lucas, H. B. (1991). Toxicological and hematological effects of sicklepod (*Cassia obtusifolia*) seeds in Sprague-Dawley rats: a subchronic feeding study. *Toxicon, 29*, 1329–1336.

Mackey, A. P., Miller, E. N., & Palmer, W. A. (1997). *Sicklepod (*Senna obtusifolia*) in Queensland*. Coorparoo, Australia: Department of Natural Resources.

Mariod, A., & Matthäus, B. (2008). Physico-chemical properties, fatty acid and tocopherol composition of oils from some Sudanese oil bearing sources. *Grasas y Aceites, 59*(4), 321–326.

Neils, J. S., Augustine, C., Mojaba, D. I., & Dazala, U. I. (2013). Blood biochemicals of broiler chickens after been fed *Cassia obtusifolia* seed meal. *International Journal of Poultry Science, 12*(11), 635–638.

Nuha, M. O., Isam, A. M. A., & Elfadil, E. B. (2010). Chemical composition, antinutrients and extractable minerals of sicklepod (*Cassia obtusifolia*) leaves as influenced by fermentation and cooking. *International Food Research Journal, 17*, 775–785.

Parsons, W. T., & Cuthbertson, E. G. (1992). *Noxious weeds of Australia*. Melbourne, Australia: Inkata Press, p. 692.

Staples, G. W., Imada, C. T., & Herbst, D. R. (2003). New Hawaiian plant records for 2001. *Bishop Museum Occasional Papers, 74*, 7–21.

Suliman, H. B., Shommein, A. M., & Shaddad, S. A. (1987). The pathological and biochemical effects of feeding fermented leaves of *Cassia obtusifolia* 'Kawal' to broiler chicks. *Avian Pathology, 16*(1), 43–49.

Teem, D. H., Hoveland, C. S., & Buchanan, G. A. (1980). Sicklepod (*Cassia obtusifolia*) and coffee senna (*Cassia occidentalis*), geographic distribution, germination and emergence. *Weed Science, 28*(1), 68–71.

Vadivel, V., & Janardhanan, K. (2002). Agrobotanical traits and chemical composition of *Cassia obtusifolia* L: a lesser-known legume of the Western Ghats region of South India. *Plant Foods for Human Nutrition, 57*(2), 151–164.

Xu, Y. L., Tang, L. Y., Zhou, X. D., Zhou, G. H., & Wang, Z. J. (2015). Five new anthraquinones from the seed of *Cassia obtusifolia*. *Archives of Pharmacal Research, 38*, 1054–1058.

Chapter 4

Cassia (Senna) occidentalis L. (Coffee Senna)

PLANT DISTRIBUTION

Cassia (Senna) occidentalis L. (coffee senna), Fabaceae (Leguminosae), belongs to the family Caesalpiniaceae. It is known by various names, for example, coffee senna, fetid cassia, and Negro Coffee (English). The plant grows throughout the tropics and subtropics including United States from Texas to Iowa eastward, Africa, Asia, and Australia (Yadav et al., 2010).

BOTANICAL DESCRIPTION

C. occidentalis is commonly known in Sudan as Soreib. The pods are 10–13 cm long and up to 0.8 cm in diameter containing dark olive green seeds. *C. occidentalis* is an erect, somewhat branched, smooth, semiwoody, fetid herb or shrub, 0.8–1.5 m tall, taproot, hard, stout, with a few lateral roots on mid section. This plant species varies from a semiwoody annual herb in warm temperate areas to a woody annual shrub or sometimes a short-lived perennial shrub in frost-free areas (Holm, Doll, Holm, Pancho, & Herberger, 1997; Yadav et al., 2010). The stem of the plant is reddish purple. The young ones are four-sided, becoming rounded with age. Leaves are alternate, even pinnately compound, each one with 4–6 pairs of nearly sessile, opposite leaflets, with a fetid smell when crushed, each leaflet 4–6 cm long, 1.5–2.5 cm wide, ovate or oblong, lanceolate with a pointed tip and fine white hairs on the margin. The rachis has a large, ovoid, shining, dark purple gland at the base. Stipules are 5–10 mm long, often leaving an oblique scar. Inflorescence is a compound of auxiliary and terminal racemes. The flower is perfect, 2-cm long with five yellowish green sepals with distinct red veins and five yellow petals. The fruit is a dry, dehiscent, transversely partitioned, faintly recurved, laterally compressed, sickle-shaped legume (pod), 7–12 cm long, 8–10 mm wide, with rounded tip, and containing 25–50 seeds. Seeds are oval shaped, 3.5–4.5 mm wide, flattened; pale to dark brown, slightly shiny, smooth, and with a round pointed tip (Roy et al., 1997; Yadav et al., 2010) (Fig. 4.1).

Unconventional Oilseeds and Oil Sources. http://dx.doi.org/10.1016/B978-0-12-809435-8.00004-4

FIGURE 4.1 *Cassia occidentalis* plant and seeds.

MEDICINAL IMPORTANCE AND BIOACTIVE POTENTIAL

S. occidentalis is used as a coffee substitute in many countries, despite reports that the seeds are toxic to cattle. The seeds contain no caffeine or tannin. A mixture of equal parts of these seeds and coffee beans was acceptable in organoleptic tests (Hassan, El-Hindawy, Bassiony, & Abd-Alla, 1974). The plant is bitter, purgative, laxative, antiinflammatory, expectorant, hepatoprotective, antimalarial, analgesic, vermifuge, and febrifuge. It is mainly used to detoxify liver, to cure internal bacterial and fungal disorders, to kill parasites and viruses; it enhances immunity and promotes perspiration. It also helps in cough, convulsions, reduces blood pressure, reduces spasms, and is used as a cardiotonic. Seeds are brewed into a coffee-like beverage for asthma and a flower infusion is used for bronchitis. Leaves are used in the treatment of gonorrhea, fevers, urinary tract disorders, and edema (http://www.motherherbs.com/cassia-occidentalis.html; https://en.wikipedia.org/wiki/Senna_occidentalis).

The *C. occidentalis* seeds are a rich source of galactomannan (Gupta, Chaudhary, Yadav, Vermab, & Dobhal, 2005). In India, it has been reported to be used as an antidote for poison, a blood purifier, an expectorant, antiinflammatory agent, and as a remedy for the treatment of liver diseases. The composition of the seed oil has also been described (Akhtar, Shahida, Khan, & Ahmad, 1988). *C. occidentalis* is used in traditional medicine in the tropical areas of the world. In Peru, the roots are used as a diuretic and the decoction is made to treat fevers. The seeds are brewed into a coffee-like beverage to treat asthma and the flower infusion is used for bronchitis in the Peruvian Amazon (Rutter, 1990). In Brazil the roots are considered to be a tonic, febrifuge, and diuretic, and are used against fevers, tuberculosis, anemia, liver complaints, and as a reconstituent for general weakness and illness (Coimbra, 1994). In southeastern Nigeria the leaves are used for fever (Chukwujekwu, Van Staden, & Smith, 2005). Chukwujekwu, Coombes, Mulholland, and van Staden (2006) examined the antibacterial activity of the ethanolic root extract of *C. occidentalis*. They isolated emodin as a biologically active component by spectroscopic analysis. They found it active against *Bacillus subtilis* and *Staphylococcus aureus*, but it was not active against two Gram-negative bacteria (*Klebsiella pneumoniae* and *Escherichia coli*) at the highest concentration tested. Different parts of this plant have been reported to possess antiinflammatory (Kuo, Chen, La, Teng, & Wang, 1996), antihepatotoxic, antibacterial (Samy & Ignacimuthu, 2000), and antiplasmodial activities (Tona et al., 2004). A wide range of the chemical constituents have been isolated from *C. occidentalis*, including sennoside, anthraquinone glycosides (Chauhan, Chauhan, Siddiqui, & Singh, 2001), fatty oils, flavonoid glycosides, galactomannan, polysaccharides, and tannins (Purwar, Rai, Srivastava, & Singh, 2003).

Cassia occidentalis SEED COMPOSITION

The seed of *S. occidentalis* was found to contain protein (19.92%), energy (3022.41 kcal/kg), crude fiber (18.52%), ether extract (10.78%), nitrogen-free extract (39.78%), dry matter (96.70%), and ash (4.33%). The seeds contain tannins (269.23 mg), alkaloids (251 mg), saponin, (176 mg), and oxalates (79.05 mg). The seed of *C. occidentalis* is a good source of cassia seed gum with molecular structure that resembles other industrial gums, such as guar gum and locust bean gum. The seeds are a rich source of galactomannan (Mirhosseini & Tabatabaee, 2012).

The results of the proximate compositions on a general view revealed that *S. occidentalis* showed higher concentration of antinutritional factors. However, the seeds may be harmful to livestock especially to monogastric animals, when consumed in an unprocessed form. It has become necessary to use effective processing methods, such as boiling, soaking, roasting, fermentation, and sprouting to reduce the toxic components before using them as feedstuff (Augustine, Abdulrahman, Masudi, & Ngiki, 2014).

SEED OIL, PHYSICOCHEMICAL PROPERTIES, FATTY ACID, AND TOCOPHEROLS

The oil content of the seeds amounted to 3.2%. This is very low compared to that of common oil seeds. Therefore, from an economical point of view, the production of oil from such seeds could not be interesting unless some genetic modifications could be applied. The free fatty acids of the obtained oil was found as 2.0%, with refractive index (30°C) as 1.4743 and specific gravity (60°C) as 0.9222 ± 0.04 while the unsaponifiables of the obtained oil was found as 0.80%, which is considered very low (Mariod & Matthäus, 2008). The major fatty acids present were palmitic, stearic, oleic, and linoleic while myristic was in trace in *Cassia tora* and *C. occidentalis* (Ayodele, Dandaso, & Olagbemiro, 2004). The oil contained 16.8% palmitic acid, 22.1% oleic acid, 45.0% linoleic acid, and traces of linolenic acid (Mariod & Matthäus, 2008). A keto fatty acid [(Z)-7-oxo-11-octadecenoic acid] has been isolated in appreciable amount (27.1%) from *C. occidentalis* seed oil. The identification was based on chemical and spectroscopic methods (Daulatabad, Bhat, & Jamkhandi, 1996). The tocopherol content of this oil amounted 32.7 mg/100 g oil. Alpha tocopherol was the predominant tocopherol in the oil of *C. occidentalis* (Mariod & Matthäus, 2008).

REFERENCES

Akhtar, M., Shahida, Z., Khan, W., & Ahmad, S. (1988). Effect of maturity on lipid class and fatty acid composition of *Cassia* seeds. *Pakistan Journal of Scientific and Industrial Research, 31*(2), 106–113.

Augustine, C., Abdulrahman, B. S., Masudi, B., & Ngiki, Y. U. (2014). Comparative evaluation of the proximate composition and anti-nutritive components of tropical sickle pod (*Senna obtusifolia*) and coffee senna (*Senna occidentalis*) seed meals indigenous to Mubi area of Adamawa State Nigeria. *International Journal of Management and Social Sciences Research, 3*(2), 2319–4421.

Ayodele, J. T., Dandaso, R., & Olagbemiro, T. O. (2004). Fatty acid composition of cassiatora, *Cassia occidentalis* and *Cassia senna* (leguminosae subfamily caesalpinoideae) seed oils. *Nigerian Food Journal, 22,* 29–32.

Chauhan, D., Chauhan, J. S., Siddiqui, I. R., & Singh, J. (2001). Two new anthroquinone glycosides from leaves of *Cassia occidentalis*. *Indian Journal of Chemistry Section B—Organic Chemistry Including Medicinal Chemistry, 40,* 860–863.

Chukwujekwu, J. C., Coombes, P. H., Mulholland, D. A., & van Staden, J. (2006). Emodin, an antibacterial anthraquinone from the roots of *Cassia occidentalis*. *South African Journal of Botany, 72,* 295–297.

Chukwujekwu, J. C., Van Staden, J., & Smith, P. (2005). Antibacterial, antiinflammatory and antimalarial activities of some Nigerian medicinal plants. *South African Journal of Botany, 71,* 316–325.

Coimbra, R. (1994). *Manual de fitoterapia* (2nd ed.). Belem, Brazil: Editora Cejup.

Daulatabad, C. D., Bhat, G. G., & Jamkhandi, A. M. (1996). A novel keto fatty acid from *Cassia occidentalis* seed oil. *Lipid/Fett., 98*(5), 176–177.

Gupta, R. S., Chaudhary, R., Yadav, R. K., Vermab, S. K., & Dobhal, M. P. J. (2005). Effect of Saponins of *Albizia lebbeck* (L.) Benth bark on the reproductive system of male albino rats. *Journal of Ethnopharmacology, 96,* 31–36.

Hassan, Y. M., El-Hindawy, S., Bassiony, S., & Abd-Alla, M. A. (1974). *Cassia occidentalis* L. as coffee substitute in Egypt. *Egyptian Journal of Horticulture, 1*(2), 137–143.

Holm, L., Doll, J., Holm, E., Pancho, J., & Herberger, J. (1997). World weeds. *Natural histories and distribution.* New York, NY: John Wiley and Sons, Inc (http://www.motherherbs.com/cassia-occidentalis.html https://en.wikipedia.org/wiki/Senna_occidentalis).

Mariod, A., & Matthäus, B. (2008). Physico-chemical properties, fatty acid and tocopherol composition of oils from some Sudanese oil bearing sources. *Grasas y Aceites, 59*(4), 321–326.

Kuo, S. C., Chen, S. C., La, C. F., Teng, C. M., & Wang, J. P. (1996). Studies on the anti-inflammatory and antiplatelet activities of constituents isolated from the roots and stem of *Cassia occidentalis* L. *Chinese Pharmaceutical Journal, 48*, 291–302.

Mirhosseini, H., & Tabatabaee, A. B. (2012). A review study on chemical composition and molecular structure of newly plant gum exudates and seed gums. *Food Research International, 46*, 387–398.

Purwar, C., Rai, R., Srivastava, N., & Singh, J. (2003). New flavonoid glycosides from *Cassia occidentalis*. *Indian Journal of Chemistry Section B—Organic Chemistry Including Medicinal Chemistry, 42*, 434–436.

Roy, L., Holm, G., Doll, J., Holm, E., Pancho, J., & Herberger, J. (1997). *World weeds: Natural histories and distribution* Cassia occidentalis *L. and* Cassia tora *L. (Syn. C. obustifolia L.)*. New York, NY: John Wiley & Sons, 1129 p.

Rutter, R. A. (1990). *Catalogo de plantas utiles de la Amazonia Peruana*. Yarinacocha, Peru: Instituto Linguistico de Verano.

Samy, R. P., & Ignacimuthu, S. (2000). Antibacterial activity of some folklore medicinal plants used by tribals in Western Ghats of India. *Journal of Ethnopharmacology, 69*, 63–71.

Tona, L., Cimanga, R. K., Mesia, K., Musuamba, C. T., De Bruyne, T., Apers, S., Hernans, N., Van Miert, S., Pieters, L., Totte, J., & Vlientinck, A. J. (2004). In vitro antiplasmodial activity of extracts and fractions from seven medicinal plants used in the Democratic Republic of Congo. *Journal of Ethnopharmacology, 93*, 27–32.

Yadav, J. P., Arya, V., Yadav, S., Panghal, M., Kumar, S., & Dhankhar, S. (2010). *Cassia occidentalis* L.: a review on its ethnobotany, phytochemical and pharmacological profile. *Fitoterapia, 81*, 223–230.

Chapter 5

Abutilon pannosum (Forst.) Ragged Mallow

BOTANICAL DESCRIPTION

Abutilon pannosum is a herb belonging to the family Malvaceae; it is an erect, 1.5–3 m tall perennial herb to shrub, leaves 4–13 cm across, broadly ovate, 7–9 nerved. The plant parts are velvet hairy. Flowers appear in racemes or panicles in terminal branches by the reduction of leaves; pedicel 1–3.5 cm long, uniformly hairy, articulate from below the middle to near the apex. Fruit globose, 10–15 mm. Seeds are three in each mericarp, 2–2.5 mm long, 2-mm broad (Mariod & Matthäus, 2008). Leaves, 4–13 cm across, are broadly ovate, irregularly toothed, blunt or sharp tipped, heart-shaped at base, 7–9 nerved, and velvety on both sides. Leaf stalk is 2–15 cm long. Yellow flowers occur solitary in leaf axils, or appear in racemes or panicles at the end of branches. Sepal cup is about 1-cm long, in fruit up to 1.2 cm, velvety on both sides. Sepals are 4–8 mm long, ovate, lance-like to broadly ovate or triangular, sharp tipped. Flowers are 3–4.5 cm across, yellow. Inverted egg-shaped petals are 1.5–2 × 1.3–1.7 cm, slightly notched. Stamen tube is 5 mm long, fruit is spherical, 1–1.5 cm across, with visible ridges and furrows. Seeds are kidney shaped (http://www.flowersofindia.net/catalog/slides/Ragged%20Mallow.html) (Fig. 5.1).

SEED COMPOSITION

The seeds of *A. pannosum* contained 13.4% oil; Mariod and Matthäus (2008) reported less oil content of 7.1%, which is very low compared with that of common oil seeds. Therefore, from an economical point of view, the production of oil from such seeds could not be interesting unless some genetic modifications could be applied. The seed contains about 23.0% protein. The respective seed oil had iodine values of 118.4 and saponification value of 194.3, and acid value of 1.9 with refractive index of 1.4652 at 30°C. The fatty acid composition (wt.%) as determined by gas liquid chromatography was: palmitic 21.3, stearic 2.8, oleic 10.2, linoleic 60.7, malvalic 2.2, and dihydrocum sterculic 1.3 (Kittur, Mahajanshetti, Rao, & Lakshminarayana, 1982). The free fatty acid of the obtained oil was found as 1.2%, with refractive index (30°C) as 1.4752 and specific

Unconventional Oilseeds and Oil Sources. http://dx.doi.org/10.1016/B978-0-12-809435-8.00005-6

FIGURE 5.1 *Abutilon pannosum* plant and seeds.

gravity as 0.9225. The oil contains 1.80% unsaponifiables, which is considered very high when compared to conventional oils (Mariod & Matthäus, 2008). The oil contained 14.3% palmitic acid, 10.9% oleic acid, 63.9% linoleic acid, and traces of linolenic acid. Total tocopherol was found as 163.5 mg/100 g; alpha tocopherol was the predominant tocopherol, it represents about 125.7 mg/100 g, followed by γ-T, which represented about 29.0 ± 0.4 mg/100 g (Mariod & Matthäus, 2008).

EDIBLE AND MEDICINAL USES

The seeds are eaten and sometimes seeds are extracted to obtain drinks. Many herbs belonging to the species *Abutilon* are documented for their various medicinal benefits. Also, the plants from *Abutilon* spp. are claimed for other medicinal properties for the treatment of different disorders, but still they are not satisfactorily exploited. Furthermore, phytochemical investigation of the plant extracts is an important tool for the determination of the phytochemicals responsible for specific pharmacological activity. The various plants of the *Abutilon* spp., because of their different phytochemicals, possess different pharmacological activities and hence are used for the treatment of associated disorders (Khadabadi & Bhajipale, 2010). *A. pannosum* was used in the treatment of dysentery, gonorrhea, and various diseases. The plant contains mucilage. It is used to treat diarrhea, dysentery, and stomach troubles (Burkil, 1985). Different parts of this plant are in use to treat various ailments in ethnomedicine, especially its leaves have been used for treating infections. Survase, Sarwade, and Chavan (2013) studied the antibacterial activity of the extracts prepared from the dried leaves of

A. pannosum, using agar well diffusion method against both Gram-positive and Gram-negative microorganisms. They found that among all the extracts, the ethanolic extract of the leaves showed significant antibacterial activity comparable to the standard penicillin potassium and streptomycin sulfate against selected Gram-positive and Gram-negative bacteria (Survase et al., 2013). Akiyama, Kazuyasu, Yamasaki, Oono, and Iwatsuki (2001) showed the presence of quercetin, kaempferol, and flavonoids derivative. The roots are medicinally used in jaundice (Elkamali, 2009).

REFERENCES

Akiyama, H., Kazuyasu, F., Yamasaki, O., Oono, T., & Iwatsuki, K. (2001). Antibacterial action of several tannins against *Staphylococcus aureus*. *Journal of Antimicrobial Chemotherapy, 48*, 487–491.

Burkil, H. M. (1985). *The useful plants of West Tropical Africa*. Kew; Charlottesville, VA: Royal Botanic Gardens; University Press of Virginia.

Elkamali, H. H. (2009). Ethnopharmacology of medicinal plants used in North Kordofan (Western Sudan). *Ethnobotanical Leaflets, 13*, 203–210.

Khadabadi, S. S., & Bhajipale, N. S. (2010). A review on some important medicinal plants of *Abutilon* spp. *Research Journal of Pharmaceutical, Biological and Chemical Sciences, 1*(4), 718–729.

Kittur, M. H., Mahajanshetti, C. S., Rao, K. A., & Lakshminarayana, G. (1982). Characteristics and composition of *Abutilon pannosum* and *Hibiscus panduriformis* seeds and oils. *Journal of the American Oil Chemists' Society, 59*, 123–124.

Mariod, A., & Matthäus, B. (2008). Physico-chemical properties, fatty acid and tocopherol composition of oils from some Sudanese oil bearing sources. *Grasas y Aceites, 59*(4), 321–326.

Survase, S. A., Sarwade, B. P., & Chavan, D. P. (2013). Antibacterial activity of various extracts of *Abutilon pannosum* (Forst.f.) Schlecht. leaves. *African Journal of Plant Science, 7*(4), 128–130.

Chapter 6

Ipomoea quamoclit
Cypress Vine (Star Glory)

BOTANICAL DESCRIPTION

Ipomoea quamoclit (Cypress vine) is a medicinal plant belonging to the family Convolvulaceae. It is an annual or perennial, herbaceous, twining vine growing 1–3 m tall. The leaves are 2–9 cm long, deeply lobed (nearly pinnate), with 9–19 lobes on each side of the leaf. The flowers are 3–4 cm long and 2 cm in diameter, trumpet-shaped with five points, and can be red, pink, or white. It flowers in summer and fall. It is widely cultivated as an ornamental plant throughout the tropics, and also outside tropical regions, where it is grown as an annual plant only, not surviving temperate zone winters. In some tropical areas, it has become naturalized. Seedlings must always be kept moist. They require full sun for good growth. This vine is one of the best plants for attracting hummingbirds and is a vigorous grower. In warmer climate, this plant can be extremely invasive (https://en.wikipedia.org/wiki/Ipomoea_quamoclit) (Fig. 6.1).

PLANT DISTRIBUTION

The plant is distributed in tropical areas; it is cultivated in different gardens as an ornamental herb.

BIOACTIVE COMPOUNDS AND SEED COMPOSITION

A comprehensive GC–MS analysis of *I. quamoclit* revealed that it is composed of two necine bases with rare necic acids resulting in unique pyrrolizidine alkaloids. *I. quamoclit* is poorly synthesized of alkaloids. However, these metabolites are apparently chemotaxonomic markers of this infrageneric taxon in general. They represent either esters of (−)-platynecine or esters of (−)-trachelanthamidine, an additional novel structural type called minalobines. The plant is characterized by section-specific rare necic acids, for example, ipangulinic/isoipangulinic acid, phenylacetic acid. The alkaloids of *I. quamoclit* were mono- and diesters of platynecine. The major alkaloid of this plant, minalobine R, has been isolated and identified as 9-O-(*threo*-2-hydroxy-2-methyl-3-phenylacetoxy-butyryl)-(−)trachelanthamidine on the basis of spectral data

Unconventional Oilseeds and Oil Sources. http://dx.doi.org/10.1016/B978-0-12-809435-8.00006-8

FIGURE 6.1 *Ipomea indica.*

(Jenett-Siems et al., 2005). Leaves and stems contain small amounts of alkaloids and cyanogenetic glycosides. Seeds have been reported to contain the resin glycosides, quamoclins I–IV and jalapin. Total alkaloid in seeds was found to be 0.012% (Ghani, 2003). The seed contains about 23.0% protein and 11.5% oil. The fatty acid composition (wt.%) as determined by gas liquid chromatography was: palmitic 19.4, stearic 9.8, oleic 27.7, linoleic 34.6, and arachidic acid 2.2. The free fatty acid content of the obtained oil was found at 1.1%, with refractive index (30°C) as 1.4685 and specific gravity (60°C) as 0.9224. The oil contains medium amount of unsaponifiables, which was found to be 0.85%. The oil showed 34.6% of polyunsaturated fatty acids and 31.8% as total saturated fatty acids; the total unsaturated fatty acids represented more than 65.1% of the total fatty acids and the ratio of unsaturated to saturated was found as 2.04. The total tocopherols content was 52.5 ± 0.2 (mg/100 g oil) with gamma tocopherol as the predominant 45.9 ± 0.3 (mg/100 g oil) (Mariod and Matthäus, 2008).

POTENTIALITY AND DIFFERENT USES

The plant is considered cooling and purgative; used in cancer and breast pain. Pounded leaves are applied to bleeding piles and as a plaster to carbuncles (Yusuf, Begum, Hoque, & Choudhury, 2009). In vitro grown untransformed adventitious roots (AR) culture of *I. quamoclit* and its endophytic fungus (EF) *Cladosporium cladosporioides* decolorized Navy Blue HE2R (NB-HE2R) at a concentration of 20 ppm up to 83.3 and 65%, respectively within 96 h.

Whereas the AR–EF consortium decolorized the dye more efficiently and gave 97% removal within 36 h. Significant inductions in the enzyme activities of lignin peroxidase, tyrosinase, and laccase were observed in roots, while enzymes like tyrosinase, laccase, and riboflavin reductase activities were induced in EF. Metabolites of dye were analyzed using UV-vis spectroscopy, FTIR, and gas chromatography–mass spectrometry. Possible metabolic pathways of NB-HE2R were proposed with AR, EF, and AR–EF systems independently. Looking at the superior efficacy of AR–EF system, a rhizoreactor was developed for the treatment of NB-HE2R at a concentration of 1000 ppm. Control reactor systems with independently grown AR and EF gave 94 and 85% NB-HE2R removal, respectively within 36 h. The AR–EF rhizoreactor, however, gave 97% decolorization. The endophyte colonization additionally increased root and shoot lengths of candidate plants through mutualism. Combined bioreactor strategies can be effectively used for future eco-friendly remediation purposes (Patil et al., 2016). The whole plant is crushed and applied externally on carbuncles. The juice is used along with other ingredients in case of blood dysentery, piles, and body weakness (Sahu & Gupta, 2014).

Three novel ipangulines were isolated from the aerial parts of *I. quamoclit* and identified as 9-O-[2-hydroxy-3-(2-acetoxy)-2-methylbutyryl]-7-O-salicyloylplatynecine (ipanguline B2), 9-O-[2-hydroxy-3-(2-methylbutyryloxy)-2-methylbutyryl]-platynecine (ipanguline D10), and 9-O-salicyloylplatynecine (ipanguline D10). Comprehensive GC–mass spectrometric analysis also revealed a great variety of >38 platynecine monoesters and diesters. They are classified as ipangulines A (diesters with one phenylacetic acid moiety), ipangulines B (diesters with one salicylic acid moiety), ipangulines C (diesters with two aliphatic moieties), and ipangulines D (either 7- or 9-monoesters). In shoot tips and young leaves, total ipangulines concentrations are up to 0.45% (on dry wt. basis). The ipangulines are stored as tertiary alkaloids and not as N-oxides. Seeds of different provenance showed similar alkaloid patterns. Taxonomically related species, such as *Ipomoea neei* and *I. × periyrenium*, also contain platynecine esters in their vegetative tissues (Jenett-Siems et al., 1998). The hydromethanol extract of *I. quamoclit*, traditionally used in different ailments, was evaluated for antioxidant potential using 1,1-diphenyl-2-picryl hydrazyl (DPPH) radical scavenging assay. The plant extract exhibited a radical scavenging, with an IC_{50} value of 25.96 $\mu g/mL$ compared to the IC_{50} value of 5.15 g/mL as shown by the reference antioxidant ascorbic acid, in a dose-dependent fashion.

REFERENCES

Ghani, A. (2003). *Medicinal plants of Bangladesh—Chemical constituents and uses* (2nd ed.). Dhaka: The Asiatic Society of Bangladesh.

Jenett-Siems, K., Ott, S. C., Schimming, T., Siems, K., Müller, F., Hilker, M., Witte, L., Hartmann, T., Austin, D. F., & Eich, E. (2005). Ipangulines and minalobines, chemotaxonomic markers of the infrageneric *Ipomoea* taxon subgenus *quamoclit*, section Mina. *Phytochemistry, 66*(2), 223–231.

Jenett-Siems, K., Schimming, T., Kaloga, M., Eich, E., Siems, K., Gupta, M. P., Witteii, L., & Hartmann, T. (1998). Pyrrolizidine alkaloids of *Ipomoea hederifolia* and related species. *Phytochemistry*, *47*(8), 1551–1560.

Mariod, A., & Matthäus, B. (2008). Physico-chemical properties, fatty acid and tocopherol composition of oils from some Sudanese oil bearing sources. *Grasas y Aceites*, *59*(4), 321–326.

Patil, S. M., Chandanshive, V. V., Rane, N. R., Khandare, R. V., Watharkar, A. D., & Govindwar, S. P. (2016). Bioreactor with *Ipomoea hederifolia* adventitious roots and its endophyte *Cladosporium cladosporioides* for textile dye degradation. *Environmental Research*, *146*, 340–349.

Sahu, P. K., & Gupta, S. (2014). Medicinal plants of morning glory: Convolvulaceae juss. of Central India (Madhya Pradesh & Chhattishgarh). *Biolife*, *2*(2), 463–469.

Yusuf, M., Begum, J., Hoque, M. N., & Choudhury, J. U. (2009). *Medicinal plants of Bangladesh—Revised and enlarged*. Chittagong, Bangladesh: Bangladesh Council of Scientific and Industrial Research Laboratory.

Chapter 7

Ipomoea indica
Blue Morning Glory

BOTANICAL DESCRIPTION AND PLANT DISTRIBUTION

Ipomoea indica is a perennial with heart-shaped leaves and with flowers produced daily from a dense clustered inflorescence. The flowers appear from the leaf axils. Plants' hairy leaves with blade ovate in outline, 4–17 cm long, 3–16 cm wide, heart-shaped base, margins entire to deeply three-lobed; leaf stalk 2–18 cm long. Capsule globe-shaped, about 10-mm wide, with three chambers (http://www.weeds.org.au). The plant grows up to 15 m height. The stems usually develop a twining habit although they occasionally spread across the ground. The alternately arranged leaves (5–18 cm long and 3.5–16 cm wide) are borne on stalks (petioles) 2–18 cm long. They range from heart-shaped (cordate) to obviously three-lobed and have pointed tips (acute apices). The flowers are bright blue or bluish-purple in color with a paler pink or whitish-pink central tube. These are large flowers (5–10 cm long and 7–10 cm across). The fruits are globular papery capsules (about 10 mm across) containing four to six dark brown– or black-colored seeds (Henderson, 2001) (Fig. 7.1).

COMPOSITION AND BIOACTIVE COMPONENTS

Some species contain alkaloids and other drugs, including strong purgatives; others have hallucinogenic properties. Plant oils are important biomolecules, mostly are in the form of triglycerides; they are good sources of usual and unusual fatty acids.

OIL PHYSICOCHEMICAL PROPERTIES, FATTY ACID, AND TOCOPHEROLS COMPOSITION

The oil content of seeds of the *Ipomoea* species ranged from 7.84% to 14.71% and the protein content in the seeds varied from 13.11% to 23.89%. The seed oil of *I. indica* showed high iodine value of 134.29. The iodine values obtained by experimental procedures are well in agreement with fatty acid composition of seed oils. The IR and UV-vis spectra of FAME exhibited no absorption band for

Unconventional Oilseeds and Oil Sources. http://dx.doi.org/10.1016/B978-0-12-809435-8.00007-X

FIGURE 7.1 *Ipomea indica.*

the presence of any *trans* unsaturation and conjugation, respectively. Linoleic acid (ω-6, essential fatty acid) was found to be the most abundant fatty acid in *I. indica* (38.72%). The second most abundant fatty acid in *I. indica* was oleic acid (ω-9, nonessential fatty acid) at 23.43%. The amount of linolenic acid (ω-3, essential fatty acid) in the seed oils of *I. indica* was 13.47%. Among the SFAs stearic was the most abundant fatty acid in *I. indica* followed by palmitic acid. This seed oil also contains lauric, myristic, and other acids in trace amount. Its SFAs content was low (24.3%), while its PUFAs content was high (52.19%). On the basis of PUFAs content the oil is categorized as semidrying oil. The *P/S* ratio of *I. indica* seed oil was found to be 2.139 (Taufeeque, Malik, & Sherwani, 2015).

The free fatty acids content of *Ipomoea* oil was found as 2.4%, with refractive index (30°C) as 1.4741 and specific gravity (60°C) as 0.9211. The oil content of unsaponifiables was high (0.81%) (Mariod & Matthäus, 2008), and its major fatty acids were: 19.9% palmitic acid, 34.0% oleic acid, and 33.5% linoleic acid, with traces of linolenic acid. The tocopherol content of this oil amounted to 30.9 mg/100 g oil and gamma tocopherol was the predominant tocopherol; it represents about 24.2 mg/100 g followed by γ-T and delta tocotrienol, which represented about 1.8 and 1.6 mg/100 g, respectively (Mariod & Matthäus, 2008).

POTENTIALITY AND DIFFERENT USES

Ipomoea species seeds show hepatoprotective activity. The dried seeds are used as an antiinflammatory, a carminative, depurative, purgative, vermifuge, and to cure inflammations, constipation, dyspepsia, bronchitis, fever, skin diseases, scabies, etc. *Ipomoea* leaves extract is administered orally for treatment of intestinal worms. Roots are proved beneficial for women having urinary retention, constipation, or gynecological disorders (Sahu & Gupta, 2014). Whole plant along with bread is eaten for healing wounds (Jagtap, Deokule, & Bhosle, 2006). *Ipomoea* seed is helpful as laxative (Rani, Pandey, & Singh, 2014); crushed root with sugar administered orally cures passing of semen with urine (Rahmatullah et al., 2009). Its leaves are beneficial in ulcer, chest pain, and along with stem are helpful in fever, diabetes. Hydroalcoholic extract of its aerial parts shows significant radical scavenger activity (Hasan et al., 2009). The plant is used for healing broken bones. Methanolic extract from the seeds of *I. indica* presented biological activity against herpes simplex-1. Methanolic and aqueous extracts from the seeds of this species were also investigated for antibacterial activity against *Streptococcus pyogenes*, *Staphylococcus aureus*, *Pseudomonas aeruginosa*, and *Escherichia coli*, but no activity was detected (Meira, da Silva, David, & David, 2012). Acetonitrile extract from the seeds of *I. indica* showed activity against *Microsporum canis* and *Epidermophyton floccosum* at a concentration of 1000 µg/mL but no growth inhibition was observed against *Trichophyton rubrum* (Locher et al., 1995). The glycoside called ipolearoside, with significant activity against Walker carcinosarcoma 256 in rats, has been isolated from ethanol extracts of the whole plants of *I. indica* (Meira et al., 2012).

REFERENCES

Hasan, S. M. R., Hossain, M. M., Raushanara, A., Mariam, J., Mazumderm, H. E. M., & Rahman, S. (2009). DPPH free radical scavenging activity of some Bangladeshi medicinal plants. *Journal of Medicinal Plants Research, 3*(11), 875–879.

Henderson, L. (2001). Alien weeds and invasive plants. *A complete guide to declared weeds and invaders in South Africa. Plant Protection Research Institute handbook Vol. 12, 300 p.* South Africa: PPR, ARC.

Jagtap, S. D., Deokule, S. S., & Bhosle, S. V. (2006). Some unique ethnomedicinal uses of plants used by the Korku tribe of Amravati district of Maharashtra, India. *Journal of Ethnopharmacology, 107*, 463–469.

Locher, C. P., Burch, M. T., Mower, H. F., Berestecky, J., Davis, H., Van Poel, B., Lasure, A., Vanden Berghe, D. A., & Vlietinck, A. J. (1995). Anti-microbial activity and anti-complement activity of extracts obtained from selected Hawaiian medicinal plants. *Journal of Ethnopharmacology, 49*, 23–32.

Mariod, A., & Matthäus, B. (2008). Physico-chemical properties, fatty acid and tocopherol composition of oils from some Sudanese oil bearing sources. *Grasas y Aceites, 59*(4), 321–326.

Meira, M., da Silva, E. P., David, J. M., & David, J. P. (2012). Review of the genus *Ipomoea*: traditional uses, chemistry and biological activities. *Brazilian Journal of Pharmacognosy, 22*(3), 682–713.

Rahmatullah, M., Mollik, M. A. H., Azam, A. T. M. A., Islam, M. R., Chowdhury, M. A. M., Jahan, R., Chowdhury, M. H., & Rahman, T. (2009). Ethnobotanical survey of the Santal tribe residing in Thakurgaon District, Bangladesh. *American Eurasian Journal of Sustainable Agriculture, 3*, 889–898.

Rani, A., Pandey, S. K., & Singh, A. N. (2014). Documentation of medicinal plants from northern coal fields areas, Singrauli, M.P. *International Journal of Basic and Applied Science Research, 1*(1), 12–18.

Sahu, P. K., & Gupta, S. (2014). Medicinal plants of morning glory: Convolvulaceae Juss. of Central India (Madhya Pradesh & Chhattishgarh). *Biolife, 2*(2), 463–469.

Taufeeque, M., Malik, A., & Sherwani, M. R. K. (2015). Screening of seed oils from four species of genus *Ipomoea. International Journal of Current Research and Review, 7*(21), 25–28.

Chapter 8

Sesamum alatum (Thonn) Winged-Seed Sesame

DISTRIBUTION AND BOTANICAL DESCRIPTION

Widespread in tropical Africa, occurring in dry regions from Ethiopia, Kenya, Namibia, South Africa, and West Africa (Senegal to Chad), Sudan, Egypt, Morocco, Swaziland, Botswana, Mauritania, Madagascar, Mali, Nigeria, Zimbabwe, Ghana, Cameroon, and India and occasionally elsewhere it has been introduced. It is sometimes cultivated around villages. *Sesamum alatum* belongs to the family Pedaliaceae; it is an erect annual herb up to 1.5 m tall, with simple or sparsely branched stem, glabrous but with mucilage glands. Leaves opposite, lower ones palmately divided or lobed, upper ones simple; stipules absent; petiole 1–7 cm long; leaflets or lobes of lower leaves lanceolate, central one longest, up to 8 × 2 cm, often with undulate margin, blade of upper leaves linear to lanceolate, 3–10 cm long. Flowers solitary in leaf axils, bisexual, zygomorphic, 5-merous; pedicel short, with a nectary at base; calyx campanulate with narrowly triangular lobes c. 3 mm long, densely glandular, deciduous; corolla obliquely campanulate, 2–3 cm long, slightly two-lipped, pink or purple, inside sometimes red-spotted, pubescent; four stamens; disk fleshy, conspicuous; ovary superior, hairy, two-celled; style filiform; stigma two-lobed. Fruit narrowly obconical, capsule up to 5 × 0.7 cm, base gradually narrowed, apex with beak up to 12 mm long, four-grooved, dehiscing longitudinally, many seeded. Seeds obconical, c. 2.5 × 1.5 mm, with a large, 2–3 mm long wing at apex and two shorter wings at base, testa with honeycomb-like structure, pale to dark brown (Jansen, 2004) (Fig. 8.1).

COMPOSITION AND BIOACTIVE COMPONENTS

There is no information on the nutritional composition of *S. alatum* leaves, but it is probably comparable to that of *Sesamum indicum* L. (cultivated sesame) leaves, which is per 100 g edible portion: water 85.5 g, energy 188 kJ (45 kcal), protein 3.4 g, fat 0.7 g, carbohydrate 8.6 g, fiber 2.4 g, Ca 77 mg, P 203 mg, riboflavin 0.3 mg. The nutritional composition of dried seeds per 100 g is: water 7.9 g, energy 1733 kJ (414 kcal), protein 10.8 g, fat 18.1 g, carbohydrate

Unconventional Oilseeds and Oil Sources. http://dx.doi.org/10.1016/B978-0-12-809435-8.00008-1

FIGURE 8.1   *Sesamum alatum.*

(including fiber) 55.2 g, fiber 30.5 g, Ca 432 mg, P 221 mg (Leung, Busson, & Jardin, 1968).

The free fatty acids content of *S. alatum* oil was found as 1.5%, with refractive index (30°C) as 1.4762 and specific gravity (60°C) as 0.9223. Its content of unsaponifiables is low (0.78%) (Mariod & Matthäus, 2008). The oil of *S. alatum* contained 10.9% palmitic acid, 45.1% oleic acid, and 36.3% linoleic acid, with traces of linolenic acid. The tocopherol content of this oil amounted to 26.4 mg/100 g oil and gamma tocopherol was the predominant tocopherol; it represents about 21.4 mg/100 g, followed by α-T and β-T, which represented about 1.9 and 1.4 mg/100 g, respectively (Mariod & Matthäus, 2008).

OIL PHYSICOCHEMICAL PROPERTIES, FATTY ACIDS, AND MINOR COMPONENTS

The seed oil has about 5% unsaponifiables (check above 0.78%), consisting of lignans (2-episesalatin 1.4%, sesamin 0.01%, sesamolin 0.01%), sterols (22%), and tocopherols (210–320 mg/kg oil). Seed oil content was 28.9. Three new saponins, alatoside, with a 18,19-secours-12-ene skeleton were isolated from the aerial parts. In addition, verbascoside and two cyclohexylethanol derivatives, rengyol (2a) and isorengyol (3a), were isolated and identified (Potterat, Kurt, Stoeckli-Evans, & Mahamane, 2004). Kamal-Eldin and Yousif (1992) isolated new sesamin-type furofuran lignan, 2-episesalatin, from the seeds of *S. alatum*; they established its structure as 2-epi-(3,4,5-trimethoxyphenyl)-6-(3,4-methylenedioxy-5-methoxy phenyl)-*cis*-3,7-dioxabicyclo[3.3.0]octane by spectroscopic methods.

The oil from *S. alatum* was characterized by higher percentages of unsaponifiables (4.9%) compared to *S. indicum* (1.4%–1.8%), mainly due to its high content of lignans. Total sterols accounted for c. 40% of the unsaponifiables of the species. The *S. alatum* was different in the relative percentages of the three sterol fractions (the desmethyl, monomethyl, and dimethyl sterols) and in the percentage composition of each fraction. Campesterol, stigmasterol, sitosterol, and Δ^5-avenasterol were the major desmethyl sterols, whereas obtusifoliol, gramisterol, cycloeucalenol, and citrostandienol were the major monomethyl sterols, and α-amyrin, β-amyrin, cycloartenol, and 24-methylene cycloartenol were the main dimethyl sterols in *S. alatum*. *S. alatum* contained less tocopherols (210–320 mg/kg oil), than did *S. indicum* (490–680 mg/kg oil). The plant was comparable in tocopherol composition, with γ-tocopherol representing 96%–99% of the total tocopherols. The four species varied widely in the identity and levels of the different lignans. *S. alatum* showed 1.37% of 2-episesalatin and minor amounts of sesamin and sesamolin (0.01% each) (Kamal-Eldin & Appelqvist, 1994).

Oil extracted from seeds of different varieties of the cultivated *S. indicum* L. and three related wild species, namely *S. alatum* Thonn, *Sesamum radiatum* Schum & Thonn, and *Sesamum angustifolium* (Oliv) Engl, were analyzed for their total unsaponifiables and for the contents and composition of the three sterol fractions (desmethyl, monomethyl, and dimethyl sterols). The sterols were analyzed after saponification by preparative TLC, capillary GC, and GC–MS of their TMS ethers. Oils from the wild species contained more unsaponifiable material (2.3%–2.4%) compared with the cultivated species (1.1%–1.3%). Considerable differences were observed in the total sterol contents and the relative proportions of the three sterol fractions in the oils from the four species studied. Sitosterol, campesterol, stigmasterol, and Δ^5-avenasterol were the major desmethyl sterols in all four species. The monomethyl sterol fraction consisted primarily of obtusifoliol, gramisterol, cycloeucalenol, and citrostadienol. Cycloartenol and 24-methylene cycloartenol were the predominant dimethyl sterols. Variations in the composition of the three sterol fractions were observed between *S. indicum* oils traditionally pressed by the camel-driven expellers and laboratory extracted oils from the same seeds (Kamal-Eldin, Appelqvist, Yousif, & Iskander, 1992).

POTENTIALITY AND DIFFERENT USES

The plant is locally known as Salalaba in Sudan and Daraba in Chad. The leaves and young shoots of *S. alatum* are used as a cooked vegetable, sometimes flavored with its pounded seeds. The seeds are occasionally cooked separately as a relish or boiled with pumpkin leaves and served with a staple food. The seed produces an edible oil, and is used as an aphrodisiac and to cure diarrhea and other intestinal disorders. A decoction of the leaves is given to cattle to promote their fertility. The plant is used as mucilaginous source; fruits are eaten in Sudan and are used for oil production. Seed cooked as a relish or boiled with pumpkin leaves and served with a staple food. They are ground into a flour and used in soups (Facciola, 1998).

S. alatum was traditionally used in the ayurvedic medicine in India. It has wider range of activities and hence its seeds (black) are medicinally preferred. Folk medicine explains the renal protective activity of *S. alatum* (Ellandala, Yasmeen, Shashank, Sujatha, & Prathapagiri, 2011).

The antioxidant effects of diet formulated from *S. alatum* and other four leafy vegetables, commonly available and extensively consumed in some rural areas in Nigeria, were studied using albino rats. The results showed that malonylaldehyde levels were significantly ($P < 0.05$) lower in all the groups treated with the experimental diets, while vitamin C and antioxidant activities were significantly higher when compared with the diabetic controls. *S. alatum* and *Ceiba pentandra* showed the highest and lowest antioxidant activities, respectively. The results suggested appreciable antioxidant potentials of the vegetables, which could be of importance in the management of oxidative stress in diabetic conditions, the results revealed the antioxidant properties of *S. alatum* and reassures the safety of its consumption as vegetables, especially in the rural communities. Therefore, it seems promising in alleviating diabetes-associated oxidative stress (Idenyi et al., 2009).

Antidiabetic activity of the methanolic leaf extract of *S. alatum* (MESA) was evaluated in the Streptozotocin (65 mg/kg)-induced male albino rats. Dose-dependent reduction in blood glucose levels was observed in treated group and it can be comparable with that of standard drug, Glibenclamide. Significant renoprotective activity was observed in the MESA-treated group. MESA (500 mg/kg) has shown the antidiabetic activity by improving the levels of non-enzymatic levels, antioxidant enzymes, and has decreased the levels of MDA in kidney. And the other blood and urine parameters were normalized in treated groups (Ellandala et al., 2011).

REFERENCES

Ellandala, R., Yasmeen, N., Shashank, T., Sujatha, K., & Prathapagiri, K. (2011). Role of *Sesamum alatum* L. on nephropathy in diabetic rats. *International Journal of Pharmaceutical Sciences and Research, 18*, 1716–1721.

Facciola, S. (1998). *Cornucopia II*. Vista, CA: Kampong Publications.

Idenyi, J., Edeogu, O., Afiukwa, C., Ugwuja, E., Aloke, C., & Nwachukwu, N. (2009). Antioxidant activity of diet formulated from selected leafy vegetables commonly available and consumed in Abakaliki, Nigeria. *The Internet Journal of Alternative Medicine, 8*(2), 1–6.

Jansen, P. C. M. (2004). *Sesamum alatum* Thonn. ex Schumach. [Internet] Record from PROTA4U. In G. J. H. Grubben & O. A. Denton (Eds.), PROTA (Plant resources of Tropical Africa/Resources végétales de l'Afrique tropicale). Wageningen, The Netherlands. Available from: http://www.prota4u.org/search.asp

Kamal-Eldin, A., & Appelqvist, L. A. (1994). Variations in the composition of sterols, tocopherols and lignans in seed oils from four *Sesamum* species. *Journal of the American Oil Chemists' Society, 71*(2), 149–156.

Kamal-Eldin, A., Appelqvist, L. A., Yousif, G., & Iskander, G. M. (1992). Seed lipids of *Sesamum indicum* and related wild species in Sudan. The sterols. *Journal of the Science of Food and Agriculture, 59*(3), 327–334.

Kamal-Eldin, A., & Yousif, G. (1992). A furofuran lignan from *Sesamum alatum. Phytochemistry, 31*, 2911–2912.

Leung, W. -T. W., Busson, F., & Jardin, C. (1968). *Food composition table for use in Africa.* Rome, Italy: FAO, 306 p.

Mariod, A., & Matthäus, B. (2008). Physico-chemical properties, fatty acid and tocopherol composition of oils from some Sudanese oil bearing sources. *Grasas y Aceites, 59*(4), 321–326.

Potterat, O., Kurt, H., Stoeckli-Evans, H., & Mahamane, S. (2004). Saponins with an unusual Secoursene skeleton from *Sesamum alatum* THONN. *African Journal of Helvetica Chimica Acta, 75*(3), 833–841.

Chapter 9

Hibiscus cannabinus L. Kenaf

DISTRIBUTION AND BOTANICAL DESCRIPTION

Kenaf (*Hibiscus cannabinus* L.) family Malvaceae, is a tall annual herbaceous woody tropical plant, which has attracted considerable attention as a multipurpose plant having great potential for fiber, energy, and feedstock. Kenaf seed has been the waste part of kenaf plant. *H. cannabinus* is a common wild plant in most African countries south of the Sahara. It may have been domesticated as a fiber plant already 6000 years ago in Sudan. Kenaf is now widespread in the tropics and subtropics. As a vegetable it is widely grown in Africa, where it is grown on a much smaller scale as a fiber crop. In the past it has been of some importance as a commercial fiber crop in Côte d'Ivoire, Burkina Faso, Togo, Benin, Niger, Kenya, Tanzania, and Malawi. India has long been the largest producer of kenaf fiber (Bukenya-Ziraba, 2011) (Fig. 9.1).

COMPOSITION AND BIOACTIVE COMPONENTS

Kenaf is composed of various active components including tannins, saponins, polyphenolics, alkaloids, essential oils, and steroids (Yazan et al., 2011). Mariod, Fathy, and Ismail (2010) used two Malaysian kenaf varieties QP3 and V36 as good protein concentrates source. The two varieties yielded protein concentrate of 13.04 and 10.56%, respectively. The two varieties showed significantly different ($P < 0.05$) water and oil absorption capacities. QP3 showed higher foaming capacity than did V38, and it was increased with increasing salt and sugar concentration. Albumin was the main fraction of protein representing 59.6 and 66.1% in QP3 and V36 varieties, respectively, followed by globulin, which represented 22.6 and 19.1%, respectively. The ratios of albumin, globulin, glutelin, and prolamin were significantly different. Based on the data obtained from sodium dodecyl sulfate polyacrylamide gel electrophoresis, the main kenaf seed proteins present in the concentrates were five proteins with molecular weights ranging from 10 to 66 kDa. From differential scanning calorimetry data, these authors found that QP3 and V36 protein concentrates had similar denaturation temperatures (82.6 and 81.8°C, respectively).

Unconventional Oilseeds and Oil Sources. http://dx.doi.org/10.1016/B978-0-12-809435-8.00009-3

FIGURE 9.1 *Hibiscus cannabinus* plant and seeds.

KENAF SEED OIL

Kenaf seed oil contains alpha-linolenic acid (ALA), the essential omega 3 fatty acid that is metabolized to eicosapentaenoic acid, a precursor of eicosanoids with antiinflammatory and antithrombotic activity (Ruiz, Castillo, Dobson, Brennan, & Gordon, 2002). The potential for using kenaf seeds as a source of edible oil is often overlooked when considering kenaf as a fiber and feed crop. Coetzeea, Labuschagnea, and Hugob (2008) studied and compared the fatty acid composition and oil content of eight commercial kenaf varieties from various countries. They found that linoleic, oleic, and palmitic acids were the predominant fatty acids in all cultivars. Percentages of fatty acids varied greatly among different kenaf varieties. The oil content was 19.84% on average and there was not much difference between the varieties. These authors concluded that the relatively high oil content and the unique fatty acid composition suggested that kenaf seed could be used as a source of edible oil. Kenaf oil can be considered nutritionally healthy because of the relatively high amount of monounsaturated and polyunsaturated fatty acids.

Mohamed, Bhardwaj, Hamama, and Webber (1995) evaluated the oil, fatty acid, phospholipid, and sterol content of nine American kenaf genotypes. They reported that the oil content ranged from 21.4% to 26.4% with a mean of 23.7%.

Total phospholipids ranged from 3.9% to 10.3% of the oil, with a mean of 6.0%. Mean sterol percent was 0.9 and ranged from 0.6% to 1.2%. These authors reported palmitic acid (20.1%), oleic (29.2%), and linoleic (45.9%) of the total fatty acids as the major fatty acids, and palmitoleic (1.6%), linolenic (0.7%), and stearic (3.5%) as the minor components. Medium (Ci2–Ci4) and long (C22–C24) chain fatty acids were less than 1%. These authors also identified different classes of phospholipids, sphingomyelin (4.42% of the total phospholipids), phosphatidyl ethanolamine (12.8%), phosphatidyl choline (21.9%), phosphatidyl serine (2.9%), phosphatidyl inositol (2.7%), lysophosphatidyl choline (5.3%), phosphatidyl glycerol (8.9%), phosphatidic acid (4.9%), and cardiolipin (3.6%) in the nine genotypes. Phosphatidyl choline, phosphatidyl ethanolamine, and phosphatidyl glycerol were the dominant phospholipids. Kenaf seed oil contains phytosterol (Kritchevsky & Chen, 2005). Mohamed et al. (1995) reported β-sitosterol (72.3% of the total sterols), campesterol (9.9%), and stigmasterol (6.07%) as the prevalent phospholipids in kenaf seed oil.

Mariod, Matthäus, and Ismail (2011) investigated and compared fatty acids, tocopherols, and sterols of kenaf seed oil extracted by supercritical fluid (carbon dioxide) and traditional solvent methods. They determined fatty acids, tocopherols, and sterols in the extracted oils as functions of the pressure (400 bar, 600 bar), temperature (40°C, 80°C), and CO_2 flow rate (25 g/min). they did not find any differences in the fatty acid compositions of the various oil extracts and the main components were found to be linoleic (38%), oleic (35%), palmitic (20%), and stearic acid (3%). Their extraction of tocopherols using high pressure (600 bar/40°C, 600 bar/80°C) gave higher total tocopherols (88.20 and 85.57 mg/100 g oil, respectively) when compared with hexane extraction, which yielded 62.38 mg/100 g oil. These authors reported that the total sterols ranged from 5870.26 to 7421.95 mg/100 g oil for kenaf seed oil extracted using different techniques. Extraction of kenaf seed oil using SFE at high temperature (80°C) gave higher amounts of sterols when compared with hexane extraction. The major component of sterols in all extracts was β-sitosterol, which represents 75.23%–76.94% of the total sterols, followed by campesterol (9.72%–10.62%) and stigmasterol (5.82%–6.54%). In kenaf seed oil extracted by solvent the major sterol fractions were β-sitosterol, which represents 75.23%, followed by campesterol 10.62% and stigmasterol 6.01% of the total oil. These authors concluded that extraction of kenaf seed oil using supercritical fluid extraction at high temperature (80°C) gave higher amounts of sterols when compared with hexane extraction.

POTENTIALITY AND DIFFERENT USES

The kenaf plant was traditionally used as fiber source for making sacs, canvases, ropes, and also carpet. Kenaf seed has been reported as a suitable source for livestock and horse feed due to its high content of protein. Kenaf seeds yield vegetable oil that is edible for human consumption (Nyam, Tan, Lai, Long, &

Che Man, 2009). The shoots or young leaves, and sometimes the flowers and young fruits, are used as vegetables. In Uganda a local delicacy is made from the seeds. These are roasted, ground, and pounded, and the flour and skin are separated in water. The flour is rejected but the floating skin parts are used for the preparation of a paste, mixed with boiled pigeon peas. Children chew the bark for its sweetness (Bukenya-Ziraba (2011). The stem is a source of fiber used in the manufacture of twine, rope, and coarse textiles for sacking and cloth for packaging; special fiber cultivars exist. The production of kenaf fiber in Africa is rather uncommon, but locally important, for example, in northern Nigeria, Niger, and Sudan, where it is used for cordage, twines, fish lines, and nets. Dried scraped ribbons are used for twine and cordage for sleeping mats. Ribbons and whole stems are raw materials for the pulp and paper industry. Seeds from the fiber crop are used for oil extraction, the rest being used as feed. The oil is suitable as a lubricant and for illumination, for manufacture of soap, linoleum, and in paints and varnishes. In local medicine in Kenya, pounded roots are administered to spider bites, and leaves are used to treat stomach disorders. In West Africa, powdered leaves are applied to sores and boils, and a leaf infusion is administered for treating cough. In India, juice from the flowers is taken against biliousness, while the seeds are considered stomachic and aphrodisiac. Whole young plants are an excellent fodder for cattle. The stem core (xylem) is used in combination with peat moss (*Sphagnum*) and fertilizers as a growth medium for plants. Kenaf plants accumulate minerals, such as selenium and boron and can be used as a bioremedial tool for removing these metals from contaminated soil. In West Africa the plants are used as boundary markers (Bukenya-Ziraba, 2011).

Chan, Khong, Iqbal, Mansor, and Ismail (2013) studied nutritional composition, phenolic content, and antioxidant activity of defatted kenaf seed meal (DKSM) in comparison to wheat, rice, and sweet potato flours. Their study revealed that DKSM was high in protein (26.19 g/100 g DKSM) and carbohydrate (57.09 g/100 g DKSM). Magnesium, potassium, and phosphorus were the major minerals (>1 g/100 g DKSM) found in DKSM through energy dispersive X-ray spectrometric analysis. DKSM also exhibited appreciably higher total phenolic (3399.37 mg gallic acid equivalent (GAE)/g defatted material) and flavonoid contents (251.00 mg rutin equivalent (RE)/g defatted material) as well as antioxidant activity as compared to all selected edible flours ($P < 0.05$). Gallic acid, (+)-catechin, 4-hydroxybenzoic acid, vanillic acid, and syringic acid were determined as the predominant phenolics in DKSM through HPLCeDAD analysis. These authors recommended commercialization of DKSM as a highly antioxidative and nutritive edible flour, which can be prospectively used in the development of natural food preservative, nutraceuticals, and functional foods.

An early study for kenaf as protein source for layer chicken was done by Rajashekher, Rao, Reddy, and Eshwaraiah (1993) who reported that the meal prepared from kenaf seed was low in protein and energy and high in fiber, compared with the composition of peanut meal. Kenaf meal replaced peanut meal in layer diets satisfactorily. The threonine, methionine, and tryptophan amino acids in kenaf meal protein were higher by approximately 14, 22, and 76%, respectively, than the values in peanut meal protein. Based on the composition of indispensable amino acids, the protein quality of kenaf meal seemed to be slightly better than that of peanut meal (Rajashekher et al., 1993).

Kenaf seed oil can be extracted out conventionally by using the organic solvents, such as *n*-hexane or petroleum ether. Nevertheless, the oil is always doubted for the safety of consumption due to the residue of the solvents. Hence, supercritical fluid extraction (SFE) is considered a better way to obtain the oil (Wang & Weller, 2006). In Malaysia recently Kenaf was introduced as a fiber crop to substitute tobacco. Kenaf has been used to treat bruises, bilious conditions, fever, and puerperium (Yazan et al., 2011). The leaves are consumed as a vegetable in certain parts of the world, and possess an erythrocyte protective activity against drug-induced oxidative stress (Agbor, Oben, & Ngogang, 2015). Noordin, Chan, Iqbal, and Ismail (2012) determined the total phenolic content and antioxidant activities of kenaf seed extracted using solvent and supercritical fluid extraction (SFE). Their results showed that hexane extract gave the highest yield (16.6%), while water extract was the highest in phenolic content, DPPH scavenging activity, and hydroxyl radical scavenging activity as compared to other solvent extracts and SFE extracts ($P < 0.05$). In total antioxidant activity test, water and methanol extraction efficiently retarded the formation of hydroperoxide and malondialdehyde ($P < 0.05$) throughout the incubation period. Total phenolic content is highly correlated with antioxidant activities. Mariod, Ibrahim, Ismail, and Ismail (2012) investigated the antioxidant activity of crude 80% methanolic extract (CME), and its fractions using ethyl acetate (EAF), hexane (HF), and water (WF) of kenaf (*H. cannabinus*) seedcake. They examined CME and its fractions for radical scavenging capacities and antioxidant activities. They reported high total phenolic compounds in kenaf seedcake. The extracts/fractions of kenaf seedcake were markedly effective in inhibiting the oxidation of linoleic acid and subsequent bleaching of β-carotene. The addition of different kenaf seedcake extracts improved the oxidative stability of corn oil at 70°C. Kenaf seed oil was known as chemopreventive agent (Williams, Verghese, & Walker, 2007). Additionally, kenaf seed oil contains phytosterol, which possesses anticancer, antioxidant, and lipid lowering cholesterol properties (Kritchevsky & Chen, 2005).

Yazan et al. (2011) investigated extensive cytotoxicity study on kenaf seed oils obtained by sonication, Soxhlet, and supercritical fluid extraction (SFE) with nine different combinations of pressure (bars) and temperature (°C) (200/40, 200/60, 200/80, 400/40, 400/60, 400/80, 600/40, 600/60, and 600/80). All the oils were found cytotoxic toward ovarian cancer (CaOV$_3$) and colon cancer (HT29) cell lines in a dose-dependent manner as detected by using the MTT assay and trypan blue dye exclusion method. Oil obtained using sonication was found the most cytotoxic toward CaOV$_3$ cell line. Treated cells exhibited characteristics of apoptosis, such as chromatin condensation and nuclear fragmentation. In considering the safety of the product, SFE technology is a better alternative extraction method that is suitable in kenaf seed oil extraction. Kenaf's relatively high oil content and its similarity to cottonseed oil suggest that the seed oil may be used as a source of edible oil (Mohamed et al., 1995).

REFERENCES

Agbor, G. A., Oben, J. E., & Ngogang, J. Y. (2015). Antioxidative activity of *Hibiscus cannabinus* leaf extract. Available from: http://foodafrica.nri.org/index.html

Bukenya-Ziraba, R. (2011). *Hibiscus cannabinus* L. [Internet] Record from PROTA4U. In M. Brink & E. G. Achigan-Dako (Eds.), PROTA (Plant Resources of Tropical Africa/Ressources végétales de l'Afrique tropicale). Wageningen, The Netherlands. Available from: http://www.prota4u.org/search.asp

Chan, K. W., Khong, N. M. H., Iqbal, S., Mansor, S. M., & Ismail, M. (2013). Defatted kenaf seed meal (DKSM): prospective edible flour from agricultural waste with high antioxidant activity. *LWT—Food Science and Technology, 53*, 308–313.

Coetzeea, R., Labuschagnea, M. T., & Hugob, A. (2008). Fatty acid and oil variation in seed from kenaf (*Hibiscus cannabinus* L.). *Industrial Crops and Products, 27*, 104–109.

Kritchevsky, D., & Chen, S. C. (2005). Phytosterols—health benefits and potential concerns: a review. *Nutrition Research, 25*, 413–428.

Mariod, A. A., Fathy, S. F., & Ismail, M. (2010). Preparation and characterization of protein concentrates from defatted kenaf seed. *Food Chemistry, 123*(3), 747–752.

Mariod, A. A., Ibrahim, R. M., Ismail, M., & Ismail, N. (2012). Antioxidant activity of phenolic rich fractions obtained from Kenaf (*Hibiscus cannabinus*) seedcake toward corn oil oxidation. *Grasas Y Aceites, 63*(2), 167–174.

Mariod, A. A., Matthäus, B., & Ismail, M. (2011). Comparison of supercritical fluid and solvent extraction methods in extracting kenaf (*Hibiscus cannabinus*) seed oil lipids. *Journal of the American Oil Chemists' Society, 88*, 931–935.

Mohamed, A., Bhardwaj, H., Hamama, A., & Webber, C. (1995). Chemical composition of kenaf (*Hibiscus cannabinus* L.) seed oil. *Industrial Crops and Products, 4*, 157–165.

Noordin, M. Y., Chan, K. W., Iqbal, S., & Ismail, M. (2012). Phenolic content and antioxidant activity of *Hibiscus cannabinus* L. seed extracts after sequential solvent extraction. *Molecules, 17*, 12612–12621.

Nyam, K. L., Tan, C. P., Lai, O. M., Long, K., & Che Man, Y. B. (2009). Some physicochemical properties and bioactive compounds of seed oils. *LWT—Food Science and Technology, 42*, 1396–1403.

Rajashekher, A. R., Rao, P. V., Reddy, V. R., & Eshwaraiah (1993). Chemical composition and nutritive value of ambadi (*Hibiscus cannabinus*) meal for layers. *Animal Feed Science and Technology, 44*, 151–166.

Ruiz, M. L., Castillo, D., Dobson, D., Brennan, R., & Gordon, S. (2002). Genotypic variation in fatty acid content of blackcurrant seed. *Journal of Agriculture Food Chemistry, 50*, 332–335.

Wang, L., & Weller, C. L. (2006). Recent advances in extraction of nutraceuticals from plants. *Trends in Food Science and Technology, 17*, 300–312.

Williams, D., Verghese, M., & Walker, L. T. (2007). Flax seed oil and flax seed meal reduce the formation of aberrant crypt foci (ACF) in azoxymethane-induced colon cancer in Fisher 344 male rats. *Food and Chemical Toxicology*, *45*, 153–159.

Yazan, L. S., Foo, J. B., Ghafar, S. A., Chan, K. W., Tahir, P. M., & Ismail, M. (2011). Effect of kenaf seed oil from different ways of extraction towards ovarian cancer cells. *Food and Bioproducts Processing*, *8*(9), 328–332.

Chapter 10

Hibiscus radiatus Roselle

DISTRIBUTION AND BOTANICAL DESCRIPTION

Hibiscus radiatus belongs to the family Malvaceae. It is native to warm-temperate, southern and Southeast Asia subtropical and tropical regions throughout the world. It has 15 cm (5.9 in.) mauve flowers that have a purple center and yellow anthers. Leaves are dentate, with upper leaves lobed into three, five, or seven parts. Leaves are mistaken as marijuana, but radiatus' stems have small thorns. It is frequently grown as a vegetable or medicinal herb (Lawton, 2004) (Fig. 10.1).

SEED COMPOSITION AND BIOACTIVE COMPOUNDS

Roselle seeds are the waste that is left behind during processing of Roselle products. Roselle seeds contain about 14.9% protein, 21.2% crude fiber, 14.6% fats and oils, 35.6% carbohydrate by weight. Roselle seed oil contains phytosterols and tocopherols (Nyam, Tan, Lai, Long, & Che Man, 2009). The nutritional compositions of *H. radiatus* seeds vary depending on the variety, location and environmental conditions where the seeds were grown. The seeds, with a 5.0% moisture content, were microwave roasted for 5, 10, and 15 min; the remaining moisture content was 3.9, 2.8, and 1.4%, respectively. Protein content of unroasted was significantly lower ($P < 0.05$) than roasted and boiled seeds. Statistical results indicated that the protein of Roselle seeds increased in the order of boiled > microwave roasted > oven roasted. Protein content of boiled, microwave roasted, and oven roasted increased significantly as compared with unroasted (Mariod et al., 2013) (Table 10.1).

POTENTIALITY AND DIFFERENT USES

Occasional use as a fiber; ornamental for showy flowers, leaves sometimes used for scouring, in solution as a laxative or as a vegetable

Unconventional Oilseeds and Oil Sources. http://dx.doi.org/10.1016/B978-0-12-809435-8.00010-X

FIGURE 10.1 *Hibiscus radiatus* **plant and seeds.**

SEED PROCESSING, OIL PROPERTIES, AND STABILITY

Mariod et al. (2013) studied the effect of roasting and boiling techniques on Indonesian Roselle (*H. radiatus*) seed constituents; their results showed that roasting and boiling temperatures can increase fat and fiber content; microwave and boiling showed higher fat content when compared with oven roasting treatment. Their results showed that the major fatty acids in Roselle seed oil were linoleic, oleic, and palmitic. Roselle oil from seeds roasted by microwave and oven and that boiled was not different from oil of untreated Roselle seeds. Fatty acid compositions of Roselle oils did not change with roasting and boiling temperatures. The total tocopherol concentration of Roselle oils decreased during processing techniques as a result of heating temperature. Tocopherol of untreated Roselle seed oil was 0.05%; this was affected by microwave temperatures and time and decreased to 0.03, 0.02, and 0.01% in seeds treated by microwave temperature for 5, 10, and 15 min, respectively. In the same manner total tocopherol concentration decreased to 0.04, 0.01, and 0.02% of samples roasted using an oven temperature of 130 and 150°C and boiling, respectively (Table 10.2).

TABLE 10.1 Proximate Analysis of Roselle Seed Powder[a]

Constituent	HU	HMR 200W/5	HMR 200W/10	HMR 200 W/15	HOR 130C/30	HOR 150C/30	HB
Moisture	5.01 ± 0.0a	3.94 ± 0.3b	2.82 ± 0.1c	1.45 ± 0.1d	3.92 ± 0.1e	2.56 ± 0.1f	4.91 ± 0.1g
Protein	20.12 ± 0.8a	20.84 ± 1.4a	22.54 ± 0.2c	21.45 ± 0.8a	21.11 ± 0.1a	20.20 ± 1.9a	22.10 ± 0.5c
Fat	15.75 ± 0.5a	18.33 ± 1.2b	17.64 ± 0.6c	19.33 ± 1.8d	15.74 ± 0.9e	16.15 ± 0.2f	17.90 ± 0.3c
Fiber	25.30 ± 02a	38.12 ± 0.3b	31.90 ± 0.3c	32.14 ± 0.2d	28.55 ± 0.2e	27.54 ± 0.2f	24.55 ± 0.2g
Ash	6.15 ± 0.1a	6.52 ± 0.1b	7.75 ± 0.1c	8.50 ± 0.1d	6.92 ± 0.1b	7.25 ± 0.1c	7.92 ± 0.1c
COH	27.80 ± 0.3a	12.75 ± 0.2b	17.55 ± 0.1c	17.37 ± 0.1c	23.95 ± 0.2d	26.55 ± 0.2e	22.74 ± 0.2f

HB, Hibiscus boiled; HMR 200W/5, hibiscus microwave roasted at 200°C for 5 min, HMR 200W/10, hibiscus microwave roasted at 200°C for 10 min; HMR 200W/15, hibiscus microwave roasted at 200°C for 15 min; HOR 130/30, hibiscus oven roasted at 130°C for 30 min; HOR 150/30, hibiscus oven roasted at 150°C for 30 min, HU, hibiscus untreated sample.

[a]Values are expressed as mean ± SEM of three measurements. Data were statistically analyzed using one-way ANOVA. Values with different letter are significantly different at $P < 0.05$ within the same row.

Source: Data taken from Mariod, A. A., Suryaputra, S., Hanafi, M., Rohmana, T., Kardono, L. B. S., Herwan, T. (2013). Effect of different processing techniques on Indonesian Roselle (Hibiscus radiates) seed constituents. Acta Scientiarum Polonorum. Technologia Alimentaria, 12(4), 359–364.

TABLE 10.2 Fatty Acid (%) and Tocopherol (%) Composition of Roselle Seeds Oil[a]

Fatty acid	HUT	HMR 200W/5	HMR 200W/10	HMR 200 W/15	HOR 130/30	HOR 150/30	HB
C14:0	0.3 ± 0.02	0.3 ± 0.02	0.2 ± 0.02	0.3 ± 0.03	0.2 ± 0.01	0.2 ± 0.02	0.3 ± 0.01
C16:0	20.21 ± 0.4	19.85 ± 0.3	20.18 ± 0.3	20.68 ± 0.4	20.15 ± 0.3	20.09 ± 0.3	20.54 ± 0.3
C16:1	0.4 ± 0.01	0.4 ± 0.03	0.4 ± 0.02	0.4 ± 0.03	0.4 ± 0.03	0.4 ± 0.02	0.4 ± 0.02
C18:0	4.58 ± 0. 2	4.5 ± 0.2	4.53 ± 0.2	4.47 ± 0.3	4.33 ± 0.3	4.46 ± 0.2	4.42 ± 0.3
C18:1	21.62 ± 0.5	21.63 ± 0.4	21.36 ± 0.4	21.32 ± 0.3	22.55 ± 0.3	21.42 ± 0.2	21.37 ± 0.3
C18:2	51.39 ± 0.7	52.02 ± 0.7	52.09 ± 0.6	51.40 ± 0.5	51.50 ± 0.6	51.89 ± 0.5	51.77 ± 0.4
C18:3	0.9 ± 0.02	0.9 ± 0.02	0.8 ± 0.1	1.05 ± 0.3	0.54 ± 0.03	1.07 ± 0.3	0.9 ± 0.1
C20:0	0.5 ± 0.02	0.4 ± 0.01	0.4 ± 0.01	0.4 ± 0.02	0.3 ± 0.01	0.4 ± 0.02	0.3 ± 0.01
Total SFA	25.57 ± 0.6a	25.08 ± 0.6a	25.38 ± 0.6a	25.53 ± 0.5a	25.02 ± 0.5a	25.22 ± 0.4a	25.53 ± 0.3a
Total PUFA	74.43 ± 0.3b	74.91 ± 0.3b	74.61 ± 0.4b	74.57 ± 0.3b	74.98 ± 0.5b	74.88 ± 0.4b	74.47 ± 0.5b
Tocopherol	0.05 ± 0.01	0.03 ± 0.01	0.3 ± 0.02	0.01 ± 0.2	0.04 ± 0.3	0.01 ± 0.2	0.02 ± 0.4

For abbreviations see Table 10.1.

[a]Values are expressed as mean ± SEM of three measurements. Data were statistically analyzed using one-way ANOVA. Values with different letter are significantly different at P < 0.05 within SFA and PUFA.

Source: Data taken from Marod, A. A., Suryaputra, S., Hanafi, M., Kardono, L. B. S., Herwan, T. (2013). Effect of different processing techniques on Indonesian Roselle (Hibiscus radiates) seed constituents, Acta Scientiarum Polonorum, Technologia Alimentaria, 12(4), 359–364.

REFERENCES

Lawton, B. P. (2004). *Hibiscus: hardy and tropical plants for the garden*. Portland: Timber Press, p. 122.

Mariod, A. A., Suryaputra, S., Hanafi, M., Rohmana, T., Kardono, L. B. S., & Herwan, T. (2013). Effect of different processing techniques on Indonesian roselle (*Hibiscus radiates*) seed constituents. *Acta Scientiarum Polonorum. Technologia alimentaria, 12*(4), 359–364.

Nyam, K. L., Tan, C. P., Lai, O. M., Long, K., & Che Man, Y. B. (2009). Some physicochemical properties and bioactive compounds of seed oils. *LWT, 42*, 1396–1403.

Chapter 11

Hibiscus sabdariffa L. Roselle

DISTRIBUTION AND BOTANICAL DESCRIPTION

Hibiscus sabdariffa is commonly known as roselle, hibiscus, Jamaica sorrel or red sorrel (English), and in Arabic, Karkadeh (Ali, Al Wabel, & Blunden, 2005). Its native distribution is uncertain, some believe that it is from India or Saudi Arabia (Ismail, Ikram, & Nazri, 2008), while Murdock (Murdock, 1959) showed evidence that *H. sabdariffa* (Hs) was domesticated by the black populations of western Sudan (Africa) sometime before 4000 BC. Nowadays, it is widely cultivated in both tropical and subtropical regions including India, Saudi Arabia, China, Malaysia, Indonesia, The Philippines, Vietnam, Sudan, Egypt, Nigeria, and México (Ismail et al., 2008; Yagoub, Mohamed, Ahmed, & El Tinay, 2004). There are two main varieties of Hs, the first being Hs var. *altissima* Wester, cultivated for its jute-like fiber and the second is Hs var. *sabdariffa*. The second variety includes shorter bushy forms, which have been described as races: *bhagalpuriensi*, *intermedius*, *albus*, and *ruber*. The first variety has green, red-streaked, inedible calyces, while the second and third race has yellow-green edible calyces (var. *ruber*) and also yields fiber (Morton, 1987).

Roselle is a species of *Hibiscus* native to West Africa, used for the production of bast fiber and as an infusion, in which it may also be known as Karkade. It is an annual or perennial herb or woody-based subshrub, growing to 2–2.5 m (7–8 ft.) tall. The leaves are deeply three- to five-lobed, 8–15 cm (3–6 in.) long, arranged alternately on the stems. The flowers are 8–10 cm (3–4 in.) in diameter, white to pale yellow with a dark red spot at the base of each petal, and have a stout fleshy calyx at the base, 1–2 cm (0.39–0.79 in.) wide, enlarging to 3–3.5 cm (1.2–1.4 in.), fleshy and bright red as the fruit matures. It takes about 6 months to mature. Stalks and leaves range from dark green to reddish color; flowers are creamy white or pale yellow. For fiber crops, seeds are sown close together, producing plants 10–16 ft. (3–5 m) high, with little branching. The stalks, cut when buds appear, are subjected to a retting process, then stripped of bark or beaten, freeing the fiber. In some areas retting time is reduced by treating only the bark and its adhering fiber. Plants for fruit crops, more widely spaced, are shorter and many branched, and their calyxes are picked when plump and fleshy. The fiber strands, 3–5 feet (1–1.5 m) long, are composed of individual fiber cells. Roselle fiber is lustrous, with color ranging from creamy to silvery white, and is moderately strong. It is used, often combined with jute, for bagging fabrics and twines. India, Java, and the Philippines are major producers (http://www.britannica.com/plant/roselle-plant). Bamgboye and Adejumo (2009) determined the physical properties of Roselle seeds at different moisture contents using ASAE standards. The mean values of the physical properties of the seeds were determined as 4.75–4.85 mm length, 4.15–4.26 mm width, 2.62–2.67 mm thickness, 3.73–3.83 mm geometric mean diameter, 648.31–619.14 kg/m^3 bulk density, 20.13°–24.85° angle of repose, 17–18.5 mm^2 surface area, 52.6–58.3% porosity, 1367.0–1487.4 kg/m^3 true density, 29.7–40 × 10–6 m^3 volume, 5.8–7.1 m/s terminal velocity, 1.36–1.44 specific gravity (Fig. 11.1).

In many tropical areas, the red, somewhat acid calyxes of *H. sabdariffa* var. *altissima* are used locally for beverages, sauces, jellies, preserves, and chutneys; the leaves and stalks are consumed as salads or cooked vegetables and used to season curries; and in Africa the oil-containing seeds are eaten (http://www.britannica.com/plant/roselle-plant).

H. sabdariffa L. (Malvaceae) has been used traditionally as a food, in herbal drinks, in hot and cold beverages, as a flavoring agent in the food industry and as a herbal medicine. In vitro and in vivo studies as well as some clinical trials provide some evidence mostly for phytochemically poorly characterized Hs

FIGURE 11.1 *Hibiscus sabdariffa* **plant and seeds.**

extracts. Extracts showed antibacterial, antioxidant, nephro- and hepatoprotective, renal/diuretic effect, effects on lipid metabolism (anticholesterol), antidiabetic and antihypertensive effects among others. This might be linked to strong antioxidant activities, inhibition of a-glucosidase and a-amylase, inhibition of angiotensin-converting enzymes, and direct vasorelaxant effect or calcium channel modulation. Phenolic acids (especially protocatechuic acid), organic acid (hydroxycitric acid and hibiscus acid), and anthocyanins (delphinidin-3-sambubioside and cyanidin-3-sambubioside) are likely to contribute to the reported effects. More well designed controlled clinical trials.

Fresh or dried calyces of *H. sabdariffa* are used in the preparation of herbal drinks, hot and cold beverages, fermented drinks, wine, jam, jellied confectionaries, ice cream, chocolates, flavoring agents, puddings and cakes (Bako, Mabrouk, & Abubakar, 2009). In Egypt, the fleshy calyces are used in making "cacody tea" and fermented drinks, while in Sudan and Nigeria, the calyces are boiled with sugar to produce a drink known as "Karkade" or "Zoborodo" (Da-Costa-Rocha, Bonnlaender, Sievers, Pischel, & Heinrich, 2014). In the West Indies the calyces can also be used as coloring and flavoring ingredient in rum (Ismail et al., 2008). The seeds are eaten roasted or ground in meals, while the leaves and shoots are eaten raw or cooked, or as a sour-flavored vegetable or condiment. In Sudan, the leaves are eaten green or dried, cooked with onions and groundnuts, while in Malaysia the cooked leaves are eaten as vegetables (Ismail et al., 2008). In Africa, the seeds are roasted or ground into powder and used in meals, such as oily soups and sauces. In China and West Africa, the seeds are also used for their oil (Atta & Imaizumi, 2002).

USE IN FOOD AND MEDICINE

Karkade has been widely used in local medicines. In India, Africa and Mexico, infusions of the leaves or calyces are traditionally used for their diuretic, choleretic, febrifugal, and hypotensive effects, decreasing the viscosity of the blood and stimulating intestinal peristalsis. It is also recommended as a hypotensive in Senegal. In Egypt, preparations from the calyces have been used to treat cardiac and nerve diseases and also to increase the production of urine (diuresis). In Egypt and Sudan, an infusion of "Karkade" calyces is also used to help lower body temperature. In Guatemala it is used for treating drunkenness. In North Africa, calyces preparations are used to treat sore throats and coughs, as well as genital problems, while the emollient leaf pulp is used for treating external wounds and abscesses (Neuwinger, 2000). Roselle is used in many folk medicines. It is valued for its mild laxative effect and for its ability to increase urination, attributed to two diuretic ingredients, ascorbic acid, and glycolic acid. Since it contains citric acid, it is used as a cooling herb, providing relief during hot weather by increasing the flow of blood to the skin's surface and dilating the pores to cool the skin. The leaves and flowers are used as a tonic tea for digestive and kidney functions. The heated leaves are applied to cracks in the feet and on boils and ulcers to speed maturation. The calyces and seeds are diuretic, laxative, and tonic. The ripe calyces, boiled in water, can be used as a drink to treat bilious attacks. A lotion made from roselle leaves is used on sores and wounds (Qi, Chin, Malekian, Berhane, & Gager, 2005).

In India, a decoction from the seeds is used to relieve pain in urination and indigestion. In Brazil, the roots are believed to have stomachic and emollient properties. In Chinese folk medicine, it is used to treat liver disorders and high blood pressure. In Iran, sour hibiscus tea is reportedly a traditional treatment for hypertension (Burnham, Wickersham, & Novak, 2002), while in Nigeria the decoction of the seeds is traditionally used to enhance or induce lactation in cases of poor milk production, poor letdown, and maternal mortality (Gaya, Mohammad, Suleiman, Maje, & Adekunle, 2009).

Many parts of roselle including seeds, leaves, fruits, and roots are used in various foods. Among them, the fleshy red calyces are the most popular. They are used fresh for making wine, juice, jam, jelly, syrup, gelatin, pudding, cakes, ice cream and flavors, and also dried and brewed into tea, spice, and used for butter, pies, sauces, tarts, and other desserts. The calyces possess pectin that makes a firm jelly. The young leaves and tender stems of roselle are eaten raw in salads or cooked as greens alone or in combination with other vegetables and/ or with meat. They are also added to curries as seasoning. They have an acid, rhubarb-like flavor. The red calyces contain antioxidants including flavonoids, gossypetin, hibiscetine, and sabdaretine. The fresh calyces are also rich in ribo-flavin, ascorbic acid, niacin, carotene, calcium, and iron that are nutritionally important. The seeds, are high in protein, can be roasted and ground into a powder then used in soups and sauces. The roasted seeds can be used as a coffee substitute. The young root is edible, but very fibrous (Qi et al., 2005).

PLANT COMPOSITION AND BIOACTIVE COMPONENTS

Early studies reported that the calyces contain protein (1.9 g/100 g), fat (0.1 g/100 g), carbohydrates (12.3 g/100 g), and fiber (2.3 g/100 g). They are rich in vitamin C (14 mg/100 g), b-carotene (300 lg/100 g), calcium (1.72 mg/100 g), and iron (57 mg/100 g) (Ismail et al., 2008). The main constituents of *H. sabdariffa* relevant in the context of its pharmacological are organic acids, anthocyanins, polysaccharides, and flavonoids. The leaves contain protein (3.3 g/100 g), fat (0.3 g/100 g), carbohydrate (9.2 g/100 g), minerals [phosphorus (214 mg/100 g), iron (4.8 mg/100 g)], thiamine (0.45 mg/100 g), b-carotene (4135 lg/100 g), riboflavin (0.45 mg/100 g), and ascorbic acid (54 mg/100 g) (Ismail et al., 2008). The seeds contained crude fatty oil (21.85%), crude protein (27.78%), carbohydrate (21.25%), crude fiber (16.44%) and ash (6.2%). In terms of minerals, the most prevalent is potassium (1329 ± 1.47 mg/100 g), followed by sodium (659 ± 1.58 mg/100 g), calcium (647 ± 1.21 mg/100 g), phosphorus (510 ± 1.58 mg/100 g), and magnesium (442.8 ± 1.80 mg/100 g) (Nzikou et al., 2011).

The seeds, regarded as by-product of roselle processing had 57.3% moisture. Hainida, Amin, Normah, and Mohd.-Esa (2008) studied the effects of sun-drying and boiling sun-drying on the nutritional composition of Roselle seeds, their results showed that, raw freeze-dried, sun-dried, and boiled sun-dried seeds contained 6.81, 9.9, and 9.8% moisture; 35.4, 33.5, and 30.6% protein; 27.2, 22.1, and 29.6% lipids; 2.3, 13.0, and 4.0% available carbohydrate; 25.5, 18.3, and 19.2 total dietary fiber; and 7.4, 7.5, and 6.6% ash, respectively. The carbohydrate, protein, lipids, and moisture of raw freeze-dried were significantly different ($P < 0.05$) from sun dried and boiled sun dried. The predominant minerals in Roselle seeds were potassium (99–109 mg/100 g), magnesium (26–28 mg/100 g), and calcium (24–31 mg/100 g). The total dietary fiber of the seeds was within the acceptable range, with soluble and insoluble fiber ratios ranging from 1.2 to 3.3. The study detected 17 essential and nonessential amino acids. The seeds were rich in lysine (14–15 g/100 g), arginine (30–35 g/100 g), leucine (15.4–18.6 g/100 g), phenylalanine (11–12 g/100 g), and glutamic acid (21–24 g/100 g). These authors indicated that Roselle seeds may serve as a potential source of functional ingredients.

PHYSICOCHEMICAL PROPERTIES, FATTY ACIDS, AND MINOR COMPONENTS OF KARKADE SEED OIL

Roselle seed oil belongs to the linoleic/oleic category, its most abundant fatty acids being C18:2 (40.1%), C18:1 (28%), C16:0 (20%), C18:0 (5.3%), and C19:1 (1.7%). Roselle seeds are a good source of lipid-soluble antioxidants, particularly gamma-tocopherol. The acidity, 2.24%; peroxide index, 8.63 mEq/kg; extinction coefficients at 232 (k(232)) and 270 nm (k(270)), 3.19 and 1.46, respectively; oxidative stability, 15.53 h; refractive index, 1.477; density, 0.92 kg/L; and viscosity, 15.9 cP. Sterols include beta-sitosterol (71.9%), campesterol (13.6%), Delta-5-avenasterol (5.9%), cholesterol (1.35%), and clerosterol (0.6%). Total tocopherols were detected at an average concentration of 2000 mg/kg, including alpha-tocopherol (25%), gamma-tocopherol (74.5%), and delta-tocopherol (0.5%). The global characteristics of Roselle seed oil suggest that it could have important industrial applications, adding to the traditional use of roselle sepals in the elaboration of Karkade tea (Mohamed, Fernández, Pineda, & Aguilar, 2007).

Akinoso and Suleiman (2011) determined the fatty acid composition of Roselle (*H. sabdariffa* L.) seed oil. They optimized the oil extraction using RSM. Five levels of the variables, roasting duration (10, 15, 20, 25, and 30 min), and temperature (80, 90, 100, 110, and 120°C) were used. Oil yield (OY), free fatty acid (FFA), anisidine value (AV), specific gravity (SG), and stability were determined using standard methods. Major fatty acids found were oleic acid (37.92%), linoleic (35.01%), palmitic (19.65%), and stearic (6.07%). Myristic (0.16), myristoleic (0.17), palmitoleic (0.56), arachidic (0.14), and eicosatrienoic (0.20) were minor fatty acids content. OY, FFA, AV, SG, and stability varied from 18% to 23%, from 0.56% to 4.11%, from 6.93 to 35.7 mg/L, from 0.90 to 0.95, and from 0.68% to 5.42%, respectively. The treatment had significant ($P < 0.05$) effect on OY, AV, and SG. However, nonsignificant effect of treatment was recorded on FFA and stability at 5% level of significance. The best desirability of 0.46 was achieved at roasting duration and temperature of 25 min and 110°C, respectively, which gave OY of 22%, initial FFA of 1.95%, AV of 31.2 mg/L, SG of 0.92, and stability of 2.67%.

The major saturated fatty acids identified in the seed oil of *H. sabdariffa* were palmitic (20.84%) and stearic (5.88%) acids and the main unsaturated fatty acids are linoleic (39.31%) and oleic acid (32.06%) (Nzikou et al., 2011). Nakpong and Wootthikanokkhan (2010) investigated Karkade oil as an alternative feedstock for biodiesel production in Thailand. They reported the following parameters; palmitic acid 18.15%, stearic acid 4.09%, oleic acid 33.31%, linoleic acid 38.17%, and linolenic acid 2.09%. The oil density at 15°C, was 919.9 kg/m^3, with free fatty acids content (as linoleic acid % w/w) of 0.67, while the kinematic viscosity at 40°C was 36.35 mm^2/s. These authors reported the possibility of producing biodiesel using *H. sabdariffa* oil. With the exception of a higher carbon residue and lower oxidation stability, the resultant methyl ester content of 99.4% w/w, plus all of the other measured properties of the roselle biodiesel, met the Thai biodiesel (B100) specifications and international standards EN 14214:2008 (E) and ASTM D 6751-07b.

REFERENCES

Ali, B. H., Al Wabel, N., & Blunden, G. (2005). Phytochemical, pharmacological and toxicological aspects of *Hibiscus sabdariffa* L.: a review. *Phytotherapy Research*, *19*(5), 369–375.

Akinoso, R., & Suleiman, A. (2011). Heat treatment effects on extraction of roselle (*Hibiscus sabdariffa* L) seed oil. *European Journal of Lipid Science and Technology*, *113*(12), 1527–1532.

Atta, M. B., & Imaizumi, K. (2002). Some characteristics of crude oil extracted from roselle (*Hibiscus sabdariffa* L.) seeds cultivated in Egypt. *Journal of Oleo Science*, *51*(7), 457–461.

Bako, I. G., Mabrouk, M. A., & Abubakar, A. (2009). Antioxidant effect of ethanolic seed extract of *Hibiscus sabdariffa* Linn (Malvaceae) alleviate the toxicity induced by chronic administration of sodium nitrate on some haematological parameters in Wistars rats. *Advance Journal of Food Science and Technology*, *1*(1), 39–42.

Bamgboye, A. I., & Adejumo, O. I. (2009). Physical properties of roselle (*Hibiscus sabdariffa* L.) seed. *Agricultural Engineering International: the CIGR Ejournal*, XI, Manuscript 1154.

Burnham, T., Wickersham, R., & Novak, K. (2002). *The review of natural products* (3rd ed.). St. Louis, MO: Facts and Comparisons.

Da-Costa-Rocha, I., Bonnlaender, B., Sievers, H., Pischel, I., & Heinrich, M. (2014). *Hibiscus sabdariffa* L.—a phytochemical and pharmacological review. *Food Chemistry*, *165*, 424–443.

Gaya, I. B., Mohammad, O. M. A., Suleiman, A. M., Maje, M. I., & Adekunle, A. B. (2009). Toxicological and lactogenic studies on the seeds of *Hibiscus sabdariffa* Linn (Malvaceae) extract on serum prolactin levels of albino wistar rats. *The Internet Journal of Endocrinology*, *5*(2), 8.

Hainida, K. I. E., Amin, I., Normah, H., & Mohd.-Esa, N. (2008). Nutritional and amino acid contents of differently treated Roselle (*Hibiscus sabdariffa* L.) seeds. *Food Chemistry*, *111*, 906–911.

Ismail, A., Ikram, E. H. K., & Nazri, H. S. M. (2008). Roselle (*Hibiscus sabdariffa* L.) seeds nutritional composition protein quality and health benefits. *Food, 2*(1), 1–16.

Mohamed, R., Fernández, J., Pineda, M., & Aguilar, M. (2007). Roselle (*Hibiscus sabdariffa*) seed oil is a rich source of gamma-tocopherol. *Journal of Food Science, 72*(3), S207–S211.

Morton, J. F. (1987). *Fruits of warm climates*. USA: Florida Flair Books.

Murdock, G. P. (1959). *Africa, its peoples and their culture history*. New York: McGraw-Hill.

Nakpong, P., & Wootthikanokkhan, S. (2010). Roselle (*Hibiscus sabdariffa* L.) oil as an alternative feedstock for biodiesel production in Thailand. *Fuel, 89*, 1806–1811.

Neuwinger, H. (2000). *African traditional medicine*. Stuttgart: Medpharm Scientific Publication.

Nzikou, J. M., Bouanga-Kalou, G., Matos, L., Ganongo-Po, F. B., Mboungou-Mboussi, P. S., Moutoula, F. E., et al. (2011). Characteristics and nutritional evaluation of seed oil from Roselle (*Hibiscus sabdariffa* L.) in Congo-Brazzaville. *Current Research Journal of Biological Sciences, 3*(2), 141–146.

Qi, Y., Chin, K. L., Malekian, F., Berhane, M., Gager, J. (2005). Biological characteristics, nutritional and medicinal value of roselle, *Hibiscus sabdariffa* circular—urban forestry natural resources and environment. 604.

Yagoub, A. G., Mohamed, B. E., Ahmed, A. H., & El Tinay, A. H. (2004). Study on furundu, a traditional Sudanese fermented roselle (*Hibiscus sabdariffa* L.) seed: Effect on in vitro protein digestibility, chemical composition, and functional properties of the total proteins. *Journal of Agricultural and Food Chemistry, 52*(20), 6143–6150.

Chapter 12

Monechma ciliatum Black Mahlab or Mahlab Balady

BOTANICAL DESCRIPTION

Monechma ciliatum (black mahlab) family Acanthaceae is an annual glabrous herb, 30–65 cm high woody below. Leaves are linear or narrowly linear-lanceolate, up to 10 cm long, 1.25 cm broad. Flowers are cream-white with purple and orange stracks, 2-lipped, in short spikes; bracts pectinate with long stiff white bristle, lanceolate, leafy, about 1.25 cm long seeds with a tuft of rigid thick hairs at the hilum (Aseel, 2008) (Fig. 12.1).

In the Sudan, autumn is reported to be the optimum sowing date. Plant spacing positively affected vegetative growth and yield components, and was inversely related to seed yield. The wider plant spacing gave higher values for both vegetative and reproductive growth. The fixed oil of black mahlab seed was found to be higher in autumn compared to winter and summer sowing dates (Mohammed & Elballa, 2015).

ORIGIN AND HABITATS

M. ciliatum is a native of tropical Africa and sometimes India. It is widely distributed from Senegal to Cameroon and eastward to the central and south provinces of the Sudan and south through East Africa, Zambia, and Nigeria (Uguru & Evans, 2000). In Sudan *M. ciliatum* was found to grow widely in Kassala state, southern Blue Nile state, Upper Nile state, Bahr el Ghazal state, and Bahr el Jabel (Aseel, 2008). *M. ciliatum* is a famous medicinal plant in western Sudan, especially in Nuba Mountains and Gabel Mara Area. Its seeds are used as an effective laxative, and contain an essential oil, which emits a sweet and pleasant odor. It is further used in traditional Sudanese fragrances, lotion, and other cosmetics used for wedding preparation and childbirth (Sharief, 2002).

USES AND MEDICINAL VALUES

The importance of black mahlab as a medicinal plant, which is used to relieve abdominal pain, and used as a laxative drug, is one of the secrets, which were hidden behind folk medicine practitioners. In spite of its values, this plant

Unconventional Oilseeds and Oil Sources. http://dx.doi.org/10.1016/B978-0-12-809435-8.00012-3

FIGURE 12.1 *Monechma ciliatum* **plant and seeds.**

species still grows as a wild plant in different areas in the world. In Botswana, *M. ciliatum* plays an important role as a medicine for remedy of general body pain, liver and bowel trouble (diarrhea), and sterility in women (Hedberg & Stangard, 1989). *M. ciliatum* is one of the important plants in western Sudan culture. It has cosmetic and nutritional value. The oil is used to manufacture traditional fragrance and lotion. The plant has medicinal importance as laxative medicine to remedy stomach troubles and also used to control scale dandruff, which form on the scalp in case of oily hair (Sharief, 2002).

The leaves of *M, ciliatum* (used in Nigerian and Sudanese traditional medicine) contain alkaloids, glycosides, proteins, tannins, and saponins (Uguru & Evans, 2000). In Blue Nile state, Sudan, seeds of *Monechma* are used for remedy of women sterility by eating seeds during menstruation period (Elnour, 2006).

The hot methanolic extract of *M. ciliatum* leaves, was previously found to have potent oxytocic effect on various species when fractionated into various solvents and the fractions analyzed chemically and also assayed on the rat uterus in an attempt to identify the chemical constituents present and in particular the oxytocic principle. Positive reaction with ninhydrin spray suggests that the oxytocic constituent is an amino acidic derivative (Uguru & Evans, 2000).

Mariod, Ibrahim, Ismail, and Ismail (2010b) evaluated the total phenolic content and the antioxidant potential of methanolic extract, ethyl acetate extract, and hexane extract from *M. ciliatum* leaves. They found that the methanolic leaves extract exhibited highest total phenolic compounds and IC_{50} values for DPPH, followed by ethyl acetate, and hexane extracts, respectively. These authors determined the peroxide value, anisidine value, conjugated dienes, and thiobarbituric acid reactive substances as parameters for evaluating the stabilization efficacy of *M. ciliatum* leaves extracts on corn oil. Their results revealed that *M. ciliatum* leave extracts are potent antioxidant for the stabilization of corn oil (Mariod et al., 2010b). The antifertility effect of the aqueous and ethanol extracts of the leaves and roots of *M. ciliatum* was studied both in vivo and in vitro. The antiimplantation and antifertility activities of the ethanol leaves extract were 37% and 20%, respectively (Uguru, Okwuasaba, Ekwenchi, & Uguru, 1995). Black mahlab is one of the useful odor plants in western Sudan. It has a pleasant aromatic odor, which is very important component of heavy Sudanese traditional fragrance like "Karkar" or as lotion, for example, "Dilka." It is very effective as medicinal plant for remedies to stomach disturbance and as a carminative for children (Aseel, 2008).

Mariod, AL-Nageeb, and Ismail (2010c) investigated the regulatory effects of *M. ciliatum* methanolic extract MCME at 10, 20, and 50 µg/mL on low-density lipoprotein receptor (LDLR) and 3-hydroxy-3-methylglutaryl-coenzyme A reductase (HMGCR) in human HepG2 cell line using quantitative real-time polymerase chain reaction. They reported that, LDLR mRNA level was increased significantly by 1.4-, 2.6-, and 4.3-fold in MCME treated cells at 10, 20, and 50, respectively, compared to untreated cells. Whereas, HMGCR mRNA level was decreased significantly by 38, 63, and 80% in MCME treated cells at 10, 20, and 50, respectively, compared to untreated cells. These authors reported that, the effect of MCME was concentration dependent, and different doses showed significant differences in regulation of both LDLR and HMGCR genes. The present study showed that MCME effectively regulated the expression of LDLR and HMGCR genes influencing the cholesterol metabolism in HepG2 cells (Mariod et al., 2010c). Oshi, Abdelkareem, and Eltohami (2013) examined and validated the folk's claim of inhalation of *M. ciliatum* seeds powder; they found that, *Klebsiella pneumoniae* was highly sensitive to the hydroalcoholic extract and moderately sensitive to the extract. *Staphylococcus aureus* was found to be moderately sensitive to the hydroalcoholic extract; *Pseudomonas aeruginosa* was resistant to all extracts used. In addition, all fungi used were found to be resistant to all extracts.

Comparative antimicrobial effects of extracts from seeds, leaves, and stems of *M. ciliatum* against standard organisms (*Bacillus subtilis, S. aureus, Escherichia coli, Proteus vulgaris, P. aeruginosa, Candida albicans,* and *Aspergillus niger*) were investigated. *M. ciliatum* stems methanolic extract was active more than chloroform extract. Water extract caused inhibition zone against *Pseudomonas vulgaris* only. Methanolic extract of *M. ciliatum* leaves was the most active extract against standard organisms (Abuelgasim, Ali, & Hassan, 2015).

SEED COMPOSITION AND BIOACTIVE COMPONENTS

The seed is black about 2-mm long, with a tuft of rigid thick hairs at the hilum. Hundred seeds weight of black mahlab was 1.88 g. The ash content was reported as 2.81%, which might be according to difference in minerals content, the protein content was 21.0%, crude fiber 2.5% (Aseel, 2008). Black mahlab contains flavonoids, tannins, triterpens, and anthraquinones alkaloids, glycosides, proteins, and saponins (Oshi et al., 2013; Uguru & Evans, 2000). Mariod, Aseel, Mustafa, and Abdel-Wahab (2009a) characterized the oil and meal from *M. ciliatum* seeds for their physicochemical properties. The oil content was found to be 13.15%. The refractive index of the oil was 1.470 (20°C), with specific gravity of 0.8167 g/cm^3. Saponification value was 180.3 mg KOH/g, peroxide value was 4.43 mEq/kg, and unsaponifiable matter was 0.66%. The major fatty acids were palmitic 4.5%, stearic 16.0%, oleic 47.3%, and linoleic 31.4%. The amount of tocopherols was found at 45.2 mg/100 g.

The concentration of major elements, such as Ca, K, and Mg in black mahlab seeds was found to be as 100.8, 80.5, and 69.13 ppm, respectively. The concentrations of minor elements Al, Pb Ni, Mn, Cu, Cr, Co, and Fe levels were at low levels in the black mahlab (Mariod et al., 2009a), and the protein content was 21%. The total amount of amino acids in black seeds was 12.55 g/16 g N. The total amount of the essential amino acids in black mahlab seed was 4.96 g/16 g N. All the essential amino acids with the exception of tryptophan, which was not analyzed, were found to be present at low levels, and the total essential amino acids were found to be lower when compared to that of four different foods (Table 12.1) broad bean (30.24 g/16 g N), wheat flour (25.6 g/16 g N), and soy flour (33.6 g/16 g N) (Aseel, 2008).

Mariod, Ibrahim, Ismail, and Ismail (2010d) investigated the antioxidant activities of phenolic rich fractions (PRFs) from crude methanolic extract (CME), and its fractions using ethyl acetate (EAF), hexane (HF), and water (WF) of black mahlab (*M. ciliatum*) seedcakes.

The antioxidant activity determined by the DPPH method revealed that black mahlab PRFs had the highest antioxidant activity. The presence of antioxidants in the black mahlab PRFs reduced the oxidation of β-carotene by hydroperoxides from these extracts/fractions. The effect of the black mahlab PRFs on the oxidative stability of corn oil at 70°C was tested in the dark and compared with butylated hydroxyanisole (BHA). The CME performed better antioxidant activity in inhibiting the formation of both primary and secondary oxidation products. These authors reported that the black mahlab seedcakes extracts/fractions can be used as an easily accessible source of natural antioxidants and as a possible food supplement mainly in edible oils. Mariod et al. (2010d) used HPLC–DAD to analyze CME of black mahlab seedcakes extracts; they revealed the presence of phenolic compounds that were monitored at 280 nm; these compounds were detected as gallic caid, (+)-catechin, chlorogenic acid, hydroxybenzoic acid, *p*-cumeric acid. Hydroxybenzoic acid was detected to be the major phenolic component in the black mahlab seedcake, contributing about 82.1% to the total

TABLE 12.1 Amino Acid Composition of Black Mahlab in Comparison With Four Different Foods (g/16 g N)

Amino acid	Black mahlab flour	White mahlab flour	Broad bean	Wheat flour	Soy flour
Aspartic acid	1.5	2.2	11.2	4.96	11.52
Threonine	0.53	0.63	3.36	2.72	3.84
Serine	0.48	0.67	4.48	5.28	5.12
Glutamic acid	1.75	3.94	15.04	27.36	18.72
Glycine	0.55	1.48	4.16	4.00	4.16
Alanine	1.00	1.30	4.16	3.68	4.32
Cystine	0.16	0.42	0.8	2.56	1.6
Valine	1.07	1.3	4.48	4.48	4.8
Methionine	0.27	0.35	0.64	1.6	1.28
Isoleucine	0.88	1.14	4.00	3.36	4.48
Leucine	1.44	1.86	7.04	6.56	7.84
Tyrosine	0.39	0.77	3.2	3.04	3.2
Phenylalanine	0.87	1.11	4.32	4.48	4.96
Histidine	0.72	1.31	2.4	2.08	2.56
Lysine	0.94	1.07	6.4	2.4	6.4
Tryptophan	—	—	0.96	1.12	1.28
Arginine	—	—	8.96	4.64	7.2
Proline	—	—	4.00	10.56	5.44
Total of amino acid	12.55	18.38	89.6	94.88	98.52
Total of essential amino acid	4.96	7.46	30.24	25.6	33.6
Essential amino acids (%)	39.5	40.5	33.75	26.9	34.1

amount, while chlorogenic, (+)-catechin, and *p*-cumeric acid were the predominant phenolics showing the levels of 1.038, 0.885, and 0.404 mg/100 g DW, respectively (Mariod et al., 2010d).

REFERENCES

Abuelgasim, A. I., Ali, M. I., & Hassan, A. (2015). Antimicrobial activities of extracts for some of medicinal plants. *International Journal of Advanced and Applied Sciences*(2), 1–5.

Aseel, K. M. (2008). Investigating two varieties of mahlab seeds: oil, fatty acids, and amino acids. *Msc. thesis*. Khartoum north, Sudan: Sudan University of Science and Technology.

Elnour, A. H. (2006). Folk medicine practitioner. Wadmedani (personal communication, May 17, 2006).

Hedberg, I., & Stangard, F. (1989). Traditional medicine in Botswana. *Traditional medicine plants*. Sweden: Ipelegeng Publisher.

Mariod, A. A., Aseel, K. M., Mustafa, A. A., & Abdel-Wahab, S. I. (2009a). Characterization of the seed Oil and meal from *Monechma ciliatum* and *Prunus mahaleb* seeds. *Journal the American of Oil Chemist Society*, *86*, 749–755.

Mariod, A. A., Ibrahim, R. M., Ismail, M., & Ismail, N. (2010b). Antioxidant activity of the phenolic leaf extracts from *Monechma ciliatum* in stabilization of corn oil. *Journal of the American Oil Chemists Society*, *87*, 35–43.

Mariod, A. A., AL-Nageeb, G., & Ismail, M. (2010c). *Monechma ciliatum* methanolic extract regulates low-density lipoprotein receptor and 3-hydroxy-3-methylglutaryl coenzyme A reductase genes expression by Hep G2 cells. *African Journal of Biotechnology*, *9*(36), 5813–5819.

Mariod, A. A., Ibrahim, R. M., Ismail, M., & Ismail, N. (2010d). Antioxidant activities of phenolic rich fractions (PRFs) obtained from black mahlab (*Monechma ciliatum*) and white mahlab (*Prunus mahaleb*) seedcakes. *Food Chemistry*, *118*, 120–127.

Mohammed, T. M. Sh., & Elballa, M. M. A. (2015). The effect of agricultural practices on growth and yield of black mahlab (*Monechma ciliatum*). *Agricultural and Biological Sciences Journal*, *1*(2), 31–36.

Oshi, M. A. M., Abdelkareem, A. M., & Eltohami, M. S. H. (2013). Phytochemical and antimicrobial activity of *Monechma ciliatum* (black mahlab) seed extracts. *Journal of Herbal Drugs*, *4*(1), 15–22.

Sharief, M. T. (2002). The effect of sowing date and plant spacing on growth and yield of black mahlab (*Monechma ciliatum*). *M.Sc thesis*. Sudan: University of Khartoum.

Uguru, M., & Evans, F. (2000). Phytochemical and pharmacological studies on *Monechma ciliatum*. *Journal of Entho Pharmacology*, *73*(1–2), 289–292.

Uguru, M. O., Okwuasaba, F. K., Ekwenchi, M. M., & Uguru, V. E. (1995). Oxytocic and oestrogenic effect of *Monechma ciliatum* methanol extract in vivo and in vitro in rodents. *Journal of Entho Pharmacology*, *9*(1), 26–29.

Chapter 13

Nigella sativa L. Black Cumin

DISTRIBUTION AND BOTANICAL DESCRIPTION

In English, *N. sativa* and its seed are variously called black caraway, black cumin, fennel flower, nigella, nutmeg flower, Roman coriander, *kalonji* (from Hindi), and *Habuta alsawda* or *habuta lbaraka* (in Arabic) (Bharat, 2016). Black-cumin *N. sativa* is native to southern Europe, northern Africa, and southern Asia. It is widely grown for its flavorful seeds and leaves, used as a culinary spice in Asian and Middle Eastern cuisines. It is an annual spice; is a member of the Ranonculaceae family (El-Dakhakhny et al., 2000). The leaves are compound (made up of two or more discrete leaflets) there is one leaf per node along the stem, the edge of the leaf blade is entire (has no teeth or lobes), there are two or more ways to evenly divide the flower (the flower is radially symmetrical), there are five petals, sepals, or tepals in the flower, both the petals and sepals are separate and not fused. The fruit is dry and splits open when ripe Up to 1.2 mm (https://gobotany.newenglandwild.org/species/nigella/sativa). Flowers are usually pale blue and white in color with 5–10 petals. In its natural form, the flowers are bluish with a variable number of sepals and are characterized by the presence of nectars. The gynoecium is composed of a variable number of multiovule carpels, developing into a follicle after pollination, with single fruits partially connected to form a capsule-like structure. The fruit is a large and inflated capsule composed of three to seven united follicles, each containing seeds. The seeds of *N. sativa* are small in size (1–5 mm), with corrugated integuments (Benkaci-Ali, Baaliouamer, Meklati, & Chemat, 2007).

SEED COMPOSITION AND BIOACTIVE COMPONENTS

N. sativa seeds are a rich source of many important nutrients that appear to have a very positive effect on human health. They constitute a good alternative source of essential fatty acids compared to common vegetable oils and could contribute to the overall dietary intake of the mineral elements studied (Cheikh-Rouhou et al., 2007). *N, sativa* seeds contain 32%–53% (w/w) of fixed oil with 85% of total unsaturated fatty acid. *N. sativa* seeds from Egypt were reported to have 34.49%–38.72% of oil (Atta, 2003), from Saudi to have 38.2% (Al-Jassir, 1992) and from Turkey to have 40.6% oil. Mariod et al. (2012)

investigated two black cumin seeds originally produced in Ethiopia and Syria germinated at 2, 4, and 6 days under light and darkness. The proximate analysis of the raw Ethiopian Nigella seeds showed that moisture content (7.9%), oil content (29.1%), crude protein (26.1%), ash content (5.7%), and total carbohydrates (31.2%), which were slightly higher than the amounts of the raw Syrian variety. In Syrian black cumin seeds, the content of moisture (7.2%), oil (13.2%), crude protein (25.8%), fiber (5.8%), ash (4.3%), and total carbohydrates (43.7%) was affected by germination process, where germination in general increase both oil and protein content. The following results (on a dry-weight basis) were obtained for Tunisian and Iranian black seed varieties, respectively: protein 26.7% and 22.6%, oil 28.48% and 40.35%, ash 4.86% and 4.41%, and total carbohydrate 40.0% and 32.7% (Cheikh-Rouhou et al., 2007). These authors attributed such variations in nutrient concentrations among species and varieties to the variations of cultivated regions, storage conditions, and maturity stage (Atta, 2003). Physicochemical properties of the oils for Tunisian and Iranian varieties, respectively, include: saponification number 211 and 217, peroxide value 5.65 and 4.35, iodine index 120 and 101, and an acidity of 22.7% and 18.6% (Cheikh-Rouhou et al., 2007).

Black cumin seed meal is high in protein also high in amino acid content and seems to be a potential source of nutrients for food and feed. The meal contains higher nonessential amino acids (60%) which are glycine, proline, alanine, serine, glutamic acid, tyrosine, cysteine, and aspartic acid compared to essential amino acids (40%) which are methionine, arginine, glutamic acid, cysteine and aspartic acid.

Tekeli (2014) reported that, black cumin meal was found to contain a total of 23 amino acids (10 of which are essential) with a total value of 202 g/1000 g. Lysine of black cumin meal content was 6.41 g/kg. Additionally, black cumin meal is a rich source of methionine and threonine (respectively 5.49 g/kg and 10.75 g/kg). During seed germination the total amino acids increased and it affected by germination conditions. Germinated seeds showed noticeable decreases in the contents of K, Na, Ca, and Fe (Maiod et al., 2012). The black seed meal is a good source of minerals as it contains the most important minerals, for example, potassium (14.78 g/kg), phosphorus (2.36 g/kg), calcium (1.76 g/kg), magnesium, (1.78 g/kg), sulfur (1.39 g/kg), zinc (26.37 mg/kg), iron (204.46 mg/kg), copper (10.20 mg/kg), manganese (1.18 mg/kg), and boron 17.95 mg/kg (Tekeli, 2014).

Mariod, Ibrahim, Ismail, and Ismail (2009) investigated the antioxidant activities of crude methanolic extract (CME) and its fractions using ethyl acetate (EAF), hexane (HF), and water (WF) of black cumin seedcake. They used DPPH radical scavenging activity, β-carotene–linoleate bleaching, and inhibition of corn oil oxidation to evaluate the antioxidant capacity. They found the total phenolics as 78.8, 27.8, 32.1, and 12.1 mg gallic acid equivalents (GAE)/g in EAF, CME, WF, and HF, respectively. The CME and EAF exhibited the highest DPPH followed by WF and HF. The extract/fractions showed high effect on reducing the oxidation of β-carotene. These authors tested the effect of extract/ fractions on the oxidative stability of corn oil at 70°C in the dark and they compared that with butylated hydroxyanisole (BHA). They concluded that the oil peroxide and anisidine values were generally lower with addition of PRFs. The predominant phenolic compounds identified by HPLC–DAD in CME and WF of black cumin seedcake were hydroxybenzoic, syringic, and p-coumaric acid.

The major fatty acids in the seed oil of *N. sativa* were myristic (C14:0), palmitic (C16:0), and stearic (C18:0) as saturated fatty acids, whereas oleic (C18:1), linoleic (C18:2), and linolenic (C18:3) were the main unsaturated fatty acids. Little change was observed in fatty acids after germination as saturated and total monounsaturated fatty acids were decreased, while the total of polyunsaturated fatty acids was increased (Mariod et al., 2012; Nergiz & Otles, 1993). The crude *N. sativa* oil and its fractions (neutral lipids, glycolipids, and phospholipids) showed potent in vitro radical scavenging activity that is correlated well with their total content of polyunsaturated fatty acids, unsaponifiables, and phospholipids, as well as the initial peroxide values of crude oils (Suboh, et al., 2004). The percentage of oil extracted using chloroform: methanol was significantly higher compared to the percentage of oil extracted using hexane and petroleum ether in the three different samples. The amount of total lipid extracted was higher in the chloroform: methanol micelle (39.2% of seed fresh weight) than in the *n*-hexane extract (37.9%).The significant percentage of oil extracted using chloroform: methanol was due to the properties of chloroform: methanol which can extract both polar and nonpolar lipids (Ramadan & Morsel, 2002). This was expected since it was reported that the composition and nutritional values vary with the country of origin, stage of maturity, type of seeds, and subsequent storage conditions (Atta, 2003). The unsaturated fatty acids were the dominant fatty acids in black cumin seed oil which make up 80.9%. Major unsaturated fatty acids were oleic (21.5%) and linolenic (58%), on the other hand the major saturated fatty acids of *N. sativa* oil were palmitic 11.56%, stearic 2.55%. In addition arachidic (C20:0), behenic (C22:0), and lignoceric (C24:0) were detected as a saturated fatty acids as well as palmitoleic (C16:1) and erucic (C22:1n–9) were detected as monounsaturated fatty acids. In an another study by Atta (2003) showed that the oils of black cumin varieties contained oleic and linoleic acids at relatively high levels (18.9–20.1 and 47.5–49.0, respectively) but these are lower than those corresponding to the Tunisian variety (25.0 and 50.3, respectively) and to the Iranian variety (23.7 and 49.2, respectively). The source of variability may be genetic (plant cultivar, variety grown), seed quality (maturity, harvesting-caused damage, and handling/storage conditions), oil processing variables, or accuracy of detection, lipid extraction method, and quantitative techniques (Ramadan and Morsel, 2002). The ratio of linoleic acid to oleic acid was more than 2.5:1. These results are supported by previous findings by Houghton, Zarka, de las Heras, and Hoult (1995), who reported that the fixed oil of *N. sativa* had antioxidant effects greater than thymoquinone, which is its active constituent.

The amount of α-tocopherol was in the range of 120–290 mg/100 g, in Yemeni black cumin samples collected from three different states; there were significant differences ($P < 0.05$) in α-tocopherol content between the three different samples. Nergiz and Otles (1993) reported content of α-tocopherol as 40 µg/100 while Ramadan et al. (2003), reported a content of 0.284 g/kg, that may be due to the different conditions and methods used. The average mean concentrations (mg/kg fresh weight) of selenium, DL-α-tocopherol, DL-γ-tocopherol, all-*trans*-retinol, thymoquinone and thymol in tested (115 samples) black cumin seeds were 0.1770.10, 9.0274.84, 5.4273.96, 0.2770.27, 2224.49, 71629.50, and 169.357100.12, respectively. Matthaus and Özcan (2011) extracted oils from the seeds of 14 black cumin obtained from Germany and Turkey, they investigated tocopherols and sterols, they found it contained between 1.70–4.12 mg/100 g α-T, 0.97–4.51 mg/100 g γ-T, and 4.90–17.91 mg/100 g β-T3. The total tocopherol content in the studied seeds varied between 9.15 and 24.65 mg/100 g. They determined the compositions of the sterol fractions by gas liquid chromatography they reported the total amounts of sterols as ranged between 1993.07 and 2182.17 mg/kg. The main component was β-sitosterol (48.35%–51.92%), followed by 5-avenasterol, campesterol, and stigmasterol.

Black cumin fixed oil contains appreciable quantity of carotenoids around 88.95 ± 3.91 mg/kg oil. Total tocopherols contents were 361.71 ± 10.23 mg/kg oil while carotenoids and tocopherols collectively were 450.66 ± 16.21 mg/kg oil. Thymoquinone as the major bioactive compound in the fixed oil its content was 201.31 ± 13.17 mg/kg oil, it represents together with dihydrothymoquinone more than 28% of the total essential oils. This compound showed good effectiveness against various maladies like oxidative stress, cancer, immune dysfunction, and diabetic complications (Sultan et al. 2009). Thymoquinone is more concentrated in black cumin essential oil ranging from 18% to 24% (Burits & Bucar, 2000) whereas extraction of fixed oil by means of *n*-hexane retains meager quantity. Some organic extracts of black cumin seed also reported to possess appreciable amounts of thymoquinone. Singh et al. (2005) estimated thymoquinone contents of 11.8% in acetone extract seed. Moreover, thymoquinone belongs to class of compounds known as terpenoids; most members are volatile in nature and their heating losses ranged from 20% to 50% (Bendahou et al., 2008) that could also be a possible reason for lower quantity of thymoquinone in fixed oil.

The concentrations of these analytes were significantly affected by the country of origin of the *N. sativa* (Al-Saleh, Billedo, & El-Doush, 2006). Oils extracted from Tunisian and Iranian Nigella seeds were analyzed for sterol composition. These components could have many health benefits, especially in lowering LDL-cholesterol and preventing heart disease. The total unsaponifiable matter was found 15.58 oil and 14.82 g/kg oil in TNS and INS seed oils, respectively). The total sterol content in the two studied oils represented 17.41% and 42.66% of the unsaponifiable matter, respectively. β-sitosterol was the major sterol in the two oils with 44%, and 54% of the total sterols in Tunisian and Iranian Nigella seed oils, respectively. The next major sterol was stigmasterol in both Nigella seed oils studied (16.57%–20.92% of total sterols). TMS 484, D7-stigmasterol, D7-avenasterol, and cholesterol were detected at lower levels (Cheikh-Rouhou et al., 2008).

USES AND MEDICINAL VALUES

The seeds of *N. sativa* are used as a spice in Indian and Middle Eastern cuisines. The black seeds taste like a combination of onions, black pepper, and oregano. They have a pungent bitter taste and smell. The dry-roasted Nigella seeds flavor curries, vegetables, and pulses. It can be used as a "pepper" in recipes with pod fruit, vegetables, salads, and poultry. In some cultures, the black seeds are used to flavor bread products. It is also used as part of the spice mixture *panch phoron* (meaning a mixture of five spices) and by itself in many recipes in Bengali cuisine and most recognizably in *naan* bread. *Nigella* is also used in Armenian string cheese, a braided string cheese called *majdouleh* or *majdouli* in the Middle East (Bharat, 2016). *N. sativa* seeds are used for edible and medicinal purposes. The seeds are sold in the markets to be used as a spice, condiment, additive on bread, for cheese and as a native medicine.

The seeds have traditionally been used in Middle Eastern folk medicine as a natural remedy for various diseases as well as a spice for over 2000 years. *N. sativa* had a rich historical and religious background. Seeds or their oils were claimed for medicinal applications dating back to the ancient Egyptians, Greeks, and Romans (Benkaci-Ali et al., 2007). In Islamic literature, *N. sativa* is considered as one of the greatest forms of healing medicine. It has been recommended for use on regular basis in *Al-Tibb-Al-Nabwi* (prophetic medicine). *N. sativa* seeds have been utilized for the treatment of asthma, cough, bronchitis, headache, rheumatism, fever, influenza, eczema, as a diuretic, lactagogue, and vermifuge (Burits & Bucar, 2000; Ramadan, 2007)(Fig. 13.1).

Recently, researchers have taken interest in the *N. sativa* seeds in different forms: the seed extract, its oil and its volatile substances. The seeds have been subjected to have a range of pharmacological, phytochemical, and nutritional investigations in recent years. Human studies and laboratory studies on the seeds and its oil have provided scientific support for the traditional uses of the seed and its oil for the treatment of rheumatism, immune stimulation, diabetic, cancer, and related inflammatory disease (Ramadan & Morsel, 2002).

Ramadan (2013) prepared healthy blends of high linoleic sunflower oil with *N. sativa* oil and he later studied the functionality, stability, and antioxidative characteristics of the prepared blends. By increasing the amounts of *N. sativa* oil in sunflower oil, linoleic acid level decreased, while tocols increased. Inverse relationships were noted between peroxide value as well as anisidine value and oxidative stability at termination of storage under accelerated oxidative conditions. Levels of conjugated dienes and conjugated trienes in sunflower oil and blends

FIGURE 13.1   *Nigella sativa* **plant and seeds.**

increased with increase in time. *N. sativa* oil blends gave about 70% inhibition of DPPH% radicals. Previously performed clinical and experimental investigations have shown that *N. sativa* has a protective effect against oxidative damage in isolated rat hepatocytes (Daba & Abdel-Rahman, 1998). In addition, it was suggested that *N. sativa* treatment increases the antioxidants defense system activity in experimentally induced diabetic rabbits. Sultan et al. (2009) showed that black cumin fixed and essential oils are rich source of phytochemicals and can be utilized against lifestyle disorders like hyperglycemia and hypercholesterolemia. Their interesting study on in vitro antioxidant capacity indicated that fixed and essential oils of black cumin seed inhibited lipid peroxidation by 25.62 and 92.56% and 2,2-diphenyl-1-picrylhydrazyl (DPPH) radical scavenging activity by 32.32 and 80.25%, respectively.

REFERENCES

Al-Jassir, M. S. (1992). Chemical composition and microflora of black cumin (*Nigella sativa* L.) seeds growing in Saudi Arabia. *Food Chemistry*, *45*, 239–242.

Al-Saleh, I. A., Billedo, G., & El-Doush, I. I. (2006). Levels of selenium, DL-a-tocopherol, DL-g-tocopherol, all-*trans*-retinol, thymoquinone and thymol in different brands of *Nigella sativa* seeds. *Journal of Food Composition and Analysis*, *19*, 167–175.

Atta, M. B. (2003). Some characteristics of nigella (*Nigella sativa* L.) seed cultivated in Egypt and its lipid profile. *Food Chemistry*, *83*, 63–68.

Bendahou, M., Muselli, A., Grignon-Dubois, M., Benyoucef, M., Desjobert, J. M., Bernardini, A. F., & Costa, J. (2008). Antimicrobial activity and chemical composition of *Origanum glandulosum* Desf. Essential oil and extract obtained by microwave extraction: comparison with hydrodistillation. *Food Chemistry*, *106*, 132–139.

Benkaci-Ali, F., Baaliouamer, A., Meklati, B. Y., & Chemat, F. (2007). Chemical composition of seed essential oils from Algerian *Nigella sativa* extracted by microwave and hydrodistillation. *Flavour and Fragrance Journal*, *22*, 148–153.

Aggarwal, B. B. (2016). *Molecular targets and therapeutic uses of spices* (p. 259). Google Books.

Burits, M., & Bucar, F. (2000). Antioxidant activity of *Nigella sativa* essential oil. *Phytotherapy Research*, *14*, 323–328.

Cheikh-Rouhou, S., Besbes, S., Hentati, B., Blecker, Ch., Deroanne, C., & Attia, H. (2007). *Nigella sativa* L.: chemical composition and physicochemical characteristics of lipid fraction. *Food Chemistry*, *101*, 673–681.

Cheikh-Rouhou, S., Besbes, S., Hentati, B., Blecker, Ch., Deroanne, C., & Attia, H. (2008). Sterol composition of black cumin (*Nigella sativa* L.) and Aleppo pine (*Pinus halepensis* Mill.) seed oils. *Journal of Food Composition and Analysis*, *21*(2008), 162–168.

Daba, M. H., & Abdel-Rahman, M. S. (1998). Hepatoprotectice activity of thymoquinone in isolated rat hepatocytes. *Journal of Toxicology Letters*, *16*, 23–29.

El-Dakhakhny, M., Barakat, M., & El-Halim, M. A. (2000). Effect of *Nigella sativa* oil on gastric secretion and ethanol-induced ulcer in rats. *Journal of Ethnopharmacology*, *72*, 299–304.

Houghton, P. J., Zarka, R., de las Heras, B., & Hoult, J. R. S. (1995). Fixed oil of *Nigella sativa* and derived thymoquinone inhibit eicosanoid generation in leukocytes and membrane lipid peroxidation. *Planta Medica*, *61*, 33–36.

Mariod, A. A., Ibrahim, R. M., Ismail, M., & Ismail, N. (2009). Antioxidant activity and phenolic content of phenolic rich fractions obtained from black cumin (*Nigella sativa*) seedcake. *Food Chemistry*, *116*, 306–312.

Mariod, A. A., Edris, Y. A., Cheng, S. F., & Abdelwahab, S. I. (2012). Effect of germination periods and conditions on chemical composition, fatty acids and amino acids of two black cumin seeds. *Acta Scientiarum Polonorum, Technologia Alimentaria*, *11*(4), 401–410.

Matthaus, B., & Özcan, M. M. (2011). Fatty acids, tocopherol, and sterol contents of some *Nigella* species seed oil Czech. *Journal of Food Science*, *29*(2), 145–150.

Nergiz, C., & Otles, S. (1993). Chemical composition of *Nigella sativa* L. seeds. *Food Chemistry*, *48*, 259–261.

Ramadan, M. F. (2007). Nutritional value, functional properties and nutraceutical applications of black cumin (Nigella sativa L.) oilseeds: an overview. *International Journal of Food Science & Technology*, *42*, 1208–1218.

Ramadan, M. F. (2013). Healthy blends of high linoleic sunflower oil with selected cold pressed oils: functionality, stability, and antioxidative characteristics. *Industrial Crops Products*, *43*, 65–72.

Ramadan, M. F., & Morsel, J. T. (2002). Characterization of phospholipid composition of black cumin (*Nigella sativa* L.) seed oil. *Nahrung/Food*, *46*, 240–244.

Ramadan, M. F., & Morsel, J. -T. (2003). Analysis of glycolipids from black cumin (Nigella sativa L.), coriander (Coriandrum sativum L.) and niger (Guizotia abyssinica Cass.) oilseeds. *Food Chemistry*, *80*, 197–204.

Singh, N., Verma, M., Mehta, D., & Mehta, B. K. (2005). Two new lipid constituents of Nigella sativa (Seeds). *Industrial Journal of Chemistry*, *44*, 742–744.

Suboh, S. M., Bilto, Y. Y., & Aburjai, T. A. (2004). Protective effects of selected medicinal plants against protein degradation, lipid peroxidation, and deformability loss of oxidatively stressed human erythrocytes. *Phytotherapy Research*, *18*, 280–284.

Sultan, M. T., Butt, M. S., Anjum, F. M., Jamil, M., Akhtar, S., & Nasir, M. (2009). Nutritional profile of indigenous cultivar of Black cumin seeds and antioxidant potential of its fixed and essential oil. *Pakistan Journal of Botany*, *41*(3), 1321–1330.

Tekeli, A. (2014). Nutritional Value of Black Cumin (Nigella Sativa) Meal as an Alternative Protein Source in Poultry Nutrition. *Journal of Animal Science Advances*, *4*(4), 797–806.

Chapter 14

Cucurbits, *Cucurbita* Species As New Oil Sources

Cucumis mello VAR. *agrestis* MUSKMELON

Distribution and Botanical Description

Distribution

Throughout the Old-World tropics, adventive in the Neotropics, the plant is cultivated in many African regions. Thin-stemmed, monoecious plants growing as weeds in African and Asian countries. Very small less than 5 cm inedible fruits with very thin mesocarp and tiny seeds (Stepansky, Kovalski, & Perl-Treves, 1999). A much branched prostrate annual or perennial climber growing up to 1.5 m. Stem generally covered with scabrous hairs, leaves triangular, ovate, 3–5(–7)-lobed, scabrid with rigid hairs: petiole 1–6 cm long (http://www.efloras.org). Flowers are small, yellow, solitary, or rarely in pairs or threes. Flower clusters are carried on stalk 5–10 mm long, densely hairy. Sepal is about 1.5 mm long, subulate. Flowers are 6–8 mm long, with petals ovate-oblong. Fruit is ellipsoid, oval-round, sometimes obscurely trigonous, smooth, and hairless, 4 × 2.5 cm, generally with dark green stripes, looks like a miniature watermelon. Fine powder of dried "Kachri" is the best natural meat tenderizer. Minced meat for *Seekh-kebabs* are mixed with Kachri powder, and left for 4–6 h to make it tender. The fresh fruits are also made into chutney, which has a tangy taste (http://www.flowersofindia.net/catalog/slides/Wild%20 Melon.html).

 C. melo L. var. *agrestis*, the snake melon, is a very popular salad plant that widely distributed in Sudan and used as a salad crop instead of cucumber in Western states where it is known as "Agoor" the fruit of which contains some amounts of carbohydrates, minerals, and vitamins, the seeds contain between 12.5% and 39.1% oil (Mariod & Matthaeus, 2008) (Fig. 14.1).

Fruit and Seed Composition and Bioactive Components

Genetic diversity of 17 wild melon (*C. melo* var. *agrestis*) genotypes was evaluated using AFLP markers. Out of 564, 386 studied primers amplified a total of bands, out of which (68.4%) were polymorphic. The study indicated differences

Unconventional Oilseeds and Oil Sources. http://dx.doi.org/10.1016/B978-0-12-809435-8.00014-7

FIGURE 14.1   *Cucumis melo* **var.** *agrestis* **plant, fruits and seeds.**

between the *C. melo* var. *agrestis*. The different estimates of genetic variability employed indicated that the wild melon genotypes had average variability. Thus, these populations were genetically similar. The study showed that the average variation observed, due to the limited number of samples of wild melons, is not accurately assessing the amount of variation within the population estimate. The relatively high genetic diversity observed may be attributed to several factors, for example, nature of melon reproduction (cross-pollination), the high differentiation ability of AFLP markers and the existence of several undistinguished subspecies (Shamasbi, Dehestani, & Golkari, 2014).

Phytochemical study of aqueous, and ethanolic extracts of the fruits reveals that carbohydrates, tannins, flavonoids, alkaloids, saponins, steroids, triterpenoids, and glycosides were present (Alagar et al., 2015). The seeds of *C. melo* var. *agrestis* contain 59.46% oil, 30.4% carbohydrate, and 3.89% protein, the seed oil is edible and nonrancid with free fatty acid value of 1.94%, peroxide value 8.0 mEq/kg, iodine value 10.5, saponification value of 193 mEq/kg (Adekunle & Oluwo, 2008). Mariod and Matthaeus (2008) investigated two Sudanese *C. melo* var. *agrestis* samples, they reported 30%–31% oil content, medium amounts of tocopherols, 39.9 mg/100 g oil. The main tocopherol of the two samples was γ-tocopherol, which represented around 80.7% of the total tocopherols, followed by α-tocopherol, which represented around

8.2 mg/100 g oil. The two samples oil contained 3879.0 mg/kg oil as sterols with β-sitosterol as the main with a remarkably high amount (10.8%) of Δ5-avenasterol, which is a well-known antioxidant and as an antipolymerization agent in frying oils (Savage, Mcneil, & Dutta, 1997). Alkaline protease was isolated, partially purified and characterized from the seeds of *C. melo* var. *agrestis*, by a four step purification process. The enzyme showed maximum activity at pH 9.0 and optimum temperature at 40°C with casein as substrate. The enzyme exhibited homogeneity as attested by a single protein band on both native PAGE and SDS PAGE. It is a monomeric enzyme and nonglycoprotein in nature. The k_m value of the isolated enzyme for casein as determined by double reciprocal plot was 2.5 mg/mL. The results indicated that the alkaline protease was a serine protease, similar to cucumisin from sarcocarp of melon fruit (Devi & Hemalatha, 2014).

Uses and Medicinal Value

This plant has medicinal use of common medicine exclusively for curing disease. The plant, fruit is used as vomitive but on a few amount of honey is a tonic for the stomach. Paste of crushed grain of the plant by *Cynodon dactylon* juice is used for curing and removing herpes grains and boils (Adekunle & Oluwo, 2008). The plant is cultivated in many African regions for its edible kernels used as a soup thickener. The plant, an annual, and romonoecious, trailing-vine species, is of high social, cultural, and economic value for local communities preferentially during popular fetes and prestigious ceremonies (Kouonon et al., 2009). Methanolic and ethanolic extracts of mature and immature fruit of *C. melo* by disc and well diffusion methods have an effect on *Aspergillus flavus*. By increasing the extract amount, antigrowth halo diameter around the disc and well was increased. The antifungal effect of ethanolic extract is more than that of methanolic one (Asadi, Azizi, & Yahyayi, 2012).

C. *melo* var. *agrestis* seed methanolic extract was found to have significant scavenging activity 75.5% using the DPPH method and 69.8% using the hydrogen peroxide method, further the extract showed maximum (70.6%) analgesic activity of acetic acid induced writhing method (Arora, Kaur, & Gill, 2011). Oral administration of *C. melo agrestis* fruit extract (CMFE) and both water fraction (CMWF) and hexane fraction (CMHF) reduced the total cholesterol, triglycerides, low-density lipoprotein cholesterol, and very low-density lipoprotein cholesterol levels in high fat diet-fed dyslipidemic hamsters. CMHF also modulated expression of genes involved in lipogenesis, lipid metabolism, and reverse cholesterol transport. Standard biochemical diagnostic tests suggested that neither of fractions cause any toxicity to hamster liver or kidneys. CMFE and CMHF also decreased oil-red-O accumulation in 3T3-L1 adipocytes (Shankar et al., 2015). The fruits of *C. melo agrestis* are used by the local people of the Nara Desert, Pakistan as a condiment for the treatment of constipation (Qureshi & Bhatti, 2008).

REFERENCES

Adekunle, A. A., & Oluwo, O. A. (2008). The nutritive value of *Cucumis melo* var. *agrestis* scrad (Cucurbitaceae) seeds and oil in Nigeria. *American Journal of Food Technology, 3*(2), 141–146.

Alagar, R. M., Sahithi, G., Vasanthi, R., Banji, D., Rao, K. N. V., & Selvakumar, D. (2015). Study of phytochemical and antioxidant activity of *Cucumis melo* var. *agrestis* fruit. *Journal of Pharmacognosy and Phytochemistry, 4*(2), 303–306.

Al-Khalifa, A. S. (1996). Physicochemical characteristics, fatty acid composition, and lipoxygenase activity of crude pumpkin and melon seed oils. *Journal of Agricultural and Food Chemistry, 44,* 964–966.

Arora, R., Kaur, M., & Gill, N. S. (2011). Antioxidant activity and pharmacological evaluation of *Cucumis melo* var. *agrestis* methanolic seed extract. *Research Journal of Phytochemistry, 5,* 146–155.

Asadi, M., Azizi, I. G., & Yahyayi, F. (2012). Antifungal effects of mature and immature fruit extracts of *Cucumis melo* L. on *Aspergillus flavus. Global Veterinaria, 8*(4), 347.

Devi, B. G., & Hemalatha, K. P. J. (2014). Isolation, partial purification and characterization of alkaline serine protease from seeds of *Cucumis melo* var. *agrestis. International Journal of Research in Engineering and Technology, 3*(6), 88–95.

Jeffrey, C. (1990). Appendix: an outline classification of the Cucurbitaceae. In D. M. Bates, R. W. Robinson, & C. Jeffrey (Eds.), *Biology and utilization of the Cucurbitaceae* (pp. 449–463). Ithaca: Comstock, Cornell University Press.

Kouonon, L. C., Jacquemart, A. -L., Arsene, I., Zoro, Bi, Bertin, P., Baudoin, J. -P., & Dje, Y. (2009). Reproductive biology of the and romonoecious *Cucumis melo* subsp. *Agrestis* (Cucurbitaceae). *Annals of Botany, 104*(6), 1129–1139.

Mariod, A., & Matthaeus, B. (2008). Investigations on fatty acids, tocopherols, sterols, phenolic profiles and oxidative stability of *Cucumis melo* var. *agrestis* oil. *Journal of Food Lipids, 15*(1), 42–55.

Qureshi, R., & Bhatti, G. R. (2008). Ethnobotany of plants used by the Thari people of Nara Desert. *Pakistan/Fitoterapia, 79,* 468–473.

Rahman, A. H. M. M. (2013). Study of species diversity on Cucurbitaceae family at Rajshahi division, Bangladesh. *Journal of Plant Sciences, 1*(2), 18–21.

Savage, G. P., Mcneil, D. L., & Dutta, P. C. (1997). Lipid composition and oxidative stability of oils in hazelnuts (*Corylus avellana* L.) grown in New Zealand. *Journal of American Oil Chemists' Society, 74,* 755–759.

Shamasbi, F. V., Dehestani, A., & Golkari, S. (2014). AFLP based genetic diversity assessment among *Cucumis melo* var. *agrestis* genotypes in Southern Caspian coastline. *International Journal of Biosciences, 4*(12), 54–61.

Shankar, K., Singh, S. K., Kumar, D., Varshney, S., Gupta, A., Rajan, S., Srivastava, A., Beg, M., Srivastava, A. K., Kanojiya, S., Mishra, D. K., & Gaikwad, A. N. (2015). *Cucumis melo* ssp. *agrestis* var. *agrestis* ameliorates high fat diet induced dyslipidemia in Syrian golden hamsters and inhibits adipogenesis in 3T3-L1 adipocytes. *Pharmacognosy Magazine, 11*(44), 501–510.

Sharma, O. P. (2004). *Plant Taxonomy.* New Delhi, India: Tata McGraw-Hill Publishing Company Limited (pp. 18–42).

Stepansky, A., Kovalski, I., & Perl-Treves, R. (1999). Intraspecific classification of melons (*Cucumis melo* L.) in view of their phenotypic and molecular variation. *Plant Systematics and Evolution, 217,* 313–333.

Chapter 15

Cucumis melo var. *flexuosus* The Armenian Cucumber or Snake Melon

DISTRIBUTION AND BOTANICAL DESCRIPTION

The origin of *Cucumis melo* var. *flexuosus* is known to lie in southeastern Anatolia, Azerbaijan, Iraq, Palestine, and Central Asia. Today snake melon is grown mostly at home and market gardens in many countries; hence, the annual production or area sown in different countries is not accurately known. *C. melo* var. *flexuosus* is a very important vegetable throughout Northern Africa, the Middle East, and India (Walters & Thieret, 1993). The Armenian cucumber is a type of long, slender fruit, which tastes like a cucumber and looks somewhat like a cucumber inside. It is actually a variety of muskmelon (*C. melo*), a species closely related to the cucumber (*Cucumis sativus*). It is also known as the yard-long cucumber, snake cucumber, snake melon. The skin is very thin, light green, and bumpless. It has no bitterness and the fruit is almost always used without peeling. It grows approximately 30–36 in. long. It grows equally well on the ground or on a trellis. The plants prefer to grow in full sun for most of the day. The fruit is most flavorful when it is 12–15 in. long (https://en.wikipedia.org/wiki/Armenian_cucumber). *C. melo* var. *flexuosus* is an annual climber growing up to 1.5 m (5 ft.). It flowers from July to September, and the seeds ripen from August to October. The flowers are monoecious (individual flowers are either male or female, but both sexes can be found on the same plant) and are pollinated by insects. The plant is self-fertile (Facciola, 1998). *C. melo* var. *flexuosus* differs from the other varieties of melon by its taxonomy. It is characterized by an important leaf area with dark green color, elongate fruits with dark green skins, green flesh with an intermediate acidity and whitish seeds (Henane, Ben Slimane, & Jebari, 2015) (Fig. 15.1).

FIGURE 15.1 *Cucumis prophetarum* fruit and seeds.

FRUIT AND SEED COMPOSITION AND BIOACTIVE COMPONENTS

Ueda, Tanaka, and Kato (1986) investigated the fruit growth of *C. melo* L. var. *flexuosus* and the plant hormones contained in its immature fruit. The fruit growth started 5 days after pollination and its rapid growth continued for about 10 days. During this period the growth rate (length) was 9 cm/day. The final size of the fruit was about 120 cm in length and 6 cm in width, 25 days after pollination. The cell number of the fruit increased to more than twice that of the fruitlet before pollination. The increase started immediately after pollination and stopped at 10 days after pollination. On the other hand, no change in cell size was observed during the first 7 days after pollination. After this period, rapid growth started and continued to the end of the fruit growth. The cell size increased to more than 7 times that of the fruitlet before pollination (Ueda et al., 1986). Mariod et al. (2009) investigated the fruit characteristics of *C. melo* var. *flexuosus*; they reported 65 g fruit weight, 30.0 cm fruit length, and 10.0 cm fruit diameter. They found that the weight of dried seeds was 19.5 g, while the weight of dried pulp was 45.5 g. These authors reported the dried pulp of *C. melo* var. *flexuosus* represented more than 70% of its original weight. These authors easily separated the seeds of *C. melo* var. *flexuosus* from the stringy pulp to which they are attached. They noticed a light fermentation for 24–72 h of the wetted seeds was useful to clean the seeds of pulp. Fresh, wet seeds of *C. melo* var. *flexuosus* are sometimes chewed without further processing. They can also be toasted, with or without light salting, or they can be cooked into soups with or without removing hull (Martin, 1984).

The *C. melo* var. *flexuosus* seeds showed 15.75% as protein content, which is slightly less than protein levels in most oily seeds, carbohydrate as 16.19, and ash as 5.69 g/100 g, respectively. The oil content of *C. melo* var. *flexuosus* is 22.33% of the seed weight, which proposed this plant as a potential source of oil. The oil obtained from the seed of *C. melo* var. *flexuosus* was odorless, of good color, and of good appearance with good physicochemical properties as refractive index (40 + 1°C) is 1.437, the relative density (30 ± 1°C) is 0.920, the unsaponifiable matter is 1.2%, and iodine value is 111.5 (Mariod et al., 2009). Seed storage proteins of *C. melo* var. *flexuosus* were analyzed for variation in polypeptide patterns and proportion of four protein fractions; these are salt-soluble globulins, glutelins, albumins, and prolamins. SDS-PAGE of total protein extracts of 11 lines of *C. melo* representing different geographic regions showed considerable variation in their polypeptides in the range of molecular weights 50–55, 35–39, 20–26, and 16–19 kDa (Singh & Matta, 2008).

The fatty acid compositions of *C. melo* var. *flexuosus* seed oil shows the principal fatty acids to be palmitic (12.9%), stearic (6.0%), oleic (19.4%), and linoleic (61.4%) acids. Palmitic and stearic acids composition in this oil are higher than their content in the commercial oils. However, oleic acid content in the *C. melo* var. *flexuosus* seed oil is lower than its content in the commercial oils (Mariod et al., 2009). Palmitic ($C16:0$), oleic ($C18:1n-9$), and linoleic ($C18:2n-6$) acids were sequentially the highest in concentration followed by stearic acid ($C18:0$) at less than 8% and myristic ($C14:0$), palmitoleic ($C16:1n-7$), margaric ($C17:0$), arachidic ($C20:0$), and erucic ($C22:1n-9$) acids at an even lower content ($<1\%$) (Kaymak, 2012).

Tocopherols and tocotrienols are fat-soluble antioxidants but also seem to have many other functions in the body. α-Tocopherol is the main source found in supplements and in the European diet, where the main dietary sources are olive and sunflower oils (Wagner, Kamal-Eldin, & Elmadfa, 2004), while γ-tocopherol is the most common form in the American diet due to a higher intake of soybean and corn oil (Wagner et al., 2004; Jiang, Christen, Shigenaga, & Ames, 2001). The crude oil of *C. melo* var. *flexuosus* had medium amounts (34.7 mg/g oil) of tocopherols when compared with other common oils, such as sesame oil, groundnut oil, or sunflower oil, in which the amount of tocopherols ranged from 17.0 to 152 mg/100 g. The main tocopherol of the *C. melo* var. *flexuosus* oil was γ-tocopherol, which represents 96.5% of the total tocopherols, followed by δ-tocopherol, which represent only 2.3% (Mariod et al., 2009). Kaymak (2012) reported that the simple correlation coefficients and stepwise multiple regression analysis showed that the low or high amount of fatty acids, such as myristic ($C14:0$), oleic ($C18:1n-9$), linoleic ($C18:2n-6$), and arachidic

(C20:0) acids in *C. melo* var. *flexuosus* seed might play a major role on germination percentage and speed.

USES AND MEDICINAL VALUE

Fruits of *C. melo* var. *flexuosus*—raw or cooked have a flavor rather like cucumber and are very refreshing when eaten raw in hot weather. They can also be added to curries, cooked, preserved, or pickled. Seaweed is rich in oil with a nutty flavor, but very fiddly to use because the seed is small and covered with a fibrous coat. The seed contains between 12.5% and 39.1% oil. An edible oil is obtained from the seed (Facciola, 1998). The fruits can be used as a cooling, light cleanser, or moisturizer for the skin. They are also used as a first aid treatment for burns and abrasions. The flowers are expectorant and emetic. The fruit is stomachic. The seed is antitussive, digestive, febrifuge, and vermifuge. When used as a vermifuge, the whole seed complete with the seed coat is ground into a fine flour, then made into an emulsion with water and eaten. It is then necessary to take a purge in order to expel the tapeworms or other parasites from the body. The root is diuretic and emetic (Duke & Ayensu, 1985).

REFERENCES

Duke, J. A., & Ayensu, E. S. (1985). *Medicinal plants of China*. Algonac, MI: Reference Publications.

Facciola, S. (1998). *Cornucopia II*. Vista, CA: Kampong Publications.

Henane, I., Ben Slimane, R., & Jebari, H. (2015). SSR-based genetic diversity analysis of Tunisian varieties of melon (*Cucumis melo* L.) and Fakous (*Cucumis melo* var. *flexuosus*). *International Journal of Advanced Research*, *3*(3), 727–734.

Jiang, Q., Christen, S., Shigenaga, M. K., & Ames, B. N. (2001). Gamma-tocopherol, the major form of vitamin E in the US diet, deserves more attention. *The American Journal of Clinical Nutrition*, *74*(6), 714–722.

Kaymak, H. C. (2012). The relationships between seed fatty acids profile and seed germination in cucurbit species. *Žemdirbyst [Agriculture]*, *99*(3), 299–304.

Mariod, A. A., Ahmed, Y. M., Matthaeus, B., Khaleel, G., Siddig, A., Gabra, A. M., & Abdelwahab, S. I. (2009). A comparative study of the properties of six Sudanese cucurbit seeds and seed oils. *Journal of the American Oil Chemists' Society*, *86*, 1181–1188.

Martin, F.W. (1984). Cucurbits seed as possible oil and protein sources, Echo Technical Note. Available from http://www.echonet.org.

Singh, N. P., & Matta, N. K. (2008). Variation studies on seed storage proteins and phylogenetics of the genus *Cucumis*. *Plant Systematics and Evolution*, *275*(3), 209–218.

Ueda, J., Tanaka, K., & Kato, J. (1986). Plant growth regulators in *Cucumis melo* L. var. *flexuosus* Naud fruit during rapid growth. *Plant and Cell Physiology*, *27*(5), 809–818.

Wagner, K. -H., Kamal-Eldin, A., & Elmadfa, I. (2004). Gamma-tocopherol—an underestimated vitamin? *Annals of Nutrition and Metabolism*, *48*(3), 169–188.

Walters, T. W., & Thieret, J. W. (1993). The snake melon (*Cucumis melo*; Cucurbitaceae). *Economic Botany*, *47*, 99–100.

Chapter 16

Cucumis sativus Cucumber

DISTRIBUTION AND BOTANICAL DESCRIPTION

The cucumber is originally from Southern Asia, but now grows on most continents. In ancient time it was produced in ancient Thrace, and it is certainly part of modern cuisine in Bulgaria and Turkey, parts of which make up that ancient state. From India, it spread to Greece and Italy, and later in China. The Ancient Greeks grew cucumbers, and there were different varieties in Italy, Africa, and Moesia (Renner, Schaefer, & Kocyan, 2007).

The cucumber originated in India, where a great many varieties have been observed from *Cucumis hystrix*. It has been cultivated for at least 3000 years and was probably introduced to other parts of Europe by the Greeks or Romans. Records of cucumber cultivation appear in France in the 9th century, England in the 14th century, and in North America by the mid-16th century (Doijode, 2001).

Cucumis sativus is an annual climber growing up to 2 m (6 ft. 7 in.). It is a widely cultivated plant in the gourd family, (Cucurbitaceae). It is a creeping vine that bears cylindrical fruits that are used as culinary vegetables. There are three main varieties of cucumber: *slicing*, *pickling*, and *burpless*. Within these varieties, several different cultivars have emerged. The cucumber is a creeping vine that roots in the ground and grows up on trellises or other supporting frames, wrapping around the supports with thin, spiraling tendrils. The plant has large leaves that form a canopy over the fruit. The fruit of the cucumber is roughly cylindrical, elongated with tapered ends, and may be as large as 60 cm (24 in.) long and 10 cm (3.9 in.) in diameter. Having an enclosed seed and developing from a flower, botanically speaking, cucumbers are classified as pipes, a type of botanical berry (Nonnecke, 1989). The flowers are monoecious and are pollinated by insects. The plant is self-fertile (Chiej, 1984). The fruit varies widely in size between cultivars, but can be up to 1-m long. The seeds are rich in oil with a nutty flavor, but very fiddly to use because the seed is small and covered with a fibrous coat. Young leaves and stems are cooked as a potherb (Usher, 1974) (Fig. 16.1).

Unconventional Oilseeds and Oil Sources. http://dx.doi.org/10.1016/B978-0-12-809435-8.00016-0

FIGURE 16.1 *Cucumis sativus* **plant, fruit and seeds.**

FOOD USES AND MEDICINAL VALUES

The cucumber is a common ingredient of salads, being valued mainly for its crisp texture and juiciness. However, it is very watery, with little flavor and is not very nutritious. Much like tomatoes and squash they are often also perceived, prepared, and eaten as vegetables. Cucumbers are usually more than 90% water. It is widely consumed fresh in salads or fermented (pickles) or as a cooked vegetable (Sotiroudis, Sotiroudis, & Chinou, 2010). The fruits are sweet, refrigerant, haemostatic, and tonic. Traditionally cucumber is used for the wide spectrum of cure in rural and urban areas to remove general debility, for treatment of skin problems, and as a cooling agent. Several pharmacological activities including the antioxidant, antiwrinkle, antimicrobial, antidiabetic, and hypolipidemic potentials have been reported with this plant. The seeds are useful for quitting burning sensation, constipation, tonic, and intermittent fevers (Warrier, 1994). The methanolic extract of *C. sativus* seeds possessed significant ulcer potential, which could be due to the antioxidant activity (Gill et al., 2009). Many people find the cucumber fruit to be indigestible; this is due to the high cellulose content (Chiej, 1984). The leaf juice is emetic; it is used to treat dyspepsia in children. The fruit is depurative, diuretic, emollient, purgative, and resolvent. The fresh fruit is used internally in the treatment of blemished skin, heat rash, etc., while it is used externally as a poultice for burns, sores, etc. and also as a cosmetic for softening the skin (Duke & Ayensu, 1985). The seed is cooling, diuretic, tonic, and vermifuge. Thoroughly ground seeds (25–50 g) (including the seed coat) are a standard dose as a vermifuge and usually needs to be followed by a purgative to expel the worms from the body (Grieve, 1998). A decoction of the root is diuretic.

Cucumber skins have been shown to repel cockroaches in laboratory experiments (Duke & Ayensu, 1985). The fruit is applied to the skin as a cleansing cosmetic to soften and whiten it. The juice is used in many beauty products (Grieve, 1998; Chiej, 1984). Smitha, Santhi, Prasad, and Manonmani (2012) studied the potential of *C. sativus* solid waste, to be a low-cost adsorbent for removing crystal violet and rhodamine B from aqueous solutions. They proved the use of *C. sativus* waste and sulfuric acid *C. sativus* waste as an eco-friendly adsorbent with an economical treatment method. These authors revealed that the adsorption was dependent on the initial pH, contact time, and dye concentration. *C. sativa* (cucumber) is a widely cultivated plant of gourd family, which is eaten in the unripe, green form. Its fruit extract has shown free radical scavenging and analgesic activities in mice (Kumar et al., 2010), carminative and antacid property (Sharma, Dwivedi, & Paliwal, 2012). Studies of Gill et al. (2009) have shown the antioxidant and antiulcer effect of *C. sativa* in rats. The acetone extract of *C. sativus* seed possessed significant antiinflammatory properties as a therapeutic agent in acute as well as chronic inflammatory conditions (Vetriselvan, Subasini, Velmurugan, Muthuramu, & Revathy, 2013). The results of Sood, Kaur, and Gupta (2012) revealed the antimicrobial activity of cucumber seed extracts against *Serratia marcescens, Escherichia coli, Streptococcus thermophilus, Fusarium oxysporum, Trichoderma reesei*, while these extracts showed no inhibition against *Aspergillus niger*.

Kaur, Aggarwal, and Javed (2014) studied the effect of addition of sodium benzoate, potassium metabisulfite, and their combination on the physicochemical and phytochemical parameters, antioxidant activity, and the shelf life of cucumber juice. They observed a very slight but nonsignificant change in the physicochemical parameters like TS, TSS, acidity, and color values. While there is a significant ($P \leq 0.05$) change in vitamin C, total phenols, and antioxidant activity.

FRUIT AND SEED COMPOSITION AND BIOACTIVE COMPONENTS

Oil from *C. sativus* seed is said to resemble olive oil; it is used in salad dressings and French cooking. The oil contains 22.3% linoleic acid, 58.5% oleic acid, 6.8% palmitic acid, and 3.7% stearic acid (Duke & Ayensu, 1985). Cucumber seed oil has four main fatty acids: linoleic acid, C18:2 (61.6%); oleic acid, C18:1 (15.7%); stearic acid, C18:0 (11.1%); and palmitic acid, C16:0 (10.7%). The chemical properties of the oil are similar to those of corn, cottonseed, sesame, and sunflower seed oils, suggesting its potential use as good table and cooking oil which can increase HDL and reduce serum cholesterol and LDL levels, hence could help prevent cardiovascular illnesses. They could also be used for making mayonnaise and soap. Cucumber seed contains more than 53.7% oil; this oil showed a saponification value of 231 mg KOH/g of oil with an acid value of 3.7 mg KOH/g (Fokou et al., 2009).

The defatted cakes of cucumber seed had high total protein content (61.9%). The trichloroacetic acid soluble fraction of this protein is 94% of total proteins, due to the postharvest treatment of the seeds. The seed cake is rich in most essential amino acids, giving protein digestibility, corrected amino acid score of 0.67 which was for lysine, indicating that in the absence of tryptophan and methionine, lysine was the limiting amino acid in these seeds. The seeds had higher levels of the essential amino acids, histidine (essential for infants), threonine, phenylalanine, and tyrosine than soybean and higher levels of valine, isoleucine, leucine, phenylalanine, and tyrosine than casein. The seeds of cucumber had low levels of phenolic compounds (0.43%) (Achu, Fokou, Kansci, & Fotso, 2013). The seeds of cucumber served as a good source of protein (33.8%), fat (45.2%), carbohydrates (10.3%), and crude fiber (2.0%); the crude fiber contains indigestible materials, which can reduce constipation by increasing bowel movement. The *C. sativus* seeds were rich in Ca content (2.0%), Mg (8.5%), Na (79.4 ppm), Mn (66.3 ppm) and Fe (21.4 ppm), K (5.1%), Cu (3.3 ppm), and Zn (11.7 ppm) (Abiodun & Adeleke, 2010). Minerals aid in digestion, formation of strong bone and teeth, and hemoglobin formation. The variation in mineral composition could be due to the climate, species, soil type, water, and the cultural practices adopted during planting (Steven, Vernon, & Michael, 1985).

Mariod et al. (2009) investigated the fruit characteristics of *C. sativus*; they reported 6.2 g fruit weight, 6.0 cm fruit length, and 3.0 cm fruit diameter. They found that the weight of dried seeds was 2.1 g, while the weight of dried pulp was 4.1 g. These authors reported that the dried pulp of *C. sativus* represented more than 66.1% of its original weight. These authors easily separated the seeds of *C. sativus* from the stringy pulp to which they are attached. The *C. sativus* seeds showed 17.5% protein content, which is slightly less than protein levels in most oily seeds, carbohydrate as 22.4 and ash as 4.0 g/100 g, respectively. The oil content of *C. sativus is* 25.8% of the seed weight, which proposed this plant as a potential source of oil. The oil obtained from the seed of *C. sativus* was odorless, of good color, and of good appearance with good physicochemical properties as refractive index (40 + 1°C) is 1.434, the relative density (30 ± 1°C) is 0.914, the unsaponifiable matter is 1.1%, and iodine value is 114 (Mariod et al., 2009).

The medicinal properties of *C. sativus* have been well described, and few bioactive compounds have been derived from this plant belonging to different chemical groups. Bitter principles (Cucurbitacins) are the characteristic properties of this species that exhibited cytotoxicity and anticancer activity. The polyphenol contents have also been reported in cucumber (Melo, Shimizu, &

Mazzafera, 2006). Sood et al. (2012) confirmed the presence of various phyto-chemicals like tannins, cardiac glycosides, terpenoids, carbohydrates, resins, saponins, and phytosterols in cucumber. They also suggested that the plant can be a source of useful drugs but they recommended further studies to isolate the active component of the crude plant extract for proper drug development. Zhao et al. (2014) studied the effects of nanoparticles (NPs) on cucumber fruit quality and its impact on carbohydrates, proteins, mineral nutrients, and an-tioxidants in the fruit of cucumber plants. The authors treated the soil where cucumber is grown with CeO_2 and ZnO NPs at 400 and 800 mg/kg. Their results showed that, none of the ZnO NP concentrations affected sugars; however, at 400 mg/kg, CeO_2 and ZnO NPs increased starch content. Conversely, CeO_2 NPs did not affect the starch content, but impacted nonreducing sugar content (sucrose). ZnO NPs did not impact protein fractionation; however, CeO_2 NPs at 400 mg/kg increased globulin and decreased glutelin. Both CeO_2 and ZnO NPs had no impact on flavonoid content, although CeO_2 NPs at 800 mg/kg significantly reduced phenolic content. ICP-OES results showed that none of the treatments reduced macronutrients in fruit. In case of micronutrients, all treatments reduced Mo concentration, and at 400 mg/kg, ZnO NPs reduced Cu accumulation. μ-XRF revealed that Cu, Mn, and Zn were mainly accumulated in cucumber seeds (Zhao, et al., 2014).

Cucumber contains a wide variety of biologically active, nonnutritive com-pounds known as phytochemicals, such as alkaloids, flavonoids, tannins, phlo-batannins, steroids, and saponins, among others (Sheetal & Jamuna, 2009). Babajide, Olaluwoye, Taofik Shittu, and Adebisi (2013) developed a new fruit drink with health benefits by blending equal portions of cucumber (50%) and pineapple (50%) juices and mixed them with clove and ginger powders at 0.25, 0.5, 0.75, and 1% (w/v), respectively. These authors analyzed the developed mix and it showed many active constituents of great health benefits, such as fumaric acid, betaxolol, benzoquinone, benzofuranone, camphene, chlorogenic acid, quinolizine, and pyridinepropanol (Babajide et al., 2013).

REFERENCES

Abiodun, O. A., & Adeleke, R. O. (2010). Comparative studies on nutritional composition of four melon seeds varieties. *Pakistan Journal of Nutrition, 9*(9), 905–908.

Achu, M. B., Fokou, E., Kansci, G., & Fotso, M. (2013). Chemical evaluation of protein quality and phenolic compound levels of some Cucurbitaceae oilseeds from Cameroon. *African Journal of Biotechnology, 12*(7), 735–743.

Babajide, J. M., Olaluwoye, A. A., Taofik Shittu, T. A., & Adebisi, M. A. (2013). Physicochemical properties and phytochemical components of spiced cucumber-pineapple fruit drink. *Nigerian Food Journal, 31*(1), 40–52.

Chiej, R. (1984). *Encyclopaedia of medicinal plants.* London, UK: MacDonald and Company Limited Maxwell House.

Doijode, S. D. (2001). *Seed storage of horticultural crops.* Binghamton, NY, USA: Haworth Press.

Duke, J. A., & Ayensu, E. S. (1985). *Medicinal plants of China.* Algonac, MI: Reference Publications.

Fokou, E., Achu, M. B., Kansci, G., Ponka, R., Fotso, M., Tchiégang, C., & Tchouanguep, F. M. (2009). Chemical properties of some Cucurbitaceae oils from Cameroon. *Pakistan Journal of Nutrition, 8*(9), 1325–1334.

Gill, N. S., Garg, M., Bansal, R., Sood, S., Muthuraman, A., Bali, M., & Sharma, P. D. (2009). Evaluation of antioxidant and antiulcer potential of *Cucumis sativum* L. seed extract in rats. *Asian Journal of Clinical Nutrition, 1*(3), 131–138.

Grieve, M. (1998). *A modern herbal—the medicinal, culinary, cosmetic and economic properties, cultivation and folklore of herbs, grasses, fungi, shrubs and trees with all their modern scientific uses.* London: Tiger Books International.

Kaur, G., Aggarwal, P., & Javed, M. (2014). Effect of chemical additives on the shelf life of cucumber juice. *International Journal of Engineering Research and Applications, 4*(1), 206–209.

Kumar, D., Kumar, S., Singh, J., Rashmi, N., Vashistha, B. D., & Singh, N. (2010). Free radical scavenging and analgesic activities of *Cucumis sativus* L. fruit extract. *Journal of Young Pharmacist, 2*(4), 365–368.

Mariod, A. A., Ahmed, Y. M., Matthaeus, B., Khaleel, G., Siddig, A., Gabra, A. M., & Abdelwahab, S. I. (2009). A comparative study of the properties of six Sudanese cucurbit seeds and seed oils. *Journal of the American Oil Chemists' Society, 86*, 1181–1188.

Melo, G. A., Shimizu, M. M., & Mazzafera, P. (2006). Polyphenoloxidase activity in coffee leaves and its role in resistance against coffee leaf miner and coffee leaf rust. *Phytochemistry, 67*, 277–285.

Nonnecke, I. L. (1989). *Vegetable production.* Springer Science & Business Media.

Renner, S. S., Schaefer, H., & Kocyan, A (2007). Phylogenetics of *Cucumis* (Cucurbitaceae): cucumber (*C. sativus*) belongs in an Asian/Australian clade far from melon (*C. melo*). *BMC Evolutionary Biology, 7*, 58.

Sharma, S., Dwivedi, J., & Paliwal, S. (2012). Evaluation of antacid and carminative properties of *Cucumis sativus* under simulated conditions. *Der Pharmacia Lettre, 4*(1), 234–239.

Sheetal, G., & Jamuna, P. (2009). Studies on Indian green leafy vegetables for their antioxidant activity. *Plant Foods for Human Nutrition, 64*, 39–45.

Smitha, T., Santhi, T., Prasad, A. L., & Manonmani, S. (2012). *Cucumis sativus* used as adsorbent for the removal of dyes from aqueous solution. *Arabian Journal of Chemistry*, Available from http://dx.doi.org/10.1016/j.arabjc.2012.07.030.

Sood, A., Kaur, P., & Gupta, R. (2012). Phytochemical screening and antimicrobial assay of various seeds extract of Cucurbitaceae family. *International Journal of Applied Biology and Pharmaceutical Technology, 3*(3), 401–409.

Sotiroudis, G., Sotiroudis, E. M., & Chinou, I. (2010). Chemical analysis, antioxidant and antimicrobial activity of three Greek cucumber (*Cucumis sativus*) cultivars. *Journal of Food Biochemistry, 34*, 61–78.

Steven, R. T., Vernon, R. Y., & Michael, C. A. (1985). Vitamins and minerals. In O. Fennema (Ed.), *Food chemistry* (2nd ed., pp. 523). New York, NY: Marcel Dekker.

Usher, G. (1974). *A dictionary of plants used by man.* London, UK: Constable and Company Ltd.

Vetriselvan, S., Subasini, U., Velmurugan, C., Muthuramu, T., & Revathy, S. J. (2013). Anti-inflammatory activity of *Cucumis sativus* seed in carrageenan and xylene induced edema model using albino wistar rats. *International Journal of Biopharmaceutics, 4*(1), 34–37.

Warrier, P. K. (1994). *Indian medicinal plants: a compendium of 500 species.* Chennai: Press Orient Longman.

Zhao, L., Peralta-Videa, J. R., Rico, C. M., Hernandez-Viezcas, J. A., Sun, Y., Niu, G., Servin, A., Nunez, J. E., Duarte-Gardea, M., & Gardea-Torresdey, J. L. (2014). CeO_2 and ZnO nanoparticles change the nutritional qualities of cucumber (*Cucumis sativus*). *Journal of Agricultural and Food Chemistry, 62*(13), 2752–2759.

Chapter 17

Cucumis prophetarum Globe Cucumber or Wild Cucumber

DISTRIBUTION AND BOTANICAL DESCRIPTION

The plant is distributed in many countries including Africa: Kenya, Tanzania, Uganda, Chad, Eritrea, Ethiopia, Somalia, Sudan, Egypt, Mali, Mauritania, Niger, Nigeria, Senegal, West-Central Rwanda; Arabian Peninsula: Oman, Qatar, Saudi Arabia, United Arab Emirates, Yemen; Western Asia: Iran, Iraq, Israel, Jordan, Syria; India; and Pakistan (Burkill, 2010). The plant belongs to the Cucurbitaceae family, which is mostly a prostate or climbing, monoecious herb. The fruits are ellipsoid, echinate, and green with white strips. It is distributed geographically in the Western India (Saldanha & Ramesh, 1984). The stem slender, sulcate, hispid, and rough; hairs whitish. Tendrils simple and short, scabrid. Leaves ovate-orbicular, 2–4 cm long. Petiole 1–4 cm long, densely scabrid with white hairs. Male flowers in fascicles of 2–3 flowers, rarely solitary; peduncle 5–10 mm long, hispid. Calyx tube 3–5 mm long, hispid, lobes linear, spreading, c. 1.5 mm long. Corolla ± villous, 3–5 mm long, lobes ovate-oblong, ± mucronate. Anthers oblong, 2–5 mm long. Female flowers solitary, peduncle 2–3 cm long, calyx and corolla as in male. Ovary muricate. Fruit ovoid or subglobose, puberulous, softly echinate, longitudinally striped green and white, yellow when ripe, 3–4 × 2–3 cm. Seeds pale ashy, c. 4 × 2 mm, ± oblong or ellipsoid (http://www.efloras.org) (Fig. 17.1).

FOOD USES AND MEDICINAL VALUES

Cucumis prophetarum Linn., the fruit is used for inflammation-related problems and is proved to be possessing anticancer and hepatoprotective effects. The fruit is used in indigenous medicine as an emetic and purgative. The aerial parts of *C. prophetarum* are used by Saudi traditional medicine practitioners as remedy for liver disorders (Al-Yahya, Al-Meshal, Mossa, Al-Badr, & Tariq, 1990). *C. prophetarum* (L.) is used in traditional Indian medicine for the treatment of inflammation-related problems. The active compound N-Trisaccharide obtained from the plant possesses the propitious effect on STZ–NA-induced type 2 diabetes, indicating its usefulness in diabetes management. N-Trisaccharide

Unconventional Oilseeds and Oil Sources. http://dx.doi.org/10.1016/B978-0-12-809435-8.00017-2

FIGURE 17.1   *Cucumis prophetarum* **fruit and seeds.**

significantly ($P \le 0.05$) increased the plasma insulin and liver glycogen levels in diabetic rats. The altered enzyme activities of carbohydrate metabolism in the liver and kidney of the diabetic rats were significantly ($P \le 0.05$) improved. Additionally, N-Trisaccharide increased glycogen synthase and decreased glycogen phosphorylase activity in diabetic rats. Histological studies confirmed that an increase in insulin level is due to stimulation of injured pancreatic cells (Kavishankar & Lakshmidevi, 2014).

Gawli and Lakshmidevi (2015) studied the effect of aqueous crude extract (CE) of *C. prophetarum* on antidiabetic and antioxidant activity. The fruits were fractionated into water-soluble fraction 1 (F1), chloroform fraction 2 (F2), basic fraction 3 (F3), and neutral fraction 4 (F4) by acid–base extraction. F1 exhibited effective antidiabetic activity ($P < 0.05$) with an IC$_{50}$ value of 20.6 and 59.9 µg/mL. The activity decreased in the order of CE > F4 > F3 > F2, according to α-amylase assay, which were the same, with the exception of the rank order of F4 and CE, as the α-glucosidase assay. Furthermore, F1 (IC$_{50}$ = 73 µg/mL) showed better reducing ability than CE > F4 > F2 > F3 (IC$_{50}$ = 78–272 µg/mL), according to the DPPH assay.

FRUIT AND SEED COMPOSITION AND BIOACTIVE COMPONENTS

Afifi, Ross, ElSohly, Naeem, and Halaweish (1999) identified cucurbitacin B, isocucurbitacin B, dihydrocucurbitacin B, cucurbitacin E, dihydrocucurbitacin E, isocucurbitacin D, dihydroisocucurbitacin D, cucurbitacin I, dihydrocucurbitacin I, cucurbitacin Q1, and dihydrocucurbitacin Q1 for the first time as constituents of *C. prophetarum* L. Ayyad, Abdel-Lateff, Basaif, and Shier (2011) isolated and elucidated the structures of dihydrocucurbitacin B (1) and cucurbitacin B (2), as major compounds from the fruit extract of *C. prophetarum*. These compounds showed potent activities toward human cancer cell lines, and could be considered as a new anticancer natural drug. The fruit is known to contain cucurbitacins B and D and traces of cucurbitacins G, Q, and H. Cucurbitacin Ql, a compound isolated from the fruit of *C. prophetarum*, has been shown to be the pure *trans* component of cucurbitacin Q (Atta-Ur-Rahman, Ahmied, Khan, & Zehra, 1973). Sadou, Sabo, Alma, Saadou, and Leger (2007) determined the chemical composition of the seeds of *C. prophetarum*. Their results showed: 26.86% protein content, 24.62% fat content, and 3.2% ash content. The mineral matter was as follows: calcium (260 mg/100 g); copper (5.8 mg/100 g); iron (15.6 mg/100 g); magnesium (239 mg/100 g); potassium (1137 mg/100 g); phosphorus (46.2 mg/100 g); sodium (12.8 mg/100 g); and zinc (2.9 mg/100 g). These authors reported that the acid value was 6.3, iodine value was 141, saponification value was 185, and refractive index was 1.4674. The nonsaponification percentage was within 1.51% and the predominant fatty acids were palmitic acid (15.9%), stearic acid (7.83%), oleic acid (10.8%), and linoleic acid (64.2%). The study of the antioxidants fraction showed: 20.4 mg/kg α-tocopherol content, 619 mg/kg γ-tocopherol content, and 0.23 mg/kg β-carotene content. Mariod et al. (2009) investigated the fruit characteristics of *C. prophetarum*; they reported 9.6 g dry fruit weight, 7.0 cm fruit length, and 4.0 cm fruit diameter. They found that the weight of dried seeds was 3.4 g, while the weight of dried pulp was 6.2 g. These authors reported that the dried pulp of *C. prophetarum* represented more than 64.5% of dry fruit weight. The *C. prophetarum* seeds showed 14.5% protein content, carbohydrate as 28.8 and ash as 8.8 g/100 g, respectively. The oil content of *C. prophetarum* is 10.9% of the seed weight. The oil obtained from the seed of *C. prophetarum* showed good physicochemical properties as refractive index (40 + 1°C) is 1.410, the relative density (30 ± 1°C) is 0.884, the unsaponifiable matter is 0.9%, and iodine value is 110.6 (Mariod et al., 2009). The total tocopherols in the seed of *C. prophetarum* oil is 1.5 mg/100 g with delta-tocopherols as the predominant one representing more than 74%, followed by gamma- and then alpha-tocopherol (Mariod et al., 2009). The fruits of *C. prophetarum* have afforded a new disecosterol characterized as 5 (10), 13 (14)-diseco-stigma-5 (6), 9 (11)-diane-3-α-ol(1) on the basis of chemical and spectral analysis, and has been designated as cucumidisecosterol (Al-Rehaily, Al-Yahya, Mirza, & Ahmed, 2002).

REFERENCES

Afifi, M. S., Ross, S. A., ElSohly, M. A., Naeem, Z. E., & Halaweish, F. T. (1999). Cucurbitacins of *Cucumis prophetarum*. *Journal of Chemical Ecology*, 25(4), 847–859.

Al-Rehaily, A. J., Al-Yahya, M. A., Mirza, H. H., & Ahmed, B. (2002). Cucumidisecosterol: a new diseco-sterol from *Cucumis prophetarum*. *Pharmaceutical Biology*, 40(2), 154–159.

Al-Yahya, M. A., Al-Meshal, I. A., Mossa, J. S., Al-Badr, A. A., & Tariq, M. (1990). *A phytochemical and biological approach*. Riyadh: King Saud University Press (pp. 142–144).

Atta-Ur-Rahman, Ahmied, V. U., Khan, M. A., & Zehra, F. (1973). Isolation and structure of cucurbitacin Ql. *Photochemistry*, *12*, 2741–2743.

Ayyad, S. E. N., Abdel-Lateff, A., Basaif, S. A., & Shier, T. (2011). Cucurbitacins-type triterpene with potent activity on mouse embryonic fibroblast from Cucumisprophetarum, cucurbitaceae. *Pharmacognosy Research*, *3*(3), 189–193.

Burkill, H. M. (2010). *The useful plants of west tropical Africa. 1985–2004*. Kew, UK: Royal Botanic Gardens.

Kavishankar, G. B., & Lakshmidevi, N. (2014). Anti-diabetic effect of a novel N-Trisaccharide isolated from *Cucumis prophetarum* on streptozotocin–nicotinamide induced type 2 diabetic rats. *Phytomedicine*, *21*, 624–630.

Gawli, K., & Lakshmidevi, N. (2015). Antidiabetic and antioxidant potency evaluation of different fractions obtained from *Cucumis prophetarum* fruit. *Pharmaceutical Biology*, *53*(5), 689–694.

Mariod, A. A., Ahmed, Y. M., Matthaeus, B., Khaleel, G., Siddig, A., Gabra, A. M., & Abdelwahab, S. I. (2009). A comparative study of the properties of six Sudanese cucurbit seeds and seed oils. *Journal of the American Oil Chemists' Society*, *86*, 1181–1188.

Sadou, H., Sabo, H., Alma, M. M., Saadou, M., & Leger, C. -L. (2007). Chemical content of the seeds and physico-chemical characteristic of the seed oils from *Citrullus colocynthis*, *Coccinia grandis*, *Cucumis metuliferus* and *Cucumis prophetarum* of Niger. *Bulletin of the Chemical Society of Ethiopia*, *21*(3), 323–330.

Saldanha, C. J., & Ramesh, S. R. (1984). UK: Oxford & IBH.

Chapter 18

Citrullus colocynthis Colocynth, Bitter Apple, Bitter Gourd

DISTRIBUTION AND BOTANICAL DESCRIPTION

Citrullus colocynthis (Cucurbitaceae) is a perennial herb usually trailing. It is commonly found wild in the sandy lands of North West, the Punjab, Sind, and Central and Southern India; also found indigenously in Mediterranean Europe, Cyprus, the Syrian Arab Republic, Lebanon, Jordan, Egypt, Kuwait, Saudi Arabia, Turkey, the Islamic Republic of Iran, Pakistan, Afghanistan, India, North Africa, Tropical Africa (Pravin, Tushar, Vijay, & Kishanchnad, 2013). In English the plant is known as colocynth, bitter apple, bitter gourd, while in Arabic it is known as hanzal. Leaves are very variable in size. Wild leaf is 3.8–6.3 cm in length and 2.5 cm in width while cultivated leaves are large in size. Leaf shows deltoid margin, pale green color above and ashy color beneath, scabrid on both surfaces, 5–7 lobed. The angular leaves are alternately located on long petioles. Each leaf is almost 5–10 cm in length and has around 3–7 lobes. Sometimes the middle lobe might have an ovate structure. The leaves have a triangular shape with many clefts. The leaves have a rough, hairy texture with open sinuses. The upper surface of the leaves is fine green color and the lower surface is comparatively pale. *C. colocynthis* shows the presence of male and female flowers. Fruit is globular, slightly depressed, 5–7.5 cm in diameter, green in color and gets white glabrous when ripe. Fruit filled with a dry, spongy very bitter pulp. Each bitter apple plant produces around 15–30 globular fruits having a diameter of almost 7–10 cm. The outer portion of the fruit is covered with a green skin having yellow stripes. The fruits may also be yellow in color. Seeds are 4–6 mm long and pale brown. The fruits have a soft, white pulp, which is filled with numerous ovate compressed seeds. The seeds are around 6 mm in size, smooth, compressed, and ovoid-shaped. They are located on the parietal placenta. The seeds are light yellowish-orange to dark brown in color (Pravin et al., 2013) (Fig. 18.1).

Unconventional Oilseeds and Oil Sources. http://dx.doi.org/10.1016/B978-0-12-809435-8.00018-4

FIGURE 18.1 *Citrullus colocynthis* plant, fruits and seeds.

FRUIT AND SEED COMPOSITION AND BIOACTIVE COMPONENTS

Bande, Adam, Jamarei, and Azmi (2012) determined basic dimensions and mass of *C. colocynthis* fruit with a digital gauge aimed to give information on loading capacity in transportation and storage of the harvested fruit prior to processing; they reported scale with accuracy 0.01 mm and 0.01 g, respectively. Maximum length, width, thickness, and mass of 100 samples were 12.86 cm, 12.53 cm, 15.52 cm, and 1031.5 g, respectively. Arithmetic and geometric mean diameters were between 5.68 and 13.63 and 5.58 and 13.22 cm while mean bulk and true densities were 404.98 and 1074.6 kg/m^3 on 3 and 6 runs, respectively. The sphericity and aspect ratio were about 1, with average packaging coefficient, on five separate runs, of 33.49. The average vertical and horizontal weight to break the fruit was 121.23 and 74.09 kg, respectively (Bande et al., 2012). The generally average mass of *C. colocynthis* fruit is 506 g and mass of pulp is almost 50% of the mass of fruit, while the seed content is 71.8 g (Aviara, Shittu, & Haque, 2007). Several bioactive chemical constituents of fruits were recorded, such as glycosides, flavonoids, alkaloids, fatty acids, and essential oils. The isolation and identification of curcurbitacins A, B, C, D, E, I, J, K, and L and colocynthosides A and B were also reported (Hussain et al., 2014).

The seeds of *C. colocynthis* are known to possess: 567.32 kcal/100 g dry weight, 17%–23% fatty acids, of which a low amount of free fatty acids (1.57% total seed weight and 7% total oil) and both triglycerides and free fatty acids comprise mostly linoleic acid (66%–76.4%), palmitic acid (6.3%–8.1%), oleic acid (7.8%–14.2%), and stearic acid (6.1%–7.3%); proteins at 11.7% of total weight; nonfibrous carbohydrates at 29.5% total weight; dietary fiber at 5.51% total weight; 7.51 ± 0.53% moisture content; 2.9%–3.2% ash content (dry weight); vitamin E at 121.85 mg/100 g with a high γ-tocopherol content (116.36 ± 0.15 mg/100 g; 95% total vitamin E) and lower α (0.72 ± 0.06 mg/100 g), β (0.57 ± 0.03 mg/100 g), and δ (4.19 ± 0.12 mg/100 g) content (Sebbagh et al., 2009; Obasi, Ukadilonu, Eze, Akubugwo, & Okorie, 2012; Nehdi, Sbihi, Tan, & Al-Resayes, 2013). Mariod et al. (2009) investigated the fruit characteristics of *C. colocynthis*; they reported 14.8 g dry fruit weight, 13.0 cm fruit length, and 13.0 cm fruit diameter. They found that the weight of dried seeds was 4.6 g, while the weight of dried pulp was 10.2 g. These authors reported that the dried pulp of *C. colocynthis* represented more than 68.9% of dry fruit weight. The *C. colocynthis* seeds showed 15.75% protein content, carbohydrate as 17.01 and ash as 4.62 g/100 g, respectively. The oil content of *C. colocynthis is* 27.10% of the seed weight. The oil obtained from the seed of *C. colocynthis* showed good physicochemical properties as refractive index (40 + 1°C) is 1.429, the relative density (30 ± 1°C) is 0.886, the unsaponifiable matter is 0.8%, and iodine value is 112.7 (Mariod et al., 2009). The total tocopherols in the seed of *C. colocynthis* oil is 34.7 mg/100 g with gamma-tocopherols as the predominant one representing more than 96%, followed by delta- and then alpha-tocopherol (Mariod et al., 2009). *C. colocynthis* is an excellent source of different amino acids, such as arginine, methionine, and tryptophan. The biological indices of its protein quality were lower than soybean, but comparable to or higher than most oilseeds. Nutritionally, the limiting amino acids are lysine and threonine. Glutamic acid and arginine were the main amino acids identified with concentrations 19.8 and 15.9 g of amino acid/100 g of protein, respectively. Other major (45 g of amino acid/100 g of protein) amino acids found were aspartic acid, serine, glycine, alanine, leucine, and phenylalanine. The literature reports suggest that its seeds have the potential for fortifying both traditional and modern food formulations (National Research Council, 2006).

The thermogravimetric analysis showed that the *C. colocynthis* oil was thermally stable up to 286.57°C, and then began to decompose in four stages. Furthermore, the seed oil absorbed strongly UV-B radiations considered to be responsible for causing skin cancer, immunosuppression, and premature skin aging (Nehdi, Sbihi, Tan, & Al Resayes, 2014). The fruit pulp contains: colocynthin (bitter component), colocynthein and colocynthetin (resin), cucurbitacin E glucose (0.05% dry weight), and I glucoside (0.01% dry weight). The fruits appear to, on the equivalent of dry mass, contain a 0.74% phenolic content [gallic acid equivalents (GAEs)] and 0.13% flavonoid content (catechin equivalents). The seeds also contain flavonoids and polyphenolics (Rahimi, Amin, & Ardekani, 2012).

DIFFERENT USES AND MEDICINAL VALUES

C. colocynthis fruit is employed as food for animals and humans (Sadou, Sabo, Alma, Saadou, & Leger, 2007). The seeds have the potential to find a place in the food industry, according to their nutritional and functional properties (National Research Council, 2006).

C. colocynthis root, bark, and leaves are medicinally used and show mild stomachic, bitter tonic, diuretic, and antilithic property. It is used as a purgative and diuretic. Fruit is bitter, pungent, and used as purgative, anthelmintic, antipyretic, carminative, cure for tumors, leukoderma, ulcers, asthma, etc. Root is useful in jaundice, ascites, urinary disease (Pravin et al., 2013). Antidiabetic; violent purge (pulp); pulp is: diuretic, antiepileptic, antiblenorrheic, strong cathartic (mixed with Arabic gum, to mitigate the effects); the emptied fruit, filled with water and honey, is drunk for rheumatism, gout, dropsy, ascites, while cataplasms are good for arthrosis; seeds are antidiabetic and anthelmintic; the fruit is used also as abortifacient (drinking water in which fruit is macerated overnight) and for poisonous bites; dog, insect, and snake bites, laxative, pain in joints, hair color (leaves, seeds, roots, and dried fruit); it is also used in many composed medicines. The colocynth is well known from ancient times. Greeks and Romans used it, and it was even cultivated in Cyprus for a long time. Common in the whole area, it is used at a popular level. Fatal cases of poisoning from overdoses of colocynth are not rare, but they represent a small percentage given its wide use. Some of its beneficial effects are, in fact, recognized by pharmacological research (Sincich, 2002). The plant has been reported to possess a wide range of traditional medicinal uses, including in diabetes, leprosy, common cold, cough, asthma, bronchitis, jaundice, joint pain, cancer, toothache, wound, mastitis, and in gastrointestinal disorders, such as indigestion, constipation, dysentery, gastroenteritis, colic pain, and different microbial infections (Hussain et al., 2014). Belsem et al. (2011), studied aqueous extracts of C. colocynthis fruit and seed at immature state for antiinflammatory activity using the carrageenan-induced paw edema assay in rats. They concluded that C. colocynthis was a potent antiinflammatory product. Aqueous extract of roots showed significant reduction in blood sugar level (58.70%) when compared with chloroform (34.72%) and ethanol extracts (36.60%) ($P < 0.01$). The aqueous extracts showed improvement in parameters like body weight, serum creatinine, serum urea, and serum protein, as well as lipid profile and also restored the serum level of bilirubin total, conjugated bilirubin, and some enzymes (Agarwal, Sharma, Upadhyay, Singh, & Gupta, 2012).

Arogundade, Akinfenwa, and Salawu (2004) investigated the effect of NaCl on some hydration and hydration-related properties of C. citrullus L. together with the possibility of replacing such salt with KCl in food formulation. Their study showed that protein solubility increased with an increase in salt concentration. Partial replacement of NaCl with KCl did not make any significant difference in protein solubility. Water absorption capacity decreased with an increase in salt concentration and neither complete nor partial replacement of NaCl with KCl made any significant difference ($P > 0.05$).

Bhattacharya (1990) studied the utilization of crude protein (CP) in *C. colocynthis* seed meal (Cit. SM) containing 18%CP and 25% fat, on a dry matter basis. Three ISO nitrogenous diet treatments consisting of 0, 10, and 20% Cit. SM replacing soybean meal protein in the control diet was allotted to three groups of four wethers each in a completely randomized block design. The results indicated that Cit. SM could be a satisfactory partial replacement of soybean meal in sheep diets.

C. colocynthis is traditionally used as a hydragogue (causes water accumulation in the colon) and catharsis (induces defecation) produced in humans when consumed at more than 2 g dry fruit weight and other traditional usages are stimulating the immune system; for the treatment of rheumatism, tuberculosis, diabetes; and as analgesic. Oral ingestion of the fruit at doses lower than this (300–800 mg daily) tends to be prescribed in some Middle Eastern locations for the treatment of diabetes to avoid the intestinal side effects (Daoudi, Aarab, & Abdel-Sattar, 2013).

C. colocynthis (100 mg) thrice daily for 2 months in type 2 diabetes has failed to significantly modify any measured parameter of cholesterol metabolism (LDL-C, HDL-C, or total cholesterol) although total cholesterol can be reduced if the same dose of the seed extract is given to hyperlipidemics (17.57 ± 39.55 mg/dL; average 8.6% reduction); the alterations in HDL-C (2.41 ± 5.63 mg/dL; average 5.4% increase) and LDL-C (7.33 ± 30.34 mg/dL; average 6.3% reduction) also showed promise, but were statistically insignificant (Daoudi et al., 2013; Rahbar & Nabipour, 2010).

C. colocynthis has been traditionally used as a medicinal plant for diabetes; since it induce insulin secretion from the pancreas after supplementation, while reducing blood glucose and improving lipid levels, this has been noted in vitro and in vivo as acute oral ingestion of 300 mg/kg of the pulp (ethanolic extraction) to rats has been noted to increase serum insulin AUC by 59.5% alongside a reduction in glucose (33%) and elsewhere this extraction has been noted to increase insulin content of pancreatic β-cells (Patel, Prasad, Kumar, & Hemalatha, 2012; Benariba et al., 2013). Type 2 diabetics given 100 mg of the dried fruit extract thrice daily (totaling 300 mg) for a period of 2 months noted significant reductions in blood glucose (8.2%) and HbA1c (13.5%) while insulin was not measured; these two parameters, however, were significantly higher at baseline in the experimental group relative to placebo (Huseini, 2009).

REFERENCES

Agarwal, V., Sharma, A. K., Upadhyay, A., Singh, G., & Gupta, R. (2012). Hypoglycemic effects of *Citrullus colocynthis* roots. *Acta Poloniae Pharmaceutica*, 69(1), 75–79.

Arogundade, L. A., Akinfenwa, M. O., & Salawu, A. A. (2004). Effect of NaCl and its partial or complete replacement with KCl on some functional properties of defatted *Colocynthis citrullus* L. seed flour. *Food Chemistry*, 84, 187–193.

Aviara, N. A., Shittu, S. K., & Haque, M. A. (2007). Physical properties of guna fruits relevant in bulk handling and mechanical processing. *International Agrophysics, 21*, 7–16.

Bande, Y. M., Adam, N. M., Jamarei, B. O., & Azmi, Y. (2012). Physical and mechanical properties of "Egusi" melon (*Citrullus colocynthis* lanatus var. lanatus) fruit. *International Journal of Agricultural Research, 7*(10), 494–499.

Benariba, N., Djaziri, R., Hupkens, E., Louchami, K., Malaisse, W. J., & Sener, A. (2013). Insulino-tropic action of *Citrullus colocynthis* seed extracts in rat pancreatic islets. *Molecular Medicine Reports, 7*(1), 233–236.

Belsem, M., Zohra, M., Ehsen, H., Manel, T., Abderrahman, B., Mahjoub, A., & Nadia, F. (2011). Anti-inflammatory evaluation of immature fruit and seed aqueous extracts from several popu-lations of Tunisian *Citrullus colocynthis* Schrad. *African Journal of Biotechnology, 10*(20), 4217–4225.

Bhattacharya, A. N. (1990). *Citrullus colocynthis* seed meal as a protein supplement for Najdi sheep in northern Saudi Arabia. *Animal Feed Science and Technology, 29*, 57–62.

Daoudi, A., Aarab, L., & Abdel-Sattar, E. (2013). Screening of immunomodulatory activity of total and protein extracts of some Moroccan medicinal plants. *Toxicology and Industrial Health, 29*(3), 245–253.

Huseini, H. F. (2009). The clinical investigation of *Citrullus colocynthis* (L.) schrad fruit in treat-ment of Type II diabetic patients: a randomized, double blind, placebo-controlled clinical trial. *Phytotherapy Research, 23*(8), 1186–1189.

Hussain, A. I., Rathore, H. A., Sattar, M. Z. A., Chatha, S. A. S., Sarker, S. D., & Gilani, A. H. (2014). *Citrullus colocynthis* (L.) Schrad (bitter apple fruit): a review of its phytochemistry, pharmacology, traditional uses and nutritional potential. *Journal of Ethnopharmacology, 155*, 54–66.

Mariod, A. A., Ahmed, Y. M., Matthaeus, B., Khaleel, G., Siddig, A., Gabra, A. M., & Abdelwahab, S. I. (2009). A comparative study of the properties of six Sudanese cucurbit seeds and seed oils. *Journal of the American Oil Chemists' Society, 86*, 1181–1188.

National Research Council. (2006). *Lost crops of Africa*. Washington, DC: National Academies Press.

Nehdi, I. A., Sbihi, H., Tan, C. P., & Al-Resayes, S. I. (2013). Evaluation and characterization of *Citrullus colocynthis* (L.) Schrad seed oil: comparison with *Helianthus annuus* (sunflower) seed oil. *Food Chemistry, 136*(2), 348–353.

Nehdi, I. A., Sbihi, H. M., Tan, C. P., Al Resayes, S. I. (2014). *Citrullus colocynthis* (L.) Schrad seed oil: a source of bioactive compounds. Abstracts of Pharma Nutrition 2013/Pharma Nutrition 2, 75–119.

Obasi, N. A., Ukadilonu, J., Eze, E., Akubugwo, E. I., & Okorie, U. C. (2012). Proximate composition, extraction, characterization, and comparative assessment of coconut (*Cocos nucifera*) and melon (*Colocynthis citrullus*) seeds, and seed oils. *Pakistan Journal of Biological Sciences, 15*, 1–9.

Patel, D., Prasad, S. K., Kumar, R., & Hemalatha, S. (2012). An overview on anti-diabetic me-dicinal plants having insulin mimetic property. *Asian Pacific Journal of Tropical Biomedicine, 2*(4), 320–330.

Pravin, B., Tushar, D., Vijay, P., & Kishanchnad, K. (2013). Review on *Citrullus colocynthis*. *Inter-national Journal of Research in Pharmacy and Chemistry, 3*(1), 46–53.

Rahbar, A. R., & Nabipour, I. (2010). The hypolipidemic effect of *Citrullus colocynthis* on patients with hyperlipidemia. *Pakistan Journal of Biological Sciences, 13*(24), 1202–1207.

Rahimi, R., Amin, G., & Ardekani, M. R. (2012). A review on *Citrullus colocynthis* Schrad.: from traditional Iranian medicine to modern phytotherapy. *Journal of Alternative and Complemen-tary Medicine, 18*(6), 551–554.

Sadou, H., Sabo, H., Alma, M. M., Saadou, M., & Leger, C. -L. (2007). Chemical content of the seeds and physico-chemical characteristic of the seed oils from *Citrullus colocynthis, Coccinia grandis, Cucumis metuliferus* and *Cucumis prophetarum* of Niger. *Bulletin of the Chemical Society of Ethiopia, 21*(3), 323–330.

Sebbagh, N., Cruciani-Guglielmacci, C., Ouali, F., Berthault, M. F., Rouch, C., Sari, D. C., & Magnan, C. (2009). Comparative effects of *Citrullus colocynthis*, sunflower and olive oil-enriched diet in streptozotocin-induced diabetes in rats. *Diabetes Metabolism, 35*(3), 178–184.

Sincich, F. (2002). *Bedouin traditional medicine in the Syrian steppe*. Rome: FAO (pp. 114–115).

Chapter 19

Cucumis melo var. *cantalupo* Cantaloupe

DISTRIBUTION AND BOTANICAL DESCRIPTION

Muskmelon, rock melon, sweet melon, honeydew, Persian melon, or spanspek (South Africa) refers to varieties of *Cucumis melo*, species in the family Cucurbitaceae. Several varieties of cantaloupes grown all over the world; however, two common varieties that named after their place of origin have become popular in the western world. The European cantaloupe (*C. melo cantalupensis*) derives its name from the Italian papal village of "Cantalup" and features lightly ribbed, pale green skin that looks quite different from the North American cantaloupe. Galia melon and charentais belong to this category. North American cantaloupe (*C. melo reticulatus*), popular in the United States and in some parts of Canada, is named *reticulatus* due to its net-like (or reticulated) skin covering. Honeydew melons have sweet, characteristic pale green succulent flesh. The cantaloupe most likely originated in a region from Iran to India and Africa. Originally, cantaloupe referred only to the nonnetted, orange-fleshed melons of Europe. However, in more recent usage, it has come to mean any orange-fleshed melon (C. melo) and is the most popular variety of melon in North America (Ensminger & Ensminger 1993).

In general, melons feature round or oblong shape, measure 4.5–6.5 in. in diameter and weigh 450–850 g, often times more than a kilo. Internally, its flesh color ranges from orange-yellow to salmon, has soft consistency and juicy texture with a sweet, musky aroma that emanates best in the completely ripe fruits. At its center, there is a hollow cavity filled with small off-white color seeds encased in a web of mucilaginous netting (Fig. 19.1).

FRUIT, SEED COMPOSITION, AND BIOACTIVE COMPOUNDS

Raw cantaloupe is 90% water, 8% carbohydrates, 0.8% protein, and 0.3% fat, providing 34 calories (kcal) and 2020 mg of the provitamin A orange carotenoid, β-carotene per 100 g amount. Fresh cantaloupe is an excellent source of vitamin C and vitamin A, content. Wonderfully delicious with rich flavor, muskmelons are very low in calories (100 g fruit has just 34 calories) and fats.

Unconventional Oilseeds and Oil Sources. http://dx.doi.org/10.1016/B978-0-12-809435-8.00019-6

FIGURE 19.1 *Cucumis melo* var. *cantalupo* plant, fruit and seeds.

Nonetheless, the fruit is rich in numerous health promoting polyphenolic plant derived compounds, and minerals that are absolute for optimum health. The fruit also contains moderate levels of B-complex vitamins, such as niacin, pantothenic acid and vitamin C, and minerals like potassium and manganese (http://ndb. nal.usda.gov/ndb/search/list). The carotenoids in freeze-dried peel and flesh of cantaloupe, were extracted and identified using HPLC. An extraction duration of 2 h and temperature of 50°C were found to be the optimum extraction conditions for the flesh while it was 2 h and 30°C for the peel. Under optimal conditions, the amount of total carotenoids in peel and flesh, measured by HPLC, was 0.33 mg/g and 0.22 mg/g of dry powder, respectively. The predominant carotenoids identified by HPLC were p-carotene and lutein. β-Carotene was the most prevalent carotenoid in the flesh, while the major carotenoid of peel was lutein (Aflaki, 2012). It is also rich in antioxidant flavonoids, such as zea-xanthin and cryptoxanthin. Total antioxidant strength measured in terms of oxygen radical absorbance capacity of cantaloupe melons is 315 μmol TE/100 g. These antioxidants have the ability to help protect cells and other structures in the body from oxygen-free radicals and hence offer protection against colon, prostate, breast, endometrial, lung, and pancreatic cancers. Zea-xanthin, an important dietary carotenoid, selectively absorbed into the retina with UV light-filtering functions.

Cantaloupe carotenoids are joined by the flavonoid luteolin, antioxidant organic acids, including ferulic and caffeic acid, and antiinflammatory cucurbitacins, including cucurbitacin B and cucurbitacin E. The nutrient diversity of cantaloupe is perhaps its most overlooked health benefit. Sulfur-containing esters and compounds containing a straight six-carbon chain were present at high concentrations in cantaloupe melons. Compounds containing a straight nine-carbon chain were in high concentrations in honeydew melons. Methyl esters were present at the highest levels in Galia melons. S-methyl 2-methylpropane-thioate and S-methyl 2-methylbutanethioate have been reported as components of melon aroma (Kourkoutas, Elmore, & Mottram, 2006).

The seeds of cantaloupe are generally treated as waste; however, medicinal effects have been reported for the seeds. Hexane-extracted seed oil of *C. melo* hybrid AF-522 was determined to contain 64 g of linoleic acid per 100 g of total fatty acids. Significant amounts of oleic, palmitic, and stearic acids were also detected in the melon seed oil (Lazos 1986).

Ismail, Mariod, Bagalkotkara, and Ling. (2010a) extracted two varieties of cantaloupe (*C. reticulatus and C. cantalupensis*) seed oil using supercritical fluid extraction (SFE) parameters in comparison with Soxhlet method. They reported the total oil content obtained from SFE for *C. reticulates* was 30.4% and for *C. cantalupensis* was 22.7%. On the other hand, the total oil content obtained from Soxhlet extraction was found to be significantly higher ($P \leq 0.05$). *C. reticulatus* was 33.5% and *C. cantalupensis* was 29.8%. The major fatty acids present in the SFE fractions from *C. reticulates* were linoleic acid (65%), followed by oleic acid (16.3%), palmitic acid (8.1%), stearic acid (5.5%), arachidic acid (2.9%), eicosanoic acid (1.1%), linolelaidic acid (0.25%), α-linolenic acid (0.13%), and γ-linolenic acid (0.06%). The major fatty acid present in the fractions from SFE for *C. cantalupensis* was linoleic acid (64.7%), followed by oleic acid (16.4%), palmitic acid (9.1%), stearic acid (5.2%), arachidic acid (1.5%) (Ismail, Mariod, Bagalkotkara, & Ling, 2010a). Ismail, Chan, Mariod, and Ismai. (2010b) determined the phenolic content and antioxidant activity of methanolic extracts from different parts of cantaloupe (leaf, stem, skin, seed, and flesh). The flesh extract afforded the highest yield ($89.6 \pm 0.3\%$) while the lowest yield was obtained from the seed ($13.7 \pm 0.5\%$) ($P < 0.05$). The leaf extract showed the highest total phenolic content (26.4 ± 0.3 mg GAE/g extract) and total flavonoid content (69.7 ± 3.37 lg RE/g extract) accompanied by best antioxidant activity through all antioxidants assays ($P < 0.05$). In addition, the stem extract also exhibited good antioxidant activity. These authors suggested that methanolic extracts of cantaloupe leaf and stem may serve as a potential source of natural antioxidant for food and nutraceutical application.

MEDICINAL AND DIFFERENT USES

The fruit is a rich source of vitamins A and C, consumption of foods rich in vitamin C helps the human body develop resistance against infectious agents and scavenge harmful oxygen-free radicals. Manganese is used by the body as a cofactor for the antioxidant enzyme, superoxide dismutase (SOD). Commercially, muskmelons are being used to extract an enzyme SOD, which plays a vital role as a strong first-line antioxidant defenses in the human body. Vitamin A is a powerful antioxidant and is essential for healthy vision. It is also required for maintaining healthy mucous membranes and skin. Consumption of natural fruits rich in vitamin A has been known to help protect from lung and oral cavity cancers (Key et al., 2004). The *C. melo* extract rich in enzyme SOD activity is able to promote, in vitro as well as in vivo, a protective innate immune response by increasing the production of the antiinflammatory cytokine, IL-10. The combination of *C. melo* extract with other vegetal extracts, such as hydrophobic gliadin biopolymers, led to the development of a 100% vegetal product that promotes the oral pharmacology of SOD (Vouldoukis et al., 2004).

Cantaloupe intake decreased risk of metabolic syndrome in individuals, since cantaloupe provides a wide range of antioxidants that help prevent oxidative stress and a wide range of antiinflammatory phytonutrients that help prevent excessive inflammation. It is also known that cantaloupe lower levels of C-reactive protein (CRP) in the bloodstream of persons who have a particularly high intake of cantaloupe, since CRP is a marker widely used to assess levels of inflammation in the body. Researchers have shown that intake of cantaloupe phytonutrients can improve insulin and blood sugar metabolism. In addition, intake of cantaloupe extracts has been shown to reduce oxidative stress in the kidneys of animals with diabetes, and to improve insulin resistance in diabetic animals (Hu & Willett 2006).

Solval, Sundararajan, Alfaro, and Sathivel. (2012) developed cantaloupe fruit juice spray-dried powder rich in vitamin C and β-carotene and evaluated its nutrition and physical properties. They found that, the powder produced at 170°C had higher β-carotene content than that produced at 180 and 190°C. The microstructure of the produced spray-dried cantaloupe powder particles was carried out in a three-dimensional characterization through scanning electron microscopy. The powder produced at 170°C had smoother surface and a more spherical shape than those powders produced at 180 and 190°C. Higher shrinkage was observed in powder produced at 180°C. This might be explained with moisture transport during the falling rate period.

REFERENCES

Aflaki, N. (2012). Optimization of carotenoid extraction in peel and flesh of cantaloupe (*Cucumis melo* L.), with ethanol solvent. *M.sc thesis*. Canada: Université Laval Québec.

Ensminger, M. E., & Ensminger, A. H. (1993). *Cantaloupe in Foods & Nutrition Encyclopedia* (2nd ed.). Boca Raton, FL, USA: CRC Press, pp. 329–331.

Hu, F. B., & Willett, W. C. (2006). Fruit and vegetable intakes, C-reactive protein, and the metabolic syndrome. *American Journal of Clinical Nutrition, 84*, 1489–1497.

Ismail, M., Mariod, A. A., Bagalkotkara, G., & Ling, H. S. (2010a). Fatty acid composition and antioxidant activity of oils from two cultivars of Cantaloupe extracted by supercritical fluid extraction. *Grasas y Aceites, 61*(1), 37–44.

Ismail, H. I., Chan, K. W., Mariod, A. A., & Ismai, M. (2010b). Phenolic content and antioxidant activity of cantaloupe (*Cucumis melo*) methanolic extracts. *Food Chemistry, 119*, 643–647.

Key, T. J., Schatzkin, A., Willett, W. C., Allen, N. E., Spencer, E. A., & Travis, R. C. (2004). Diet, nutrition and the prevention of cancer. *Public Health Nutrition, 7*(1A), 187–200.

Kourkoutas, D., Elmore, J. S., & Mottram, D. S. (2006). Comparison of the volatile compositions and flavor properties of cantaloupe, Galia and honeydew muskmelons. *Food Chemistry, 97,* 95–102.

Lazos, E. S. (1986). Certain functional properties of defatted pumpkin seed flour. *Journal of Food Science, 51,* 1382–1383.

Solval, K. M., Sundararajan, S., Alfaro, L., & Sathivel, S. (2012). Development of cantaloupe (*Cucumis melo*) juice powders using spray drying. *LWT—Food Science and Technology, 46,* 287–293.

Vouldoukis, I., Lacan, D., Kamatea, C., Coste, P., Calenda, A., Mazier, D., Conti, M., & Dugasa, B. (2004). Antioxidant and anti-inflammatory properties of a *Cucumis melo* LC. extract rich in superoxide dismutase activity. *Journal of Ethnopharmacology, 94,* 67–75.

Chapter 20

Luffa echinata Bitter Luffa, Bristly Luffa

DISTRIBUTION AND BOTANICAL DESCRIPTION

Luffa echinata is a climber with bifid, bristly or smooth tendrils. The stem is selected, slightly hairy to smooth. Leaves are kidney-shaped, round, and shallowly or deeply 5-lobed. Tip is rounded or rarely pointed. *L. echinata* is widely distributed in Sudan and is locally known as "Umshwaika" (Mariod et al., 2009). Leaves are bristly on both surfaces, margin is minutely toothed. Leaf stalk is stout, bristly, up to 12-cm long. Flowers are white, stalked, about 2.5 cm across. Male flowers are borne in 5–12-flowered, up to 15 cm long raceme. Sepal tube is about 5.6-mm long, hairy; sepals lance-shaped. Petals are ovate, 1–1.2 cm long, blunt, hairy at the base. Stamens are three, with filaments united, 3–9 mm long, anthers entire or ± bifid. Fruit is ashy, oblong, ovoid, 2–5 cm long, densely covered with 4–7 mm long bristles. Seeds are ovate, black, 4–5 mm long, 3–5 mm broad, and 2-mm thick (Fig. 20.1). The plant is widely distributed in Mauritania, Mali, Niger, NE-Nigeria, Chad, Sudan, Ethiopia, Somalia, Pakistan, India, Bangladesh, and Nepal (http://www.gbif.org).

DIFFERENT USES AND MEDICINAL VALUES

Bristly luffa (*L. echinata*), a member of the Cucurbitaceae family is a medicinal plant, which has been used in the traditional system of medicine for variety of symptoms; the dried fruits of the plant are used by the traditional Indian Ayurvedic system of medicine, for various illnesses like chronic bronchitis, dropsy, nephritis, intestinal and biliary colic, fever and jaundice, etc. It has many proposed systemic effects, such as hepatoprotective, digestive stimulant, diuretic, purgative, antiinflammatory, antipyretic, cough expectorant, etc. (Giri, Lokesh, Sahu, & Gupta, 2014). All cucurbits produce varying quantities of a complex substance known as cucurbitacin. This is responsible for the bitter quality of the plant, which produces a characteristic aroma and protects it from insects and animals. Cucurbitacin B and C are cytotoxic; they cause the G2 cell cycle arrest and apoptosis of malignant cells via reactive oxygen species–dependent mechanism. They have been tried in many cancers, such as

Unconventional Oilseeds and Oil Sources. http://dx.doi.org/10.1016/B978-0-12-809435-8.00020-2

FIGURE 20.1 *Luffa echinata* **plant, fruits and seeds.**

pancreatic cancer, melanoma, breast cancer, colon cancer (Duangmano, Sae-Lim, Suksamrarn, Patmasiriwat, & Domann, 2012). Cucurbitacin also have hepatoprotective, antiinflammatory, and cardiovascular effects. The antiinflammatory activities of some of the cucurbitacin are linked to inhibition of cyclooxygenase enzyme (COX). Saponins of *L. echinata* reduce serum cholesterol and improve immunity. Its β-sitosterol was used to treat hypercholesterolemia, benign prostatic hyperplasia, and as an immunomodulator and anticancer agent. Excess consumption of the fruits of *L. echinata* may be toxic to humans, due to different active ingredients that it contains (Giri et al., 2014). Yadav and Kumar (2013) used *L. echinata* successfully to treat 500 people who were victims of dog bite. The petroleum ether, acetone, and methanolic extracts of *L. echinata* showed a significant hepatoprotective activity against CCl_4-induced hepatotoxicity in albino rats (Ahmed, Alam, & Khan, 2001).

FRUIT AND SEED COMPOSITION AND BIOACTIVE COMPONENTS

The active constituents of *L. echinata* include saponins, hentriacontane, gypsogenin, sapogenin, cucurbitacin B and E, β-sitosterol, echinatol-A and B, oleanolic acid, elaterin glycoside, chrysoeriol-7-glucoside, sitosterol glycosides, etc. These active constituents are known to have a wide variety of pathophysiological effects on human body. Leaf flavonoids in *L. echinata* contain flavone 7-glucosides (Giri et al., 2014). The fruit of *L. echinata* contains flavonoid, chrysoeriol, apigenin, and luteolin in minor amounts as glycosides, besides β-sitosterol glucoside and apiose. Datiscacin, a cucurbitacin 20-acetate, 2-O-β-D-glucopyranosyl cucurbitacin B, and 2-O-β-D-glucopyranosyl cucurbitacin S have been isolated from the fruits of *L. echinata* (Ahmad, Enamul, & Sutradhar, 1994). *L. echinata* oil is found to deposit the hydrocarbon, hentriacontane, $C_{31}H_{64}$. Saponin has been isolated in partially purified form and the sapogenin is identified as gypsogenin (Khorana & Raisinghani, 1961). Mariod et al. (2009) investigated the fruit characteristics of *L. echinata*; they reported 1.2 g dry fruit weight, 2.7 cm fruit length, and 1.9 cm fruit diameter. They found that the weight of dried seeds was 0.45 g, while the weight of dried pulp was 0.75 g. These authors reported the dried pulp of *L. echinata* represented more than 62.5% of dry fruit weight. The *L. echinata* seeds showed 15.75% as protein content, carbohydrate as 20.1, and ash as 3.4 g/100 g, respectively. The oil content of *L. echinata is* 23.8% of the seed weight. The oil obtained from the seed of *L. echinata* showed good physicochemical properties as refractive index (40 + 1°C) is 1.409, the relative density (30 ± 1°C) is 0.918, the unsaponifiable matter is 0.8%, and iodine value is 100.1 (Mariod et al., 2009). The principal fatty acids are palmitic (14%), stearic (8.6%), oleic (32%), and linoleic (43.6%). The total tocopherols in the seed of *L. echinata* oil is 1.7 mg/100 g with delta-tocopherols as the predominant one representing more than 52.3%, followed by gamma and then alpha tocopherol (Mariod et al., 2009).

REFERENCES

Ahmad, M. U., Enamul, H. M., & Sutradhar, R. K. (1994). Bitter principles of *Luffa echinata*. *Phytochemistry, 36*(2), 421–423.

Ahmed, B., Alam, T., & Khan, S. A. (2001). Hepatoprotective activity of *Luffa echinata* fruits. *Journal of Ethnopharmacology, 76*, 187–189.

Duangmano, S., Sae-Lim, P., Suksamrarn, A., Patmasiriwat, P., & Domann, F. E. (2012). Cucurbitacin B causes increased radiation sensitivity of human breast cancer cells via G2/M cell cycle arrest. *Journal of Oncology, 2012*, 1–8.

Giri, S., Lokesh, C. R., Sahu, S., & Gupta, N. (2014). *Luffa echinata*: healer plant or potential killer? *Journal of Postgraduate Medicine, 60*, 72–74.

Khorana, M. L., & Raisinghani, K. H. (1961). Studies of *Luffa echinata* III: the oil and the saponin. *Journal of Pharmaceutical Sciences, 50*, 687–689.

Mariod, A. A., Ahmed, Y. M., Matthaeus, B., Khaleel, G., Siddig, A., Gabra, A. M., & Abdelwahab, S. I. (2009). A comparative study of the properties of six Sudanese cucurbit seeds and seed oils. *Journal of the American Oil Chemists' Society, 86*, 1181–1188.

Yadav, U., & Kumar, M. (2013). Luffa echinata: A valuable medicinal plant for the victims of dog bite. *Octa Journal of Environmental Research, 1*(1), 01–04.

Chapter 21

Luffa clynderica Loofah, Sponge Gourd

DISTRIBUTION AND BOTANICAL DESCRIPTION

Luffa cylindrica is a subtropical vegetable and widely cultivated in tropics, subtropics in Asia, India, Brazil, and USA. It is widely distributed as a cultivated and/or neutralized plant. In many countries the plant grows as a weed. *L. cylindrica* (Linn) (Cucurbitaceae) is the genus name of a group of gourds also known as vegetable sponges, dishcloth gourds, running okra, strainer vine, Chinese okra, California okra, and loofah. Two species are commonly known angled luffa is *Luffa acutangula*, while the smooth-fruited version is *L. cylindrica* or *Luffa aegyptiaca* (Stephens, 2015). *Luffa* is a large climber with a stout, 5 angled pubescent stem tendrils usually 2–3 branched. Leaves are large, 10–20 cm long, and 9–21 cm wide. Orbicular to reniform orbicular in outline, base cordate, 5–7 lobed, lobes broadly triangular, acute, margins irregularly, shallowly dentate, hispid, scab rid above, finely pubescent-hispid beneath, petioles 2–8 cm long. The male and female flowers deep, bright yellow and male flowers borne in racemes on peduncles 6–17 cm long. It is usually aggregated near apex, stamens usually 15–30 cm long and 6–10 cm in diameter, smooth with 10 dark green longitudinal lines, fibrous within, seeds 1–2 cm long and 0.8 cm wide, including the narrow, smooth, marginal wing, broadly ellipsoid, longitudinally compressed, rough on the surface (Balakrishnan & Huria, 2011).

The two species of *Luffa* are somewhat similar in appearance. Both are vigorous, climbing, annual vines with several lobed cucumber-like leaves. When crushed, the leaves give off a rank odor (Stephens, 2015). Both male and female flowers occur on the plant with a much greater number of male flowers. The rather large male flowers are bright yellow and occur in clusters. The female flowers are solitary and have the tiny, slender ovary attached. Angled luffa flowers appear later in the day than the smooth type and stay open through the night. Bees pollinate the flowers. Leaves are covered with short, stiff hairs. Smooth luffa fruits are shaped like cucumbers, but are larger, 1–2 ft. in length and 4–5 in. thick. The exterior is green, sometimes mottled, and smooth with longitudinal lines. Fruits of the angled luffa are characterized by sharp, elevated ridges running the length of the pods. The interiors of both are cucumber-like when

Unconventional Oilseeds and Oil Sources. http://dx.doi.org/10.1016/B978-0-12-809435-8.00021-4

immature, but quickly develop a network of fibers surrounding a large number of flat blackish seeds. Smooth luffa seeds are a bit narrower and have a pitted appearance (Stephens, 2015) (Fig. 21.1).

DIFFERENT USES AND MEDICINAL VALUE

Loofah sponge is fast becoming an indispensable crop because of its very wide industrial applications. In the context of the morphosynthesis, the capability of replication of the loofah sponge opens the possibility of the use of biodiversity in obtaining new materials. Effort is being made toward the possibility of

FIGURE 21.1 *Luffa cylindrica* **plant and seeds.**

harnessing, converting, and recycling waste seeds from edible fruits and those regarded as weeds (nonedible ones), like *L cylindrical* into industrial, domestic, or technological resources. Loofah sponge is a suitable natural matrix for immobilization of microorganisms and has been successful in the process of absorption of heavy metals from wastewater. This emerging cash crop will improve the economies of many nations in the near future because of its numerous potentials (Oboh & Aluyor, 2009).

Both luffas species have value as food items, but are seldom eaten. In Florida, USA most gardeners grow them in their fibrous interior, which is useful as a rough cloth or sponge for cleaning and scouring. Small (less than 6 in. long) young luffa gourds are used as a vegetable either prepared like squash or eaten raw like cucumbers. Some varieties are sweeter than others, particularly of the smooth type. Bitter types should not be eaten, as some poisonous properties have been reported. However, best sponges come from mature-green fruits, although dry fruits may be used. The fruits should be soaked for several days and then peeled. Once cleaned, the sponges should be bleached and then dried in the sun. Sponges have been used for cleaning, filtering, and bathing (Stephens, 2015).

Balakrishnan and Huria (2011) reported pharmacological activities of *L. cylindrica,* which include antiplasmodial and antibacterial activity. These authors investigated the preliminary phytochemical and various pharmacological activities (analgesic, antipyretic, and antidiabetic) activities of aqueous (AELC) and ethanol (EELC) extracts of *L. cylindrica* fruit in a scientific manner in rats. In their studies the conclusively damage of the pancreas in alloxan treated diabetic control rats and regeneration of beta cells by glibenclamide was observed. The comparable regeneration was also shown by ethanolic and aqueous extracts of the fruit of *L. cylindrica.* Antihyperglycemic and antihyperlipidemic activities of ethanol and aqueous extracts of the fruit of *L. cylindrica* may be attributed to the presence of various phytochemical, such as alkaloid, steroid, triterpinoid, flavonoid, and glycoside. Photomicrographical data in their studies reinforce healing of pancreas by *L. cylindrica* fruit extracts as a plausible mechanism of their antidiabetic activity. Their study has validated the medicinal potential of the fruit extracts of *L. cylindrical* (Balakrishnan & Huria, 2011).

Seeds of *L. cylindrica* are famous for their emetic and cathartic properties. Young fruits have demulcent and appetizers properties and are exertive of mind, bile, and phlegm. Fruits are used as vegetable in early stages whereas its medicinal uses increase with its maturity. *L. cylindrica* is famous for its fruit, eaten as vegetable, and is good for health. It has been used for the treatment of cynocytous, flu, and other diseases in Asia. It is used as an anthelmintic, stomachic, and antipyretic phytomedicinal drug (Kou, Zhuang, Tang, Tong, & Yan, 2001). The *n*-hexane fraction of *L. cylindrical* exhibited good (64%) and crude methanolic extract, moderate (58%) antibacterial activity against *Bacillus subtilis.* Butanol (BuOH) fraction presented moderate activity (58%) against *Shigella flexneri.* The BuOH fraction of *L. cylindrica* showed significant antifungal activity against *Fusarium solani* (85%) and *Trichophyton longifusus* (80%). The crude methanolic extract and ethyl acetate fraction (EtOAc) presented good linear growth inhibition against *Microsporum canis* (70%) (Ahmad & Khan, 2013). Fruit peel was evaluated for antiemetic and antiinflammatory effect using chick emesis model and carrageenan induced rat paw edema. The ethanol extract produced antiemetic activity by receptor antagonism and has peripheral antiemetic action. The extract was able to effectively prevent its effect and has a peripheral antiemetic action. Rats pretreated with *L. cylindrica* ethanol extract showed a significant edema inhibitory response at 5th hour following carrageenan injection. This result suggests that *L. cylindrica* extracts may act by suppressing the later phase of the inflammatory process by the inhibition of cyclooxygenase (Kanwal, Waseemuddin, Salman, & Mohtasheem, 2013). *L. cylindrica* seed oil is used in mineral based makeup to give a softer and smoother appearance. The seeds are pressed to produce the oil. The oil from the seed is considered an excellent lubricant, and externally used for shingles and boils, leprosy, and skin diseases. The kernel of the seed is an expectorant, and used in dysentery. In cosmetics, luffa cylindrical seed oil is used for its antifungal, antiinflammatory, and antitumor properties. Since it prevents synthesis of certain proteins, it is also considered toxic to skin cancer cells. It is high in the essential amino acid, arginine. The seeds have laxative properties, when eaten, due to their high oil content. The kernels provide essential, valuable, and useful minerals needed for good body development. Over half of the seed is oil. The luffa cylindrical seed oil has been used in sunscreens, sunless tanning products, antiaging products, facial moisturizers and body treatments, oils and facial cleansers. It is used in mineral based make up because of the healing powers of its antifungal and inflammatory traits. Its high toxicity to skin cancer cells is why it is used in sunscreen products. The high content of arginine, which is an essential amino acid that plays an important role in cell division and healing of wounds, provides healthy cell benefits when used in facial makeup. The seed oil is a rich emollient that adds smoothness and softness to the skin. Since it is a natural vegetable oil, it contains moisturizing elements for use in antiaging cosmetics (http://www.essence-of-mineral-makeup.com/luffa-cylindrica.html).

Different chemical treatments were conducted on the fibers of *L. cylindrica* with aqueous solutions of NaOH 2%, or methacrylamide (1%–3%) at distinct treatment times. *L. cylindrica* was characterized via chemical analysis and analytic techniques, such as FTIR. Different chemical treatments were effective to modify *L. cylindrica* fibers. The extent of their impact on the fiber was found to be dependent on the chemical agent, its concentration, and treatment time. Treatment with NaOH was found to adequately modify the fiber surface in preparation for its use as reinforcement of composite materials, whilst not causing such severe damage to the fiber as the methacrylamide treatment (Tanobe, Sydenstricker, Munaro, & Amico, 2005). The chloroform extract of whole plant of *L. cylindrica* (Linn) was found significant antibacterial and antifungal activity which might be helpful in preventing the progress of various diseases and can be used in alternative system of medicine (Indumathy, Kumar, Pallavi, & Devi, 2011).

FRUIT AND SEED COMPOSITION AND BIOACTIVE COMPONENTS

The fibers of *L. cylindrica* are mainly composed of cellulose (54.2%) and lignin (15.10%) and have been evaluated for the use in strengthening resin matrix compound materials (Kaufman, Csake, Warber, Duke, & Brielmann, 1999). Flavonoids, saponins, tannins, and cardiac glycoside compounds were found in the aqueous extract of *L. aegyptiaca* leaves with a wide range of biological activities with pharmaceutical importance (Mhya & Mankilik, 2014). The luffin, which was reported to be a ribosome-inactivating protein, isolated from *Luffa* seed. Many polyphenols including *p*-coumaric acid, 1-O-feruloyl-*b*-D-glucose, 1-O-*p*-coumaroyl-*b*-D-glucose, 1-O-caffeoyl-*b*-D-glucose, 1-O-[4-hydroxybenzoyl]-glucose, diosmetin-7-O-*b*-Dglucuronide methyl ester, apigenin-7-O-*b*-D-glucuronide methyl ester, and luteolin-7-O-*b*-D glucuronide methyl ester have been discovered in *Luffa* pulp (Ng et al., 2011). Also, sponge gourd was shown to contain 20.74 mg/g of total phenolics, 17.94 mg/g of flavonoids, 0.5 mg/g of total anthocyanins, and 1.2 mg/g of ascorbic acid (Reddy, Reddy, Reddy, Mohan, & Sarma, 2010). While in *Luffa* seed, both oleanolic acid and echinocystic acid were the major triterpene acids present (Khajuria, Gupta, Garai, & Wakhloo, 2007). The variety and amount of triterpene acids in *Luffa* pulp and peel remain unknown (Kao, Huang, & Chen, 2012). Fresh *Luffa* pulp was found to contain vitamin C at 60 g/g. The *Luffa* extracts contain various chlorophylls (chlorophyll A and chlorophyll B) and carotenoids (9- or 90-*cis* neoxanthin, all-*trans*-lutein, 9- or 90-*cis*-lutein, all-*trans*-β-carotene, and 9- or 90-*cis*-β-carotene) (Azeez, Bello, & Adedeji, 2013). *Luffa* peel, a major waste produced during preparation of *Luffa* prior to cooking, also proved to be a vital source of functional components.

Adewuyi and Oderinde (2012) evaluated oil extracted from seeds of *L. cylindrica* in order to study their chemical compositions and lipid profile. Oil content of seed of *L. cylindrica* was 390.10 ± 0.20 g/kg. Iodine value of *L. cylindrica* was 147.01 ± 0.70 cg iodine/g. Seed as well as oils of *L. cylindrica* were found rich in K and Ca. C18:2 was the dominant fatty acid found in the oils while neutral lipids were the dominant lipid class. Sitosterol, beta-tocopherol, phytol, stigmasterol, and hydrocarbons were identified in the unsaponifiable matter of the oil using GC–MS. HPLC results revealed presence of glycolipids, such as monogalactosyldiacylglycerol, digalactosyl diacyl glycerol, and digalactosyl monoacyl glycerol. Phosphatidyl ethanol amine was the dominant phospholipids in the oils. Molecular speciation of triacylglycerol revealed presence of C_{38} (LLnLn) and C_{40} (LLLn/LnLnP/LnLnO) to be dominantly present in the oil.

The essential mineral elements (Na, K, Mg, and Ca) were found in high amounts with K being the highest in the seeds and oils of *L. cylindrical*. K was found to be 1216.66 mg/kg in the seed. The oil of *L. cylindrica* had K as 866.66 mg/kg. Calcium was also found relatively high in the seeds (933.87 mg/kg and oils) it and is higher than the oils (858.73 mg/kg). The concentration of Na was found higher in the seed (384.72 mg/kg) is higher than the oils 348.40 ± 0.04 mg/kg. The accumulation of Mg and Fe also varied in the seeds and their oils. Mg was found higher in the seed (239.13 mg/kg) and 228.50 mg/kg in the oil, while Fe of *L. cylindrica* was 31.40 ± 0.01 mg/kg in the seed and 24.17 ± 0.01 mg/kg in the oil. The concentration of Pb was found as 4.57 ± 0.03 mg/kg in the seeds, while it was found to be 2.90 ± 0.05 mg/kg in the oil. Cd was detected in the seed (1.17 ± 0.01 mg/kg) and oil (0.73 ± 0.01 mg/kg) (Adewuyi & Oderinde, 2012).

REFERENCES

Adewuyi, A., & Oderinde, R. A. (2012). Analysis of the lipids and molecular speciation of the triacylglycerol of the oils of *Luffa cylindrica* and *Adenopus breviflorus*. *CyTA—Journal of Food, 10*(4), 313–320.

Ahmad, B., & Khan, A. A. (2013). Antibacterial, antifungal and phytotoxic activities of *Luffa cylindrica* and *Momordica charantia*. *Journal of Medicinal Plants Research, 7*(22), 1593–1599.

Azeez, A. M., Bello, O. S., & Adedeji, A. O. (2013). Traditional and medicinal uses of *Luffa cylindrica*: a review. *Journal of Medicinal Plants Studies, 1*(5), 102–111.

Balakrishnan, N., & Huria, T. (2011). Protective effect of *Luffa cylindrica* L. fruit in paracetamol induced hepatotoxicity in rats. *International Journal of Pharmaceutical and Biological Archives, 2*(6), 176–1764.

Indumathy, R., Kumar, S. D., Pallavi, K., & Devi, G. S. (2011). Antimicrobial activity of whole plant of *Luffa cylindrica* (Linn) against some common pathogenic micro-organisms. *International Journal of Pharmaceutical Sciences and Drug Research, 3*(1), 29–31.

Kanwal, W., Waseemuddin, S. A., Salman, A., & Mohtasheem, H. M. (2013). Anti-emetic and anti-inflammaotry activity of fruit peel of *Luffa cylindrica* (L.) Roem. *Asian Journal of Natural and Applied Sciences, 2*(2), 175–180.

Kao, T. H., Huang, C. W., & Chen, B. H. (2012). Functional components in *Luffa cylindrica* and their effects on anti-inflammation of macrophage cells. *Food Chemistry, 135*(2), 386–395.

Kaufman, P. B., Csake, L. J., Warber, S., Duke, J. A., & Brielmann, H. L. (1999). *Natural products from plants*. FL: Boca Raton Press.

Khajuria, A., Gupta, A., Garai, S., & Wakhloo, B. P. (2007). Immunomodulatory effects of two sapogenins 1 and 2 isolated from *Luffa cylindrica* in Balb/C mice. *Bioorganic and Medicinal Chemistry Letters, 17*, 1608–1612.

Kou, J., Zhuang, S., Tang, X., Tong, C., & Yan, Y. (2001). Preliminary studies of *Luffa* extract on allergy and acute inflammation. *Journal of Chinese Pharmacology, 23*, 293–296.

Mhya, D. H., & Mankilik, M. (2014). Phytochemical screening of aqueous extract of *Luffa aegyptiaca* (Sponge gourd) leave sample from Northern Nigeria. *International Journal of Pharma Sciences and Research, 5*(7), 344–345.

Ng, Y. M., Yang, Y., Sze, K. H., Zhang, X., Zheng, Y. T., & Shaw, P. C. (2011). Structural characterization and anti-HIV-1 activities of arginine/glutamate-rich polypeptide Luffin P1 from the seeds of sponge gourd [*Luffa cylindrica*]. *Journal of Structural Biology, 174*, 164–172.

Oboh, I. O., & Aluyor, E. O. (2009). Review *Luffa cylindrica*—an emerging cash crop. *African Journal of Agricultural Research*, 4(8), 684–688.

Reddy, B. P., Reddy, A. R., Reddy, B. S., Mohan, S. V., & Sarma, P. N. (2010). Apoptosis inducing activity of *Luffa acutangula* fruit in leukemia cells [HL-60]. *International Journal of Pharmaceutical Research and Development*, 2, 109–122.

Stephens, J. M. Gourd, Luffa—*Luffa cylindrica* (L.) Roem., *Luffa aegyptica* Mill., and *Luffa acutangula* (L.) Roxb. (2015). http://edis.ifas.ufl.edu.

Tanobe, V. O. A., Sydenstricker, T. H. D., Munaro, M., & Amico, C. S. (2005). A comprehensive characterization of chemically treated Brazilian sponge-gourds (*Luffa cylindrica*). *Polymer Testing*, 24, 474–482.

Chapter 22

Trigonella foenum-graecum
Fenugreek, Bird's Foot, Greek Hayseed

BOTANICAL DESCRIPTION AND DISTRIBUTION

Trigonella foenum-graecum (Linn.) belonging to the family Papilionaceae commonly known as fenugreek is an aromatic, 30–60 cm tall, annual herb, cultivated throughout the country (Toppo, Akhand, & Pathak, 2009). The name comes from *foenum-graecum*, meaning Greek Hay, the plant being used to scent inferior hay. Fenugreek is an erect annual herb, growing about 2 ft. high. A nearly smooth erect annual. Leaflets 2–2.5 cm long, oblanceolate-oblong, toothed. Flowers 1–2, calyx-teeth linear, corolla much exerted, pod 5–7.5 cm long, with a long persistent beak, often falcate, 10–29 seeded, without transverse reticulations (Toppo et al., 2009). The seeds are brownish, about 1/8 in. long, oblong, rhomboidal, with a deep furrow dividing them into two unequal lobes. They are contained, 10–20 together, in long, narrow, sickle-like pods. The plant is indigenous to the countries of the Mediterranean. Cultivated in India, Africa, Egypt, Morocco, and occasionally in England (https://www.botanical.com). Major producing countries are Afghanistan, Pakistan, India, Nepal, Iran, Bangladesh, Argentina, Egypt, France, Spain, Turkey, and Morocco (Parthasarathy, Kandinnan, & Srinivasan, 2008).

DIFFERENT USES AND MEDICINAL VALUE

Fenugreek is used as an herb (dried or fresh leaves), spice (seeds), and vegetable (fresh leaves, sprouts, and microgreens). It is the famous spices in human food. The seeds and green leaves of fenugreek are used in food as well as in medicinal application that is the old practice of human history. It has been used to increase the flavoring and color, and also modifies the texture of food materials (Wani & Kumar, 2016). Cuboid-shaped, yellow- to amber-colored fenugreek seeds are frequently encountered in the cuisines of many countries, used both whole and powdered in the preparation of pickles, and vegetable dishes. They are often

Unconventional Oilseeds and Oil Sources. http://dx.doi.org/10.1016/B978-0-12-809435-8.00022-6

roasted to reduce bitterness and enhance flavor (https://en.wikipedia.org/wiki/Fenugreek). The leaves contain seven saponins, known as graecunins. These compounds are glycosides of diosgenin. Leaves contain about 86.1% moisture, 4.4% protein, 0.9% fat, 1.5% minerals, 1.1% fiber, and 6% carbohydrates. The mineral and vitamins present in leaves include calcium, zinc, iron, phosphorous, riboflavin, carotene, thiamine, niacin, and vitamin C (Rao, 2003). Fresh leaves of fenugreek contain ascorbic acid of about 220.97 mg per 100 g of leaves and β-carotene is present about 19 mg/100 g. On the other side, it was reported that 84.94% and 83.79% ascorbic acid were reduced in the sun and oven-dried fenugreek leaves, respectively. Fresh leaves are used as vegetables in the diets. It was found that there was a better retention of nutrients in the leaves of fenugreek. The leaves of fenugreek should be stored in either in refrigeration conditions, or dried in oven, or blanched for some time (about 5 min) and should be cooked in pressure cooker (Wani & Kumar, 2016).

Sprouted seeds and micro greens are used in salads, and said to be equal to quinine in preventing fevers; and have been utilized for diabetes. The seeds are used in the preparation of emollient cataplasms, ointments, and plasters. Externally the fine ground seeds paste is used as a poultice for abscesses, boils, carbuncles, etc. For neurasthenia, gout, and diabetes it can be combined with insulin (https://www.botanical.com). Fenugreek is known for its pleasantly bitter, slightly sweet seeds. The seeds are available in any form, whether whole or ground form is used to flavor many foods, mostly curry powders, teas, and spice blend. Fenugreek seed has a central hard and yellow embryo which is surrounded by a corneous and comparatively large layer of white and semi-transparent endosperm. The extracts of endosperm husk, and fenugreek seed at about 200 µg concentration exhibited antioxidant activity 72%, 64%, and 56%, respectively by free-radical scavenging method (Naidu, Shyamala, Naik, Sulochanamma, & Srinivas, 2010) (Fig. 22.1).

Fenugreek is one of the primary supplements used to support noninsulin-dependent diabetes mellitus (type II diabetics). Fenugreek seed helps to reduce blood sugar levels with its high concentrations of phytochemicals, and also reduce low-density cholesterols and triacylglycerols. The plant was found to possess different activities, such as anticancer, antiinflammatory, antiseptic, aphrodisiac, astringent, bitter, demulcent, emollient, expectorant, anthelmintic, wound healing, and gastroprotective. Fenugreek seeds are a rich source of the polysaccharide galactomannan. They are also a source of saponins, such as diosgenin, yamogenin, gitogenin, tigogenin, and neotigogens. Other bioactive constituents of fenugreek include mucilage, volatile oils, and alkaloids, such as choline and trigonelline (Toppo et al., 2009). Fenugreek seed is used in Ethiopia as a natural herbal medicine in the treatment of diabetes. When the seed kernels are ground and mixed with water they greatly expand, to which hot spices, turmeric, and lemon juice are added to produce a frothy relish eaten with a soup. It is recommended by lactation advisers and used by breastfeeding mothers to help stimulate milk production and supply and it is also supposed to help either

FIGURE 22.1 *Trigonella foenum-graecum* (Fenugreek) plant and seeds.

gain or reduce weight (http://ethnomed.org/clinical/pharmacy/ethiopian-herb-drug-interactions). Dietary fiber of fenugreek seed is about 25%, which change the texture of food. These days it is used as food stabilizer, adhesive, and emulsifying agent due to its high fiber, protein and gum content. The protein of fenugreek is found to be more soluble at alkaline pH (Meghwal & Goswami, 2012). The seeds of fenugreek have been used medicinally all through the ages and were held in high repute among the Egyptians, Greeks, and Romans for medicinal and culinary purposes (https://www.botanical.com).

The plant and seeds are hot and dry, suppurative, aperient, diuretic, emmenagogue, useful in dropsy, chronic cough, enlargement of the liver and the spleen. The leaves are useful in external and internal swellings and burns; prevent the hair falling off. Fenugreek seeds are considered carminative, tonic, and aphrodisiac. The plant used in dyspepsia with loss of appetite, in the diarrhea of puerperal women, and in rheumatism. The toasted infused seeds are used by native practitioners in southern India for dysentery. The leaves are used both externally and internally, on account of their cooling properties. An infusion of the seeds is given to smallpox patients as a cooling drink (Toppo et al., 2009; Prajapati, Purohit, Sharma, & Kumar, 2003). Fenugreek has been found to stimulate sweat production as it contains hormone precursor to increase milk formation. Some scientists reported that fenugreek can increase a nursing mother's milk supply within 24–72 h after first taking the herb (Snehlata & Payal, 2012). Fenugreek flour has been incorporated up to a 10% level in the formulation of biscuits without affecting their overall quality. The physical, sensory, and nutritional characteristics generally revealed that biscuits containing 10% germinated fenugreek flour were the best among all the composite fenugreek flour biscuits. Hence, development and utilization of such functional foods will not only improve the nutritional status of the general population, but also helps those suffering from degenerative diseases (Hooda & Jood, 2005).

Fenugreek seed (raw, soaked, and germinated) significantly reduced total lipids, serum total cholesterol, and LDL cholesterol. It can be recommended that fenugreek may be used for lipid lowering purposes (Hussein, Abd El-Azeem, Hegazy, & Abeer, 2011).

Supplementation of basal diets with fenugreek leaves, seeds (dry and germinated), and wheat flour supplemented with germinated fenugreek powder at 5%–10%. Nabila, Mahmoud, Salem, and Amal (2012) recommended that intake of fenugreek dietary supplements may be beneficial for patients who suffer from iron deficiency anemia owing to their nutritive and restorative values. These dietary supplements increased the total proteins, fibers, iron, zinc, calcium, vitamin B_2, carotene, vitamin E, and vitamin C contents and also improve the blood picture of anemic rats so they have nutritive and restorative properties. These authors confirmed the use of fenugreek dietary supplement as safe and healthy.

Fenugreek has been used as food stabilizer, food adhesive, food emulsifier, and gum. Fenugreek has been used to produce various types of bakery products and extruded product. Fenugreek can be recommended and must be taken as a part of our daily diet as its liberal use is safe and various health benefits can be drawn from this natural herb. The functional, nutritional, and therapeutic characteristics of fenugreek can be exploited further in the development of healthy products (Wani & Kumar, 2016).

FRUIT AND SEED COMPOSITION AND BIOACTIVE COMPONENTS

The seeds of fenugreek contain about 0.1%–0.9% of diosgenin and are extracted commercially. The plant tissue cultures from the seeds of fenugreek when grown under optimal conditions have been found to produce as much as 2% diosgenin with smaller amounts of trigogenin and gitongenin. Seeds also contain the saponin (fenugrin B). Fenugreek seeds have been found to contain several coumarin compounds as well as a number of alkaloids (e.g., trigonelline, gentianine, carpaine). The large amount of trigonelline is degraded to nicotinic acid and related pyridines during roasting (Wani & Kumar, 2016). Two alkaloids, trigonelline and choline, and a yellow coloring substance were extracted from the plant. The chemical composition of seed oil resembles that of cod-liver oil, as it is rich in phosphates, lecithin, and nucleoalbumin, with considerable quantities of iron in an organic form, which can be readily absorbed. Trimethylamine, neurin, and betaine are found; these substances stimulate the appetite by their action on the nervous system, or produce a diuretic or ureo-poietic effect. Sotolon is the chemical responsible for fenugreek's distinctive sweet smell (https://www.botanical.com).

Fenugreek seeds contain 26% mucilage, 22% protein comprising of globulin, histidine, and albumin with a good amount of phosphorus, sulfur, and also lecithin, it contains also 50% of soluble and insoluble fiber found essential for good health (Al-Jasass & Al-Jasser, 2012). Fenugreek seeds contain a low amount of fat 4.51%, crude fiber 13.1%, and ash content 4.2% and 57.5% total carbohydrate. The oil contains ω-3, ω-6, and ω-9 fatty acids along with many saponins, alkaloids, and sterols that serve as a source of proestrogens and inhibit intestinal cholesterol absorption (Mathur & Choudhry, 2009). The seeds contain high levels of potassium (603.0 mg/100 g), followed by calcium (75.0 mg/100 g), magnesium (42.0 mg/100 g), and iron (25.8 mg/100 g), with low levels of zinc, manganese, and copper. The major fatty acids were linoleic acid (48.43%) and linolenic acid (30.8%), and oleic (10.76%). Total unsaturated fatty acids were 92.99% (Al-Jasass & Al-Jasser, 2012). Kochhar, Nagi, and Sachdeva (2006) reported that fenugreek seeds contain 11.8% moisture, 25.8% crude protein, 6.53% oil, 3.26% ash, and 6.28% crude fiber and 58.13% total carbohydrates on dry basis. However, El-Nasri and El-Tinay (2007) found that the protein content of fenugreek was found to be 28.4%, crude fiber content was 9.3%, and crude fat was 7.1%. The fatty acid profile was dominated by unsaturated acids, namely, oleic, linoleic, and linolenic acids accounting for 16.3%, 50%, and 24.4%, respectively of the total fatty acids. However, El-Sebaiy and El-Mahdy (1983) reported that the fatty acids C18:2 and C18:3 were the most abundant fatty acids in the lipids of the fenugreek seeds. Thus fenugreek seeds may serve to be beneficial health food if consumed.

REFERENCES

Al-Jasass, F. M., & Al-Jasser, M. S. (2012). Chemical composition and fatty acid content of some spices and herbs under Saudi Arabia conditions. *The Scientific World Journal, 2012,* 859892.

El-Nasri, N. A., & El-Tinay, A. H. (2007). Functional properties of fenugreek (*Trigonella foenum graecum*) protein concentrate. *Food Chemistry, 103,* 582–589.

El-Sebaiy, L. A., & El-Mahdy, A. R. (1983). Lipid changes during germination of fenugreek seeds (*Trigonella foenum graecum*). *Food Chemistry, 10*(4), 309–319.

Hooda, S. H., & Jood, S. (2005). Effect of fenugreek flour blending on physical, organoleptic and chemical characteristics of wheat bread. *Nutrition & Food Science, 35*(4), 229–242.

Hussein, A. M. S., Abd El-Azeem, A. S., Hegazy, A. M., & Abeer, A. (2011). Physiochemical, sensory and nutritional properties of corn-fenugreek flour composite biscuits. *Australian Journal of Basic Applied Sciences, 5,* 84–95.

Kochhar, A., Nagi, M., & Sachdeva, R. (2006). Proximate composition available carbohydrates, dietary fiber and anti-nutritional factors of selected traditional medicinal plants. *Journal of Human Ecology, 19,* 195–199.

Mathur, P., & Choudhry, M. (2009). Consumption pattern of fenugreek seeds in Rajasthani families. *The Journal of Human Ecology*, 25(1), 9–12.

Meghwal, M., & Goswami, T. K. (2012). A review on the functional properties, nutritional content, medicinal utilization and potential application of fenugreek. *Journal of Food Processing & Technology*, 3(9), 1–10.

Nabila, Y., Mahmoud, R., Salem, H., & Amal, A. (2012). Mater nutritional and biological assessment of wheat biscuits supplemented by fenugreek plant to improve diet of anemic rats. *Academic Journal of Nutrition*, 1(1), 01–09.

Naidu, M. M., Shyamala, B. N., Naik, P. J., Sulochanamma, G., & Srinivas, P. (2010). Chemical composition and antioxidant activity of the husk and endosperm of fenugreek seeds. *Food Science Technology*, 44, 451–456.

Parthasarathy, V. A., Kandinnan, K., & Srinivasan, V. (2008). "Fenugreek." *Organic spices* (p. 694). New India Publishing Agencies.

Prajapati, Purohit, Sharma, & Kumar (2003). *A Handbook of medicinal plants—a complete source book (p. 523)*. India: Agrobios.

Rao, A. V. (2003). *Herbal Cure for Common Diseases*. New Delhi, India: Fusion Books.

Snehlata, H. S., & Payal, D. R. (2012). Fenugreek (*Trigonella foenumgraecum* L.): an overview. *International Journal of Current Pharmaceutical Review and Research*, 2(4), 169–187.

Toppo, F. A., Akhand, R., & Pathak, A. K. (2009). Pharmacological actions and potential uses of *Trigonella foenum-graecum*: a review. *Asian Journal of Pharmaceutical and Clinical Research*, 2(4), 29–38.

Wani, S. A., & Kumar, P. (2016). Fenugreek: a review on its nutraceutical properties and utilization in various food products. *Journal of the Saudi Society of Agricultural Sciences*, Available from http://dx.doi.org/10.1016/j.jssas.2016.01.007.

Chapter 23

Guizotia abyssinica (L.f.) Cass. Niger

BOTANICAL DESCRIPTION AND DISTRIBUTION

Niger is an annual dicotyledonous herb. The root system is well developed, with a central tap-root and its lateral branching. The stem is smooth to slightly rough and the plant is usually moderately to well branched. Niger stems are hollow and break easily. The number of branches per plant varies from 5 to 12 and in very dense plant stands fewer branches are formed. The color of the stem varies from dark purple to light green and the stem is about 1.5 cm in diameter at the base. The leaves are arranged on opposite sides of the stem; at the top of the stem leaves are arranged in an alternate fashion. Leaves are 10–20 cm long and 3–5 cm wide. The leaf margin morphology varies from pointed to smooth and leaf color varies from light green to dark green, the leaf surface is smooth. The plant height of niger is an average of 1.4 m, but can vary considerably as a result of environmental influences and heights of up to 2 m have been reported from the Birr valley of Ethiopia. The niger flower is yellow and, rarely, slightly green. The heads are 15–50 mm in diameter with 5–20 mm long ray florets. Two to three capitula (heads) grow together, each having ray and disk florets. The receptacle has a semispherical shape and is 1–2 cm in diameter and 0.5–0.8 cm high. The receptacle is surrounded by two rows of involucral bracts. The capitulum consists of six to eight fertile female ray florets with narrowly elliptic, obovate ovules. The stigma has two curled branches about 2 mm long. The hermaphrodite disk florets, usually 40–60 per capitulum, are arranged in three whorls. The disk florets are yellow to orange with yellow anthers, and a densely hairy stigma. The achene is club-shaped, obovoid, and narrowly long. The head produces about 40 fruits. The achenes are black with white to yellow scars on the top and base and have a hard testa. The embryo is white (Getinet & Sharma, 1996). The fruit is an achene, small, 3–5 mm in length and 1.5 mm in width, almost lanceolate in shape, without pappus. There are usually between 15 and 30 mature seeds/head; occasionally more, and a varying number of immature seeds or pops at the center (Yadav et al., 2012) (Fig. 23.1).

The genus *Guizotia* has six species. All species except *Guizotia abyssinica* are wild and are endemic to East Africa especially Ethiopia. Center of origin for this crop is Ethiopia from where it was introduced to India (Yadav et al., 2012).

Unconventional Oilseeds and Oil Sources. http://dx.doi.org/10.1016/B978-0-12-809435-8.00023-8

FIGURE 23.1 *Guizotia abyssinica* plant, flowers and seeds.

India is the most important country accounting for more than 50% of world Niger area and production. Niger constitutes about 50%–60% of Ethiopian oilseed production, with a productivity level of 500 kg/ha. In Ethiopia, it is cultivated on water logged soils where most crops and all other oilseeds fail to grow and contributes a great deal to soil conservation and land rehabilitation. In India, niger is grown in many states.

DIFFERENT USES AND MEDICINAL VALUE

The seed is eaten fried, used as a condiment, or dried then ground into a powder and mixed with flour to make sweet cakes. It is used as a substitute for olive oil, can be mixed with linseed oil, and is used as an adulterant for rape oil, sesame oil etc. The oil is used in cooking as a ghee substitute and can be used in salad dressings, etc. (Facciola, 1990). In Ethiopia, niger seed is used at house hold level to make oil, to make paste known as "litlit" and also paste mix with snacks, which is commonly given to a large family to subdue appetite at extra meal hour. Niger seed paste mixed with roasted cereals or sandwiched with flat bread or "injera" is used during holidays in Ethiopia, especially in Northern Ethiopia. The niger seed roasted and pounded into flour is boiled and inhaled, and then drunk as a remedy for common cold. The flour or the meal from the oil press is also used to oil/smoothen heated baking clay pan before baking as "masesha" (Melaku, 2013).

In North America, the production of niger is mainly for bird-feed and is being sold under various names, such as thistle or niger seed. Niger seed oil is being used for various industrial purposes, such as soaps, paints, illuminants, and lubricants and its potential as a source of biodiesel was also reported (Devi, Kumar, Naidu, & Rao, 2006; Sarin, Sharma, & Khan, 2009). In addition to its oil, the crop offers an important source of seed protein that significantly contributes to the human dietary protein intake. Niger seeds are also used in human consumption in southern parts of India. Niger seeds are used to make dry chutney, which is used as an accompaniment with breads. They are also used as a spice in some curries. In Ethiopia, an infusion made from roasted and ground Niger seeds, sugar, and water, is used in treating common colds. The oil from the seeds is used in the treatment of rheumatism. It is also applied to treat burns. A paste of the seeds is applied as a poultice in the treatment of scabie (Chopra, Nayar, & Chopra, 1986). The oil from the seeds is used in the treatment of rheumatism. Niger seed can be used to treat the rheumatoid by application of oil on the parts where the pain is present. By periodically consumption of the niger seed oil, the risk of the disorder can be avoided (Adarsh, Poonam, & Shilpa, 2014). The oil is also used in birth control and to treat syphilis. The oil is applied topically to treat burns. A medical test for the identification of the fungus *Cryptococcus neoformans*, which causes a serious brain disease, is carried out on a niger seed-based agar medium. Niger seed sprouts, mixed with garlic and honey, are taken to treat cough (Manandhar, 2002). A drying oil is obtained from the seed. It is used medicinally, for burning, in making soap, paints, etc. The oil readily absorbs smells and appears to offer some scope in cosmetics. In Ethiopia, the straw is used as fuel for cooking (Burkill, 2010). Niger seed is a novel biologically approaches to overcome against nematodes and arthropods. Niger seed can be used as poultices which can be applied to the surface of the body to relieve pain, itching, swelling, and inflammation, abscesses, boils, etc. A niger seed poultice is the most effective treatment available for many types of disorders. Even today, they may provide effective, economical benefits (Adarsh et al., 2014).

The high oleic acid genotypes make niger oil suitable for cooking purposes while the high linoleic acid content makes the oil of niger nutritionally valuable. Thus, niger could be nutritionally considered as a new potential source of edible seed oils. It also has potential for industrial uses as in paint industries and soap making factories. It can be used as alternative fuel options for biodiesel. The seeds are thus valuable not only in meeting demands for food and food supplements with functional, health-promoting properties but also to industrial uses (Yadav et al., 2012).

FRUIT AND SEED COMPOSITION AND BIOACTIVE COMPONENTS

The success of oilseed crops is partly measured by their high seed oil content and quality and high-value protein meal after oil extraction. Previous reports on the oil content of niger seeds were within the range of 27%–47% of dry seed weight (Asilbekova, Ulchenko, Rakhimova, Nigmatullaev, & Glushenkoval, 2005; Ramadan & Morsel, 2003a) with an overall mean of about 35%.

The analysis of 153 niger populations revealed a twofold variation in seed oil content (27%–56%) with 7% of the populations having more than 50% oil. There was a high variation in oleic acid content between populations (3.3%–31.1%). Breeding of selected genotypes for two generations revealed a highly significant positive correlation between the parents and their progenies both in oil and oleic acid contents (Geleta, Stymne, & Bryngelsson, 2011).

Niger seed grown in the USA exhibited the best visual characters (glossiest and darkest brown seeds), with lower levels of moisture and foreign matter. The USA samples contain 36.6% oil content, while imported seeds varied from 36.5% to 38.2%. The USA grown seeds showed higher levels of proteins (28.2%), while imported seeds exhibited lower levels (18.2%–26.6%). Fatty acid composition showed that the USA sample was high in linoleic acid (71.3%), one of the imported seeds showed a similar profile (70.9% of linoleic acid) (Fatimaa, Villania, Komar, Simon, & Juliania, 2015). The high C18:2 types can preferably be used for fresh consumption, especially by roasting the seeds and processing in different ways, which is common, for example, among Ethiopian farmers whereas the oil from the high C18:1 types are preferable for high temperature cooking, for use in processed and fortified foods, and as a biodiesel (Geleta et al., 2011).

The meal remaining after the oil extraction contains about 24% protein and 24% crude fiber. Niger meal from India contains higher protein (30%) and lower crude fiber (17%) levels than meal from Ethiopia. Niger cake replacing linseed cake at levels of 0, 50, and 100% was fed as a nitrogen supplement for growing calves (Getinet & Sharma, 1996).

The predominant fatty acid in common niger seed oil is linoleic acid (C18:2), which is within the range of 65%–80% of the total fatty acids in Ethiopian niger and 45%–66% of the total fatty acids in Indian niger. The major monounsaturated fatty acid in niger seed oil is oleic acid (C18:1), which commonly accounted for less than 13% of the total fatty acids in Ethiopian niger (Alemaw & Teklewold, 1995).

The oil, protein, and crude fiber contents of niger are affected by the hull thickness and thick-hulled seeds tend to have less oil and protein and more crude fiber. Most of the total lipid was triacylglycerides and polar lipids accounted for 0.7%–0.8% of the total lipid content. The amount of total tocopherol was 720–935 µg/g oil of which approximately 90% was α-tocopherol, 3%–5% was gamma-tocopherol and approximately 1% was β-tocopherol. As α-tocopherol is an antioxidant, high levels of α-tocopherol could improve stability of niger oil. The total sterol consists of β-sitosterol (38%–43%), campesterol ($\sim$14%), stigmasterol ($\sim$14%), Δ5-avenasterol (5%–7%), and Δ7-avenasterol ($\sim$4%) (Getinet & Sharma, 1996).

Ramadan and Morsel (2003b) used column chromatography with solvent of increasing polarity to fractionate niger lipids; they reported a yield of 93.0% as neutral lipids and 4.90% as glycolipids, and 0.60% as phospholipids (PL). They separated PL subclasses via HPLC. The major individual PL subclasses were found to be phosphatidylcholine followed by phosphatidylethanolamine, phosphatidylinositol, and phosphatidylserine, respectively.

Phosphatidylglycerol and lysophosphatidylcholine, on the other hand, were detected in smaller quantities.

Ramadan and Morsel (2003) separated total glycolipids (TGL) from niger (*G. abyssinica* Cass.) seed oils by silica gel chromatography. Different GL subclasses were then identified and separated using high-performance liquid chromatography with ultraviolet adsorption (HPLC/UV). Among the TGL from niger oilseed, acylated steryl glucoside (ASG), steryl glucoside (SG), and glucocerebroside (CER) were detected. The linoleic acid C18:2n–6 was the dominating fatty acid, followed by oleic acid C18:1n–9. The fractions from niger oilseed showed only three distinct sugar and sterols peaks. As component sugar, glucose was the only sugar detected in all samples.

Bhagya and Shamanthaka (2003) dehulled niger seeds using hot lye treatment. They found that, the protein and fat contents were increased from 24 to 35 and 31–53 g/100 g, and the crude fiber was decreased from 16.9 to 2.2 g/100 g, respectively. They evaluated the defatted flour prepared from dehulled niger seeds for chemical, functional, and nutritional properties and compared to the undehulled flour. They reported that protein content of the dehulled flour was increased from 44 to 63 g/100 g. Dehulling resulted in inactivation of trypsin inhibitor activity. The dehulled flour had higher water and fat absorption capacities. However, the nitrogen solubility, emulsification capacity, and foaming properties were decreased. Chemical score of the dehulled flour was higher; threonine was the first limiting amino acid followed by lysine and isoleucine. The in vitro digestibility of the protein increased due to dehulling (85.5%) compared to that of undehulled flour (76%). These authors calculated the nutritional indices, essential amino acid index, biological value, nutritional index, and C-PER; they found it higher in dehulled flour compared to undehulled flour. The available lysine content was unchanged. The amino acid composition of niger protein was deficient in tryptophan. The protein quality of Ethiopian niger was evaluated using chemical score and essential amino acid requirement score. Using chemical score and whole egg protein as a standard, methionine, lysine, cysteine, isoleucine, and leucine were considered as limiting amino acids. When essential amino acids were used as a reference, lysine was the limiting amino acid. A lipoprotein concentrate was isolated from niger seed using hot water/ ethanol sodium chloride solution extraction. The lipoprotein contained 4% moisture, 12% ash, 46% protein, 20% fat, 7% crude fiber, and 11% soluble carbohydrate. From the amino acid composition, nitrogen to protein conversion ratio of 5.9 was calculated. The energy content of the niger lipoprotein concentrate was 400 kcal/100 g (Getinet & Sharma, 1996).

Sarin et al. (2009) synthesized biodiesel from crude niger seed oil with more than 3% FFA in a single step base catalyzed transesterification, with approximately 99% ester conversion. They also studied the effect of various reaction parameters and effect of impurities, like FFA and moisture, on the product conversion and purity. They reported that, the fuel properties of synthesized biodiesel meet specifications, except oxidation stability. They improved the stability of the produced biodiesel by using phenolic and aminic antioxidants.

Further they blended niger biodiesel with diesel at different dosages and they tested it according to IS 1460 specifications. Their result confirms the possibility of utilizing *G. abyssinica* seed oil for biodiesel production as a potential alternative feedstock.

REFERENCES

Adarsh, M. N., Poonam, Kumari, & Shilpa, Devi. (2014). A review of *Guizotia abyssinica*: a multipurpose plant with an economic prospective. *Journal of Industrial Pollution Control, 30*(2), 277–280.

Alemaw, A., & Teklewold, A. (1995). An agronomic and seed-quality evaluation of niger (*Guizotia abyssinica* Cass.) germplasm grown in Ethiopia. *Plant Breeding, 114*, 375–376.

Asilbekova, D. T., Ulchenko, N. T., Rakhimova, N. K., Nigmatullaev, A. M., & Glushenkoval, A. I. (2005). Seed lipids from *Crotalaria alata* and *Guizotia abyssinica*. *Chemistry of Natural Compounds, 41*, 596–597.

Bhagya, S., & Shamanthaka, M. C. (2003). Chemical, functional and nutritional properties of wet dehulled niger (*Guizotia abyssinica* Cass.) seed flour. *LWT—Food Science and Technology, 36*, 703–708.

Burkill, H. M. (2010). *The useful plants of West Tropical Africa*. UK: Royal Botanic Gardens, Kew (1985-2004).

Chopra, R. N., Nayar, S. L., & Chopra, I. C. (1986). *Glossary of Indian medicinal plants (including the supplement)*. New Delhi, India: Council of Scientific and Industrial Research.

Devi, A. N., Kumar, C. M., Naidu, Y. R., & Rao, M. N. (2006). Production and evaluation of biodiesel from sunflower (*Helianthus annus*) and niger seed oil (*Guizotia abyssinica*). *Asian Journal of Chemistry, 18*, 2951–2958.

Fatimaa, A., Villania, T. S., Komar, S., Simon, J. E., & Juliania, H. R. (2015). Quality and chemistry of niger seeds (*Guizotia abyssinica*) grown in the United States. *Industrial Crops and Products, 75*, 40–42.

Facciola, S. (1990). *Cornucopia - A Source Book of Edible Plants*. CA, USA: Kampong Publications.

Geleta, M., Stymne, S., & Bryngelsson, T. (2011). Variation and inheritance of oil content and fatty acid composition in niger (*Guizotia abyssinica*). *Journal of Food Composition and Analysis, 24*, 995–1003.

Getinet, A., & Sharma, S.M. (1996). Niger. *Guizotia abyssinica* (L.f.) Cass. *Promoting the conservation and use of underutilized and neglected crops. 5*. Rome, Italy: Institute of Plant Genetics and Crop Plant Research, Gatersleben/International Plant Genetic Resources Institute.

Manandhar, N. P. (2002). *Plants and people of Nepal*. Oregon: Timber Press.

Melaku, E. T. (2013). Evaluation of Ethiopian niger seed (*Guizotia abyssinica* Cass.) production, seed storage and virgin oil xxpression. *M.sc thesis*. Germany: Humboldt-Universität zu Berlin.

Ramadan, M. F., & Morsel, J. -T. (2003). Analysis of glycolipids from black cumin (*Nigella sativa* L.), coriander (*Coriandrum sativum* L.) and niger (*Guizotia abyssinica* Cass.) oilseeds. *Food Chemistry, 80*(2003), 197–204.

Ramadan, M. F., & Morsel, J. -T. (2003a). Determination of the lipid classes and fatty acid profile of niger (*Guizotia abyssinica* Cass.) seed oil. *Phytochemical Analysis, 14*, 366–370.

Ramadan, M. F., & Morsel, J. -T. (2003b). Phospholipid composition of niger (*Guizotia abyssinica* Cass.) seed oil. *Lebensmittel-Wissenschaft und -Technologie, 36*, 273–276.

Sarin, R., Sharma, M., & Khan, A. A. (2009). Studies on *Guizotia abyssinica* L. oil, biodiesel synthesis and process optimization. *Bioresource Technology, 100*, 4187–4192.

Yadav, S., Kumar, S., Hussain, Z., Suneja, P., Yadav, Sh. K., Nizar, M. A., & Dutta, M. (2012). *Guizotia abyssinica* (L.f.) cass: an untapped oilseed resource for the future. *Biomass Bioenergy, 43*, 72–78.

Part B

Unconventional Oils from Tree Sources

Chapter 24

Acacia tortilis Umbrella Thorn Tree Seed Oil

INTRODUCTION

Acacia tortilis is small- to medium-sized slow growing tree with an umbrella-shaped canopy, belonging to the subfamily Mimosoideae of the family Fabaceae. It has both straight and hooked thorns to help protect its highly nutritious leaves and pods from herbivores. These drought-resistant trees occur in deciduous woodland, bushveld, and grassveld. The tree is also known as Savanna Israeli babool. The tree carries leaves that grow to about 1 in. in length with between 4 and 10 pairs of pinnae each has about 15 pairs of leaflets. The flowers are white, small, aromatic, and occur in close-fitting clusters. Seeds are produced in pods that are flush and looped and assembled into a spring-like structure. The plant is known to tolerate drought, high temperatures, sandy and stony soils, strongly sloped rooting surfaces. In addition, elder plants have been observed to be frost resistant (Trees, 1994).

Acacias are shrubs and trees found in warm and tropical regions, from the Americas and Europe to Asia and Africa, however, the vast majority of species are native to Australia. Belonging to the pea family, acacias have pods, and acacia beans are on the menu in Mexico (Fig. 24.1).

EXTRACTION OF OIL

A. tortilis seeds were shade dried to constant weight, grinded, and passed through sieves to obtain powder of 40 mesh size. The fatty acid methyl esters (FAMES) references were palmitic, oleic, stearic, linoleic, and linolenic acid (99% pure). For oil extraction, 5.0 g of air-dried powdered seeds were extracted with petroleum ether (40–60°C) in a Soxhlet apparatus and dried over anhydrous sodium sulfate in vacuum at 40°C for removal of solvents and percentage yield was calculated (Srivastava & Kumar, 2013).

Unconventional Oilseeds and Oil Sources. http://dx.doi.org/10.1016/B978-0-12-809435-8.00024-X

FIGURE 24.1 *Acacia tortilis* tree, seeds and flowers.

METHYL ESTERIFICATION

The oil was hydrolyzed with 0.5 N KOH in ethanol (15 mL) for 4–6 h. Fatty acids were separated from unsaponified part of the oil after hydrolysis. The hydrolysate was extracted with diethyl etherand allowed to stand for few minutes. The upper layer containing unsaponified part was separated from the lower aqueous layer containing fatty acids. $HCl:H_2O$ (1:2) was added to make the aqueous layer acidic and finally extracted with diethyl ether for complete separation of fatty acids.

The fatty acids were derivatized into their methyl esters after complete evaporation of diethyl ether adding 25 mL of 2% methanolic H_2SO_4 (both methanol and H_2SO_4, AR grade) which was refluxed on a water bath for 4 h. Washing of the methyl esters was carried out with 1% aqueous potassium hydroxide and distilled water 3–4 times, finally transferred to Eppendorf tube after removing moisture through anhydrous sodium sulfate (if any) for injection in gas chromatograph.

FATTY ACID ANALYSIS

Gas chromatography (GC) was performed to determine the fatty acid composition of *A. tortilis* seeds oil by Hewlett Packard Gas Chromatograph Model 5890 Series II using a stainless steel column coated with diethylene glycol succinate (DEGS) on Chromosorb W (HP) (180 × 0.3 cm), equipped with flame ionization detector attached to Pentium III computer. The column, injector, and detector temperatures were maintained at 170, 230, and 250°C, respectively using nitrogen as carrier gas (Banerji, Shukla, Bajpai, & Dixit, 2007). Different commercial indices were established to calculate monounsaturated fatty acids (MUFA), polyunsaturated fatty acid (PUFA), total unsaturation (TU = PUFA + MUFA), total saturation (SFA), polyunsaturated to saturated ratio (PUFA/SFA), and omega-6/omega-3 ratio (n-6/n-3) (Pellew, 1980; Hui, 1996; Banerji et al., 2007).

ORGANOLEPTIC CHARACTERISTICS AND ANALYTIC COMPARISON WITH OTHER OIL

The oil extracted was viscous and cream to light yellow in color with the total oil content of 3.8%. The quality of oil is mainly governed by the fatty acid composition, hence the standardization of oil on the basis of fatty acid composition is mandatory. Table 24.1 shows that seed oil consists mainly of essential saturated and unsaturated fatty acid. In plant tissues, the most abundant saturated fatty acids are palmitic and stearic acids and the most common unsaturated fatty acids are oleic, linoleic, and linolenic (Muralidharudu & Virupakshappa, 1998). In *A. tortilis* seed oil, linolenic acid (50.43%) was found to be maximum, followed by linoleic acid (36.75%), palmitic acid (6.41%), oleic acid (3.96%), and stearic acid (0.63%). Higher content of linoleic acid and linolenic acid that is, 36 and 50%, respectively in analyzed oil is noteworthy. Content of linoleic acid in seed that is, 36.75% is similar to soybean, sunflower, and safflower vegetable oils having specified codex standard range of 48–55, 48–74, and 67.6%–73.2%, respectively. Thus such linoleic acid–rich oils as grape seed oil, sunflower oil, and sunflower oil including *A. tortilis* oil can play a significant role in reducing blood cholesterol levels when consumed regularly as part of the diet (Isay & Busarova, 1984; Hui, 1996; Srivastava & Kumar, 2013). Linolenic acid was found to be significantly high that is, 50.43%, which is comparable (Table 24.2) to various commercially available edible oils (Trees, 1994). In addition to this omega-3 fatty acid lowers the blood cholesterol, reduces the symptoms of depression, dementia, anxiety, and possess antiinflammatory action. It is also

TABLE 24.1 Fatty Acid Composition of *A. tortilis* Seed Oil

S. No.	Fatty acid	Values (%)
1	Palmitic acid (16:0)	6.4174
2, 3	Stearic acid (18:0)	0.6304
4	Epoxy (18:1)	1.8013
5	Oleic acid (18:1)	3.9630
6	Linoleic acid (18:2)	36.7484
7	Linolenic acid (18:3)	50.4390

TABLE 24.2 Comparison of Linoleic and Linolenic Content in *A. tortilis* Seed Oil With Some Other Commercial Oils

		Fatty acid	
No.	Plant oil	C18:2	C18:3
1	Matthiola	11.0	65.0
2	Chis	19.8	63.4
3	Kiwi	17.0	62.0
4	Perils	17.6	65.8
5	Flax	18.0	55.0
6	Lingonberry	35.0	49.0
7	Purslane	34.6	35.3
8	Sea buckthorn	40.4	32.3
9	Hemp	60.0	20.0
10	Canola	20.0	11.0
11	Walnut	53.9	10.4
12	Soybean	53.4	7.0
13	*A. tortilis*	36.7	50.43

vital for infants' health, pregnant women, seems to improve rheumatoid arthritis, diabetes, and improves the life span of patients with autoimmune diseases (Banerji et al., 2007). The high content of linoleic and linolenic acid of the oil analyzed may have a significant role in clinical use (Kang & Leaf, 1996) and contribute significantly in cholesterol-lowering effect also (Chan, Bruce, & McDonald, 1991; Srivastava & Kumar, 2013). Moreover, the higher content of saturated fatty acids like palmitic acid (6.41%) and stearic acid (0.63%) support the fact that this genus belongs to the family Fabaceae.

REFERENCES

Banerji, R., Shukla, P., Bajpai, A., & Dixit, B. (2007). *Review article: omega-3 fatty acid.* Paper presented at the Applied Botany Abstracts.

Chan, J. K., Bruce, V. M., & McDonald, B. E. (1991). Dietary alpha-linolenic acid is as effective as oleic acid and linoleic acid in lowering blood cholesterol in normolipidemic men. *The American Journal of Clinical Nutrition, 53*(5), 1230–1234.

Hui, Y. (1996) *Bailey's industrial oil and fat products. Edible oil and fat product: processing technology* (Vol. 4). (5th ed.). USA: John Wiley & Sons.

Isay, S., & Busarova, N. (1984). Study on fatty acid composition of marine organisms—I. Unsaturated fatty acids of Japan Sea invertebrates. *Comparative Biochemistry and Physiology Part B: Comparative Biochemistry, 77*(4), 803–810.

Kang, J. X., & Leaf, A. (1996). Protective effects of free polyunsaturated fatty acids on arrhythmias induced by lysophosphatidylcholine or palmitoylcarnitine in neonatal rat cardiac myocytes. *European Journal of Pharmacology, 297*(1), 97–106.

Muralidharudu, Y., & Virupakshappa, K. (1998). Extent of variation in oil, protein and fatty acid composition of high oleic hybrids, wild species and some cultivars of sunflower. *Journal of the Oil Technologists Association of India, 30*, 159–161.

Pellew, R. A. (1980). *The production and consumption of Acacia browse and its potential for animal protein production.* Paper presented at the Browse in Africa: The current state of knowledge. Papers presented at the International Symposium on Browse in Africa, Addis Ababa.

Srivastava, M., & Kumar, S. (2013). Fatty acid compositional studies of lesser known *Acacia tortilis* seed oil. *Food and Nutrition Sciences, 4*, 59–62.

Trees, S. (1994). Forest gaps and isolated Savanna trees. *BioScience, 44*(2), 77–84.

Chapter 25

Annona squamosa L. Sugar Apple Seed Oil

INTRODUCTION

The sugar apple tree (*Annona squamosa* L.) is a small tree or shrub of the Annonaceae family, native to the tropics. The tree ranges from 10 to 20 ft. (3–6 m) in height with open crown of irregular branches, and somewhat zigzag twigs. Deciduous leaves, alternately arranged on short, hairy petioles, are lanceolate or oblong, blunt tipped, 2–6 in. (5–15 cm) long and 3/4–2 in. (2–5 cm) wide; dull green on the upper side, pale, with a bloom, below; slightly hairy when young; aromatic when crushed. *A. squamosa* is the most widely grown of all the species of *Annona*; the sugar apple has acquired various regional names: *anon* (Bolivia, Costa Rica, Cuba, Panama); *anon de azucar, anon domestico, hanon, mocuyo* (Colombia); *anona blanca* (Honduras, Guatemala, Dominican Republic); *anona de castilla* (El Salvador); *anona de Guatemala* (Nicaragua); apple bush (Grenadines); *ata, fruta do conde, fruta de condessa, frutiera deconde, pinha, araticutitaia,* or *ati* (Brazil); *atte* (Gabon); *chirimoya* (Guatemala, Ecuador); *cachiman* (Argentina); *cachiman cannelle* (Haiti); *kaneel apple* (Surinam); *rinon* (Venezuela); *saramulla, saramuya, ahate* (Mexico); *scopappel* (Netherlands Antilles); sweetsop (Jamaica, Bahamas); *ata, luna, meba, sharifa, sarifa, sitaphal, sita pandu,* custard apple, scaly custard apple (India); *bnah nona, nona, seri kaya* (Malasyia); *manonah, noinah, pomme cannelle du Cap* (Thailand); *qu a na* (Vietnam); *mang cau ta* (Cambodia); *mak khbieb* (Laos); *fan-li-chi* (China); *ates* or *atis* (Philippines); and *pomme cannelle* (Guadeloupe, French Guiana, Africa).

The fruit is dark brown in color and marked with depressions giving it a quilted appearance; its pulp is reddish yellow, sweetish, and very soft. It had also been known as sweetsop (English) (KV, 2013). This chapter focuses on the characteristics and proposed uses of sugar apple seed oil (Fig. 25.1).

FIGURE 25.1 *Annona squamosa* **tree, fruit and seeds.**

ANALYTICAL AND TEST METHODS

The collected seeds were dried and crushed in a mechanical expeller cold press; others were solvent extracted. For complete extraction of oil, the seeds were passed 4 times through the expeller. The neat oil is allowed to settle for 48 h and after that oil is stored in an airtight container to avoid oxidation.

The mean molecular weight, saponification number (SN), iodine value (IV), acid value (AV), free fatty acids (FFAs), and other parameters of oil were analyzed. The oil extracted from the mechanical expeller was weighed after filtering; it was found that the sugar apple seeds contain a moderate quantity of oil, 24.5 w/w% oil, and thus it could be suitable for many applications including feedstock for the production of biodiesel.

PHYSICOCHEMICAL PROPERTIES OF SUGAR APPLE SEED OIL

According to the results by Moreno Luzia and Jorge (2012), the sugar apple seeds constitute significant sources of lipids, proteins, and carbohydrates and can therefore be used in food and feed, and offer relevant antioxidant activity of phenolic compounds. The oil seeds are a good source of unsaturated fatty acids, especially oleic and linoleic acids and they have significant amounts of total tocopherol.

TABLE 25.1 Physicochemical Properties of Sugar Apple Seed Oil

Property	Sugar apple seed oil
IV (mg I_2/g)	118
Saponification value	192
AV (mg KOH/g)	1.93
Flash point (°C)	235
Density (kg/m^3)	910
Kinematic viscosity (mm^2/s) at 40°C	42.63
Calorific value (MJ/kg)	37.95
Molecular weight (g/mol)	872.5

AV, Acid value; IV, iodine value.

The physical and chemical properties of sugar apple seed oil were determined as per the standard test procedures and tabulated in Table 25.1. The IV of the oil is 118 mg I_2/g; as the IV is higher, it indicates the unsaturation of the oil. The heating of these higher fatty acids results in polymerization of glycerides, which necessitates the limitation of unsaturated fatty acids; otherwise, it leads to the formation of deposits and deterioration of the lubricating oil. The saponification value of sugar apple seed oil was 192, which indicates that the oil is normal triglyceride and useful in the production of soaps (Sukasih, 2015). The FFA content of oil is 0.965%; it is less than the biodiesel that can be produced using single-stage transesterification process, that is, by base-catalyzed transesterification process. The flash points of sugar apple seed oil were 235°C. The flash points of sugar apple seed oil are found to be much higher in comparison with diesel, which helps in safe storage and transportation. The densities of sugar apple seed oil were higher than that of diesel, which may be due to the presence of higher molecular weight triglycerides. The kinematic viscosity of sugar apple seed oil was found to be 42.63 mm^2/s, which is much higher than that of diesel, hence the direct use of sugar apple seed oil may lead to poor combustion, untimely wear of fuel pumps, and injector. The viscosity of sugar apple seed oil was reduced by converting it to biodiesel and it was found to be 5.90 mm^2/s that is within the limits of the international standards specifications for biodiesel fuel. Calorific values of sugar apple seed oil and diesel were found to be 37.95 and 43.00 MJ/kg, respectively. The calorific values of studied oil and biodiesel are found to be lower than that of diesel, which may be due to the difference in chemical composition or presence of oxygen molecule in the molecular structure of the oil (Sravanthi, Waghrey, & Daddam, 2014).

FATTY ACID COMPOSITION OF SUGAR APPLE SEED OIL

The quality of the oil will be affected by its fatty acid composition; it is always desirable if the vegetable oil has low saturated and low polyunsaturated fatty acid in its composition, and the composition of monounsaturated fatty acid should be high. Table 25.1 shows the fatty acid composition of two samples of sugar apple (*A. squamosa*) seed oils and both contain predominant amount of monounsaturated fatty acid composition, that is, 39.72 and 50.5%, followed by 31.56 and 22.7% of polyunsaturated fatty acid and 24.07%–28.6% of saturated fatty acids. The major fatty acids present in the sugar apple seed oils were oleic acid, 39.72 and 50.5%; linoleic acid, 29.13 and 22.7%; palmitic acid, 17.79 and 15.5%; and stearic acid, 4.29 and 10.6%. Thus the oil can be classified as oleic–linoleic oil (Mariod, Elkheir, Ahmed, & Matthäus, 2010).

The vegetable oil extracted from a plant is composed of triglyceride, which is an ester derived from three fatty acids and one glycerol. The fatty acid composition of sugar apple (*A. squamosa*) seed oil samples is given in Table 25.2.

TABLE 25.2 Fatty Acid Composition of the Sugar Apple (*A. squamosa*) Seed Oil

Fatty acid	A. squamosa seed oil		
	Sample from India[a]	Sample from Indonesia[a]	
	Sample 1	Sample 2, ASCE	Sample 3, ASSE
Capric acid	—	0.3	0.2
Lauric acid	0.08%	—	—
Myristic acid	—	0.7	0.6
Palmitic acid	17.79%	15.5	15.2
Stearic acid	4.29%	10.6	9.3
Oleic acid	39.72%	49.2	50.5
Linoleic acid	29.13%	22.3	22.7
Linolenic acid	1.37%	—	—
Arachidic acid	—	1.3	1.5
Arachidonic acid	1.06%	—	—
Behenic acid	2.01%	—	—
Saturated	24.89	28.4	26.8
Unsaturated	71.28	71.5	73.2
Sat./unsat. ratio	0.35	0.40	0.37
Oleic/linoleic ratio	1.36	2.2	2.2

[a]*Sources are: KV (2013), Hotti and Hebbal (2015), and Mariod et al. (2010).*

TRANSESTERIFICATION REACTION

The transesterification reaction was carried out in a laboratory-scale batch reactor equipped with thermometer and condenser; the heating and stirring were done with a hot plate magnetic stirrer system. In each set of experiment, 50 g of oil was heated to the predefined temperature and after attainment of predefined temperature the mixture of catalyst and methanol was transferred to a reactor and all the predefined sets of transesterification reaction conditions were measured from this point for each set of experiment. Stoichiometrically 3:1 molar ratio of alcohol to oil is needed for completion of the transesterification reaction, but many researchers reported that biodiesel yield is maximum with an excess molar ratio of alcohol to oil. Hence in the present investigation, in each set of experiment, 6:1 molar ratio of alcohol to oil and constant stirrer speed were maintained (Doddabasawa, 2014; Doddabasava, 2014).

After the completion of a predefined set of transesterification reaction conditions, the reaction mixture was transferred into a separating funnel left for 60 min to separate into biodiesel and glycerol. The lower layer of glycerol was removed and the upper layer of crude biodiesel was washed several times with hot water at 50°C to remove the impurities, such as residual catalyst, methanol, soap, and glycerol. The removal of impurities was confirmed by measuring the pH of water. The biodiesel was dried by heating it to a temperature of 110°C and allowing overnight for evaporation and cooling. The final product was weighed to determine the biodiesel yield.

EFFECT OF CATALYST CONCENTRATION ON BIODIESEL PRODUCTION

The catalyst concentration is one of the most significant parameters, which affect the biodiesel yield. Figure 25.1 shows the effect of NaOH catalyst concentration on the yield of biodiesel; during the process, the other parameters are kept constant. It is observed that the biodiesel yield increases when catalyst concentration is increased from 0.25 to 0.50 w/w% and decreases when increased from 0.50 to 0.75 w/w%. The decrease in yield at 0.25 w/w% of catalyst concentration may be due to incomplete reaction and at 0.75 w/w% of catalyst concentration may be due to the fact that the higher amount of catalyst concentration favors the saponification reaction; thus a further increase in catalyst concentration was not studied. From the present investigation, the optimum amount of catalyst concentration was found to be 0.50 w/w% with 90.69% biodiesel yield.

The apple seed oil is a potential raw material for the production of biodiesel and to assess its feasibility for the replacement of petroleum fuel. The sugar apple seed oil was converted into biodiesel successfully by transesterification process and following conclusions were drawn: (1) the sugar apple oil was converted to biodiesel by single-stage base-catalyzed transesterification process without any pretreatment as the FFA content is found to be less than 1%. (2)

The optimized process parameters are catalyst concentration of 0.5 w/w%, the reaction time of 75 min and a reaction temperature of 60°C with alcohol to oil molar ratio of 6:1, and constant stirrer speed. The biodiesel yield was found to be 95.15% at the optimized process parameters. (3) The physical and chemical properties of biodiesel produced were found to be close to those of diesel fuel and also they meet the ASTM standard specifications for biodiesel (Sit & Tep, 2013).

The oil extracted from the sugar apple seeds is a good source of unsaturated fatty acids, especially oleic and linoleic acids and they have significant amounts of total tocopherol and offer relevant antioxidant activity of phenolic compounds. The seed also contains proteins and carbohydrates and can, therefore, be used in food and feed (Moreno Luzia & Jorge, 2012).

REFERENCES

Doddabasava, R. P. (2014). Biodiesel production cost analysis from the *Pongamia pinnata*: a case study in Yadagiri district of Karnataka, India. *International Journal of Scientific Research, 3*, 128–131.

Doddabasawa, R. P. (2014). Physico-chemical analysis of biodiesel derived glycerin and saponification of crude glycerin with different concentrations of sodium hydroxide lye. *International Journal of Science and Research, 3*, 210–212.

Hotti, S. R., & Hebbal, O. D. (2015). Biodiesel production process optimization from sugar apple seed oil (*Annona squamosa*) and its characterization. *Journal of Renewable Energy, 2015*(1–6).

KV, Y. (2013). *Biodiesel production from custard apple seed (Annona squamosa) oil and its characteristics study*. Paper presented at the International Journal of Engineering Research and Technology.

Mariod, A. A., Elkheir, S., Ahmed, Y. M., & Matthäus, B. (2010). *Annona squamosa* and *Catunaregam nilotica* seeds, the effect of the extraction method on the oil composition. *Journal of the American Oil Chemists' Society, 87*(7), 763–769.

Moreno Luzia, D. M., & Jorge, N. (2012). Soursop (*Annona muricata* L.) and sugar apple (*Annona squamosa* L.) antioxidant activity, fatty acids profile and determination of tocopherols. *Nutrition & Food Science, 42*(6), 434–441.

Sit, T., & Tep, M. (2013). Biodiesel production from custard apple seed (*Annona squamosa*) oil and its characteristics study. *International Journal of Engineering, 2*(5), 31–36.

Sravanthi, T., Waghrey, K., & Daddam, J. R. (2014). Evaluation of organolepictic properties and microbial studies of custard apple pulp. *International Journal of Analytical, Pharmaceutical and Biomedical Sciences, 3*(3), 6–12.

Sukasih, E. (2015). Total phenol and antioxidant from seed and peel of ripe and unripe of Indonesian sugar apple (*Annona squamosa* L.) extracted with various solvents. *IOSR Journal of Pharmacy, 5*, 26–32.

Chapter 26

The Potential of Sodom Apple (*Calotropis procera* and *Calotropis gigantea*) Seed Oil

INTRODUCTION

There are two varieties of Sodom apple, *Calotropis procera* and *Calotropis gigantea*, from the plant family Apocynaceae that are native to north and tropical Africa, western and south Asia. The two types of the plant are also known as the apple of Sodom, desert apple, giant milkweed or Sodom's milkweed, rubber bush, swallow-wort, Dead Sea apple, and *Ushar* (Arabic). Both plants are green throughout the year, have green globes that are hollow, but the flesh contains a milky sap that is extremely bitter and turns into a gluey coating resistant to soap. The fruits are green, turn to light yellow, and dry upon maturity and fruiting takes place throughout the year (Little & Mirrlees, 1974). The fruit contains numerous seeds, which are light, flat, brown, and attached to long white silky hair (pappus) at one end that help the seeds to fly from one place to another spreading the plant. Husain, Malik, Javaid, and Bibi (2008) reported that the leaves are smoked to cure asthma and cough. It is poisonous. The latex is commonly used for ringworm, dog bitten wounds, and skin diseases. Latex is used in the tanning industry. It is also reported as one of the ranges of garden plants that are considered poisonous (Johnson, Johnson, & Wales, 2006). In the traditional herbal remedies within the rural communities in the district in India, the roots and flower are used in nocturnal enuresis (Ghosh, Guria, Sarkar, & Ghosh, 2013). Seeds are numerous (350–500 per fruit), flat, obovate, with silky white pappus of 3 cm or longer (Orwa, Mutua, Kindt, Jamnadass, & Simons, 2015). The showy milkweed, *Asclepias speciosa* was studied and stated as a potential new semiarid land crop for energy and chemicals (Adams, Balandrin, & Martineau, 1984). The *C. gigantea* seeds and oil specifications are shown in Table 26.1 as reported by Rao, Pantulu, and Lakshminarayana (1983).

Unconventional Oilseeds and Oil Sources. http://dx.doi.org/10.1016/B978-0-12-809435-8.00026-3

TABLE 26.1 Physicochemical Characteristics of *C. gigantea* Seed and Oil

Seed		Seed oil	
Weight of 100 seeds (g)	0.8	Refractive index at 25°C	1.4639
Volume of 100 seeds (mL)	0.4	Acid value	26.0
Moisture (%)[a]	7.5	SV	191.3
Oil (%)[a]	30.8	IV	89.3
Protein[a]	19.0	Unsaponifiable matter (%)	1.2
Ash[a]	4.6		

IV, Iodine value; SV, saponification value.
[a]*Dry bases.*

APPLE OF SODOM SEED OIL

The seed was collected, the silky hairs were removed, and the remaining seeds were prepared for oil extraction. The oil content is 24%–34%, having a bright yellow color. The oil was characterized for fatty acid composition, iodine value (IV), saponification value (SV), free fatty acids (FFAs), and other parameters. The oil has been used in traditional medicine and cosmetics; however, no study has been found in literature about medicinal uses. In this study, the oil was successfully used to produce soap and shampoo. It is also proposed for biodiesel production (Kumar, 2013). The remaining cake was tested for protein content. Also, cake extracts were tested for their potential antimicrobial activity using a gram positive and gram negative as well as fungus (Fig. 26.1; Tables 26.2 and 26.3).

The seeds were found to contain 33.3% oil and several previously unreported minor fatty acids were identified (Rao et al., 1983). As the acid value of the oil was 27.5 wt.%, acid-catalyzed esterification has to be conducted to esterify FFAs and alkali-catalyzed transesterification will follow to produce biodiesel. The triglyceride content, diglyceride content, monoglyceride content, free glycerol, methanol, ester content, carbon residue, acid value, oxidation stability, tocopherol, water content, kinematic viscosity, density, cloud point, and flash point of the prepared biodiesel were determined with the exception of oxidation stability.

POTENTIAL USES OF SODOM APPLE SEED OIL

A report (Vázquez, Palazon, & Navarro-Ocaña, 2012) describes that the roots of *C. gigantea* (Linn.) R. Br, traditionally used in India to treat asthma, possess antilipoxygenase activity; it was found that intraperitoneal administration of indomethacin did not block edema formation, but edema was inhibited by montelukast and methanolic extracts of *C. gigantea* roots. This result indicates that

FIGURE 26.1 *Calotropis procera* **plant, fruits, and seeds.**

TABLE 26.2 Sodom Apple (*C. procera* and *C. gigantea*) Seed

Parameter	C. procera	C. gigantea
Oil content (%)	24.35	33.30
Appearance	Clear bright yellow	Clear bright yellow
IV	138	124
SV	183	195
Specific gravity	0.926	0.917
Unsaponifiable matter	2.10	1.94
FFA (%)		27
Acid value (mg KOH/g)		55
Peroxide value (mEq./kg)	06.8	09.2

FFA, Free fatty acid.

the extract from *C. gigantea* was responsible for the inhibition of the lipoxygenase pathway in the arachidonate metabolism. Therefore, it can be concluded that *C. gigantea* may have a similar mechanism of action as dexamethasone, as well as antioxidant and antilipoxygenase effects, possibly due to the presence of α- and β-amyrin (Bulani et al., 2011).

TABLE 26.3 Fatty Acid Profile of Sodom Apple (*C. procera*) Seed Oil

Fatty acid	Number of carbon atoms and bonds	*C. procera*	*C. gigantea*
Lauric acid	C12:0	—	—
Myristic acid	C14:0	Traces–5	Traces
Palmitic acid	C16:0	9–12	15.5
Palmitoleic acid	C16:1	—	0.3
Stearic acid	C18:0	11–17	10.5
Oleic acid	C18:1	30–34	30.3
Linoleic acid	C18:2	17–25	36.3
Linolenic acid	C18:3	09–12	0.8
Arachidic acid	C20:0	0.55	0.6
Behenic acid	C22:0	0.76	0.1
Others		05–7	0.4

The oil of *C. gigantea* had been reported to impart benefits to skin and/ or improve skin conditions resulting from aging or damaged skin, so it could be used as a massage oil (Mei & Lyga, 2015). The oil had been reported to be used in dental caries (Singh, Kumar, Budhiraja, & Hooda, 2007). Sodom apple (*C. procera* and/or *C. gigantea*) seed oil is a new unconventional vegetable oil found to be useful for soap making, cosmetics, and other industrial applications, such as tanning industry. The plant is evergreen although it's in the middle of the desert, and is easily spread in wide areas in Africa and the Middle East countries; if fully utilized, it could be a good source for biodiesel production as a renewable energy. Evaluation of Indian milkweed (*C. gigantea*) seed oil as alternative feedstock for biodiesel was reported by Phoo et al. (2014) who reported that all fuel properties conformed to four standards (Philippine National Standard PNS2020:2003, Japanese Automotive Standards Organization JASO M360, European Standard EN 14214, American Society for Testing Materials ASTM D6751). However, it was found that this biodiesel can be only used in tropical countries due to the poor cold flow properties.

REFERENCES

Adams, R. P., Balandrin, M. F., & Martineau, J. R. (1984). The showy milkweed, *Asclepias speciosa*: a potential new semi-arid land crop for energy and chemicals. *Biomass*, 4(2), 81–104.

Bulani, V., Biyani, K., Kale, R., Joshi, U., Charhate, K., Kumar, D., & Pagore, R. (2011). Inhibitory effect of *Calotropis gigantea* extract on ovalbumin-induced airway inflammation and arachidonic acid induced inflammation in a murine model of asthma. *International Journal of Current Biological and Medical Science*, 1(2), 19–25.

Ghosh, S. K., Guria, N., Sarkar, A., & Ghosh, A. (2013). Traditional herbal remedies for various ailments within the rural communities in the district of Bankura and Purulia, West Bengal, India. *International Journal of Pharmacy and Pharmaceutical Sciences, 5*(4), 195–198.

Husain, S. Z., Malik, R. N., Javaid, M., & Bibi, S. (2008). Ethonobotanical properties and uses of medicinal plants of Morgah biodiversity park, Rawalpindi. *Pakistan Journal of Botany, 40*(5), 1897–1911.

Johnson, A., Johnson, S., & Wales, N. S. (2006). *Garden plants poisonous to people.* New South Wales: NSW Department of Primary Industries.

Kumar, A. (2013). Biofuels utilisation: an attempt to reduce GHG's and mitigate climate change. *Knowledge systems of societies for adaptation and mitigation of impacts of climate change.* Berlin, Heidelberg: Springer, pp. 199–224.

Little, I. M., & Mirrlees, J. A. (1974). *Project appraisal and planning for developing countries.* New York, NY: Basic Books.

Mei, B. C., & Lyga, J. W. (2015). *Calotropis gigantea extracts and methods of use.* US Patent 20,150,366,788.

Orwa, C., Mutua, A., Kindt, R., Jamnadass, R., & Simons, A. (2015). Agroforestree database: a tree reference and selection guide version 4.0. 2009. Available from: http://www.worldagroforestry.org/af/treedb/

Phoo, Z. W. M. M., Razon, L. F., Knothe, G., Ilham, Z., Goembira, F., Madrazo, C. F., & Saka, S. (2014). Evaluation of Indian milkweed (*Calotropis gigantea*) seed oil as alternative feedstock for biodiesel. *Industrial Crops and Products, 54,* 226–232.

Rao, K. S., Pantulu, A., & Lakshminarayana, G. (1983). Analysis of *Calotropis gigantea, Acacia caesia* and *Abelmoschus ficulneus* seeds. *Journal of the American Oil Chemists' Society, 60*(7), 1259–1261.

Singh, J., Kumar, A., Budhiraja, S., & Hooda, A. (2007). Ethnomedicine: use in dental caries. *Brazilian Journal of Oral Sciences, 6,* 21.

Vázquez, L. H., Palazon, J., & Navarro-Ocaña, A. (2012). The pentacyclic triterpenes α-, β-amyrins: a review of sources and biological activities. *Phytochemicals—A Global Perspective of Their Role in Nutrition and Health, 23,* 487–502.

Chapter 27

Balanites aegyptiaca Seed Oil

INTRODUCTION

Balanites aegyptiaca is a species tree that classified either as a member of the Zygophyllaceae or the Balaniteceae. This species, known as desert date, Hoba call it Boha, Hausa call it Adu'a and the Fulanis call it Tanni. This tree has 25 known species of the plant widely distributed from tropical Africa to Burma (Usher, 1984). *B. aegyptiaca*, Hegleig tree is indigenous to all dry lands south of the Sahara and extending southward (Sands et al., 2001; Levey, Gassman, Hall, & Walker, 1991; Shanks & Schwerdtner, 1991; Sidiyene, 1996). This plant also natives of Arabian Peninsula (Fregon, 2015), India, Iran and Pakistan (Amalraj and Shankamarayan, 1988). In Sudan it is more likely the species with widest natural range, occur in all zones, except in very high latitudinal areas or when the rainfall exceeds 1100 mm/annum Badi, Ahmed, & Bayoumi, 1989). It makes up to one-third of the total free population in the central region of the Sudan (NRC, 2008) (Okia et al., 2011). This plant can found in many types of habitat, tolerating a wide variety of soil types, from sand to heavy clay and climatic moisture levels (Elfeel, 2010).

MORPHOLOGICAL STUDY OF *Balanites aegyptiaca* DEL.

The plant is multibranched, spiny shrub or trees up to 10 m tall. It has a crown spherical, in one or several distinct masses. Its short trunk and often branching from near the base. Their bark is dark brown to gray, deeply fissured. Its branches armed with stout yellow or green thorns up to 8 cm long. Their leaves consist of two separate leaflets, which are leaflets obovate, asymmetric, 2.5–6 cm long, bright green, leathery, with fine hairs when young. Flowers in fascicles in the leaf axils, and are fragrant, yellowish-green (Omer, 2015). *B. aegyptiaca* consists of different biological active compounds which contribute a vital role in its nutraceutical applications.

The leaves contain saponin, furanocoumarin, and flavonoid namely quercetin 3-glucosidase, quercetin-3-rutinoside, 3-glucoside, 3-rutinoside, 3-7-diglucoside, and 3-rhamnogalactoside of isorhamnetin.

Unconventional Oilseeds and Oil Sources. http://dx.doi.org/10.1016/B978-0-12-809435-8.00027-5

FLOWERING, FRUIT, AND SEED DESCRIPTION

Flowers are small, inconspicuous, hermaphroditic, and pollinated by insects. The seeds of this plant are dispersed by ingestion by birds and animals. The tree begins to flower and fruit at 5–7 years of age and maximum seed production is when trees are 15 years old (Chothani & Vaghasiya, 2011).

The fruit is a rather long, narrow drupe, 2.5–7 cm long, 1.5–4 cm in a diameter. The young fruits of this are green and tomentose, turning yellow and glabrous when to mature. Its pulp is bittersweet and edible. Their seed is the pyrene (stone), 1.5–3 cm long, light brown, fibrous, and extremely hard. It also makes up 50%–60% of the fruit. There are 500–1500 dry, clean seeds per kg (Okia et al., 2011, Mohamed, 2016) (Fig. 27.1; Table 27.1).

The kernels have been reported that contain 45.0%–46.1% oil and protein (32.4%), the oil contains mainly palmitic, stearic, oleic, and linoleic acids which were the main fatty acids (Fregon, 2015). The oil exhibited anticancer activity against lung, liver, and brain human carcinoma cell lines. It also had antimutagenic activity against *Fasciola gigantica*-induced mutagenicity besides anthelmintic activity against hepatic worms (*Schistosoma mansoni* and *F. gigantica*). Preliminary screening showed that the oil had antiviral activity against Herpes simplex virus. It also had antimicrobial activity against selected strains of Gram-positive bacteria, Gram-negative bacteria, and *Candida*.

DISTRIBUTION AND HABITAT

Natural distribution is obscured by cultivation and naturalization. It is believed indigenous to all dry lands south of the Sahara, extending southward to Malawi in the Rift Valley, and to the Arabian Peninsula, introduced into cultivation in Latin America and India. It has wide ecological distribution but is mainly found on level alluvial sites with deep sandy loam and free access to water. After the seeding stage, it is intolerant to shade and prefers open woodland or savannah for natural regeneration. It is a lowland species; growing up to 1000 m altitude in areas with means the annual temperature of 20–30°C and mean annual rainfall of 250–40 mm.

PHARMACOLOGICAL ACTIVITY

Antiinflammatory Activity

The aerial part of *B. aegyptiaca* (L.) Del. was investigated the first time for their antiinflammatory activity using ethanol extract. The activity has been evaluated in the rat by using a carrageenan-induced paw edema method. When compared to the reference standard, the ethanolic extract was identified to exhibit potent antiinflammatory activity at the concentration of 200 mg/kg 3 h. The paw volume was measured plethysm metrically at 0 and 3 h after injection and analgesic activity by using Eddy's hot plate method and tail-flick method in albino rats (Chothani & Vaghasiya, 2011). From the results, it was found that the ethanol extract did not show any behavioral changes or mortality even at a dosage of

FIGURE 27.1   *Balanites aegyptiaca* **tree and fruits.**

2000 mg/kg. Thus, the ethanolic extract of the aerial part of the plant can be used for the treatment of the inflammatory disorders.

Toxicity Study

B. aegyptiaca seed oil has used as the ingredient and substitute for the ground-nut oil in Nigeria region for the preparation of the local foods. The 4-week repeated dose toxicity study was done in male Wister strain rats to evaluate the

TABLE 27.1 Physical Properties of *B. aegyptiaca* Del. Kernels

No.	Property	Mean (SD, n)[a]
1	Weight of 100 kernels (g)	68.5 (0.02, 4)
2	Volume of 100 kernels (cm^3)	68.9 (0.01, 4)
3	Density (g/cm^3)	1.0 (0, 1)
4	Hardness (N/m^2)	10.4 (1.38, 20)
5	Length (mm)	17.9 (1.35, 50)
6	Width (mm)	8.7 (0.95, 50)

[a]n, Number of replicates; SD, standard deviation.

toxicity of crude *B. aegyptiaca* seed oil. The result has indicated that the dietary exposure of the crude *B. aegyptiaca* seed oil in the rats did not show any remarkable changes in the toxicological parameter that have been assayed. Thus, consumption of the crude oil at a present level of exposure may be no serious safety concern, especially in liver and kidney injury (Anto, Aryeetey, Anyorigiya, Asoala, & Kpikpi, 2005; Gnoula et al., 2008; Obidah, Nadro, Tiyafo, & Wurochekke, 2009).

Antidiabetic Property

The water and the ethanolic extracts obtained from the fruits of the plant have been investigated for their hypoglycemic and hypolipidemic effect in the normal senile diabetic rats in addition to the hormones that are associated with diabetes mellitus (Zaahkouk, Rashid, & Mattar, 2003). The changes in the serum total protein, albumin, and globulin level during the experimental period were found to be insignificant. The fruit flesh was found to attribute at least in part to the increased glucose metabolism and produces an increase in serum insulin concentration. From another study show that an aqueous extracts of mesocarps of the fruits *B. aegyptiaca* exhibited a prominent antidiabetic activity by oral administration in streptozotocin-induced diabetic mice. From one of the active fractions of this extract, two new steroidal saponins are isolated and their structures are determined as 26-O-beta-D-glucopyranosyl-(25R)-furost-5-ene-3beta,22,26-triol 3-O-(alpha-L-rhamnopyranosyl1-(1-4))-beta-D-glucopyranoside and its methyl ether were isolated and identified (Gad, El-Sawalhi, Ismail, & El-Tanbouly, 2006). It was revealed that the individual saponins did not show antidiabetic activity, while the recombinant of these saponins resulted in significant activity. From an ethanolic extract of the epicarps, two known flavonol glycosides, isorhamnetin-3-O-rutinoside were isolated and identified (Zaahkouk et al., 2003).

Antimicrobial Activity

Flavonoid extracts obtained from the callus tissue of *B. aegyptiaca* were screened for their antimicrobial effect against *Escherichia coli, Proteus vulgaris, Pseudomonas aeruginosa, Citrobacter amalonaticus, Staphylococcus aureus, Micrococcus lylae, Bacillus subtilis,* and *Sporolactobacillus.* The evaluation was done by using the adopted disc diffusion method. The results obtained were compared with the zone of inhibition that is produced by commercially available standard antibiotics. The free flavonoid fraction of callus tissue was found to exhibit more antimicrobial activity against Gram-positive bacteria that the ethyl ether and ethyl acetate bounded flavonoid extracts showing moderate activity against both the Gram-positive bacteria as well as Gram-negative bacteria. The antimicrobial activity of the ethanolic extract of the plant materials shows that *B. aegyptiaca* demonstrates higher activity (16 mm zone of inhibition) at 100 mg/mL. The result showed that the organic extract (acetone and ethanol) had a higher activity compared to the aqueous extract, it has been reported that different solvent have different solvent extract capabilities and different spectrum of solubility of phytoconstituents (Karuppusamy, Rajasekaran, & Karmegham, 2002).

Pesticidal Effect

B. aegyptiaca's active component, saponin was investigated for its biopesticidal value against *Tribolium castaneum* (red flour beetle or bran bugs). The saponins were extracted by using Soxhlet and maceration extraction methods with aqueous and different solvents. The study was conducted at different concentrations from 1%–5% and to higher 2.5%–17.5%. The results obtained were found to be dozing dependent because, at 5% concentration 100% mortality has been seen in the period of 8 days whereas 17.5% concentration showed 100% mortality in the period of 24 h itself. Thus, *B. aegyptiaca* was proved to contain the insecticidal property that was evaluated against *T. castaneum* (Ufodike & Omoregie, 1994).

Antioxidant Activity

The plant acts as an antioxidant against Adriamycin-induced cardiotoxicity in experimental mice. Adriamycin is an anthracycline antibiotic that is widely used as chemotherapeutic agent. However, the usefulness of this agent is limited due to its cytotoxic effects. Increase oxidative stress and antioxidant deficit have been suggested to play a major role in Adriamycin The result showed that, Adriamycin elevated the activities of LDH, CPK, GOT, GPT, LPO, and total NO content in mouse heart tissue (El Mastry, Ebeed, El Sayed, Nasr, & El Halafawy, 2010). Also, Adriamycin drug reduced the activities of SOD, GPx, and CAT. So, these findings demonstrate the cardio protective effect *B. aegyptiaca* on the antioxidant tissue defense system during Adriamycin-induced cardiac damage tissue (Speroni et al., 2005).

Anthelmintic Activity

The crude aqueous extract of root bark of *B. aegyptiaca* was showing a dose-dependent inhibition of spontaneous motility (paralysis) in adult earthworms. Balanitin-7 is isolated from aqueous extract of *B. aegyptiaca* seed and reported as an anthelmintic agent when tested by in vitro means of an original anthelmintic assay, using *Caenorhabditis elegans* as a biological model. The methanolic extract of *B. aegyptiaca* fruits are reported to have anthelmintic action against different stages of *Trichinella spiralis* in rats compared with anthelmintic drug albendazole. The aqueous extract of *B. aegyptiaca* also has molluscicidal agent to juvenile and adult *Bulinus globosus,* and *Bulinus truncatus*. From the result, the paralysis occurs when the worms do not revive even in normal saline while the death was considered worms lost their motility, and their color are fading away. Thus, it can concluded that, the aqueous extract showed marked and potent anthelmintic activity than the standard drug albendazole (Dubey et al., 2011).

Larvicidal Activity

The effect of aqueous extract of the fruit pulp, seed kernel, roots, barks, and leaves of *B. aegyptiaca* against the larvae of the *Culex pipiens* mosquito was investigated. Early fourth instars larvae of *C. pipiens* mosquitoes were exposed, for up to 3 days, to a dilution of 0, 0.1, 0.25, 0.5, 1.0, and 2.0% aqueous root extract. The lowest concentration of root extract (0.1%), showed 100% larval mortality (Bagavan, Rahuman, Kamaraj, & Geetha, 2008; Chapagain & Wiesman, 2011). Aqueous extract of fruit pulp, seed kernel, and leaves showed less larval mortality compared the root and/or bark extracts. A saponin extract and water extract from fruit kernel of *B. aegyptiaca* was investigated as a mosquito larvicide. Both extracts were tested against second and fourth instar larvae of the three mosquito species, namely *Anopheles arabiensis, Culex quinquefasciatus*. Second instar larvae were more susceptible than fourth instar larvae in all cases. The larvae of *A. arabiensis* were more susceptible than *C. quinquefasciatus*. The saponin was more active than the water extract (Wiesman & Chapagain, 2006; Chapagain, Saharan, & Wiesman, 2008).

OIL EXTRACTED FROM *Balanites aegyptiaca* SEED KERNEL

B. aegyptiaca seed kernel is used for oil extraction and the oil is used for human consumption and cosmetics. However, the oil cake is regarded as unsuitable for feeding because of the presence of many toxic substances. *B. aegyptiaca* naturally grows, contained the highest level of diosgenin as well as oil percentage. The study also showed that there is a strong positive correlation between diosgenin level and oil percentage in the *B. aegyptiaca* seed kernel. The oil content in various samples was found to be ranged between 39% and 50 % (Chapagain & Wiesman, 2005). The oil contained 54.53% unsaturated fatty acids and 1.14% sterols. The oil exhibited anticancer activity against lung, liver, and brain human carcinoma cell lines. It also had antimutagenic activity against *F. gigantica*-induced mutagenicity besides anthelmintic activity against hepatic worms (*S. mansoni* and *F. gigantica*) (Folarin, Eromosele, & Eromosele, 2011).

Seed kernel from three distinct ecological zones in Sudan and individual trees within each zone were analyzed for minerals (N, P, K, Ca, Mg, and Fe), oil, and protein contents. The results showed a very high significant variation in all seed kernel chemical contents analyzed among and within locations. The oil content was found to be ranged between 20% and 50%, while protein varied from 27% to 37%. The study revealed that *B. aegyptiaca* seed kernel chemical content was remarkably variable among and within locations that should be considered in conservation, domestication, and improvement plans for the tree (Elfeel, 2010). The fatty acid composition as methyl esters is presented in Table 27.2 where gas chromatographic (GC) results matched with standard samples (Al Ashaal, Farghaly, El Aziz, & Ali, 2010).

The kernels of *B. aegyptiaca* seed produce edible oil used for cooking. The oil remains stable when heated and has a high smoking point, and therefore its free fatty acid content is low. Its scent and taste are acceptable (Booth & Wickens, 1988). The effect of pressing temperature on fatty acid composition was insignificant (Table 27.3). Total unsaturated fatty acids at 50°C and 115°C were 74.8 and 75.1 g/100 g oil, respectively. There was agreement between these results and published data (Hall & Walker, 1991; Abu-Al-Futuh, 1983;

TABLE 27.2 Gas Chromatographic (GC) Data of Methyl Esters of Fatty Acid Constituents of *B. aegyptiaca* Seed Oil

No.	Fatty acid	Area (%)	Molecular formula
1	Lauric C12:0	05.17	$C_{12}H_{24}O_2$
2	Myristic C14:0	00.73	$C_{14}H_{28}O_2$
3	Palmitic C16:0	20.03	$C_{16}H_{32}O_2$
4	Palmitoleic C16:1	06. 64	$C_{16}H_{30}O_2$
5	Stearic C18:0	04.13	$C_{18}H_{36}O_2$
6	Oleic C18:1	22.47	$C_{18}H_{34}O_2$
7	Linoleic C18:2	10.97	$C_{18}H_{32}O_2$
8	Arachidic C20:0	02.10	$C_{20}H_{40}O_2$
9	Arachidonic C20:1	01.76	$C_{20}H_{32}O_2$
10	Behenic C22:0	13.29	$C_{22}H_{44}O_2$
11	Erucic C22:1	12.69	$C_{22}H_{42}O_2$

GC conditions: For sap—initial temperature 70°C for 2 min then with flow rate 40°C/min temperatures rising to 120°C for 2 min then with flow rate 4°C/min temperature raising to 220°C for 20 min inlet temperature 250°C, detector temperature 300°C (FID) (Al Ashaal et al., 2010).

TABLE 27.3 Fatty Acid Composition of *B. aegyptiaca* Del. Kernel Oil as a Function of Pressing Temperature (Mohamed et al., 2002)

Fatty acid	50°C (mg/g oil)	115°C (mg/g oil)
C16:0	95.6	89.7
C18:0	94.5	89
C18:1	287.6	270.5
C18:2	263.9	259.9
C18:3	12.6	9.7
UFA (%)[a]	74.8	75.14
Total	754.2	718.8

[a] Unsaturated fatty acids $= \dfrac{g\,of\,unsaturated\,fatty\,acids}{g\,of\,total\,fatty\,acids} \times 100$

Nour, Ahmed, & Abdel-Gayoum, 1985; Abdel-Rahim, El-Saadany, & Wasif, 1986). It could be concluded that a pressing temperature up to 115°C did not affect the fatty acid composition of *B. aegyptiaca* Del. kernel oil. However, the after effects on oil quality (i.e., the physical constant) should be investigated (Mohamed, Wolf, & Spiess, 2002).

CONCLUSION

The *B. aegyptiaca* is popular from present research because the tree is valued for its fruits and seeds. The seed kernel is rich in oil, protein, and minerals and edible as snacks after boiling. Also, many studies highlight the potential of *Balanites* for oil production as an agro-industrial material and the potential of the cake for animal feed.

The potent biological activities and also their active compounds were obtained from the extracts that were used in many pharmacological products because of the side effects and the immense curation potential. In the previous review, the biological activities of *B. aegyptiaca* were briefly summarized. They exhibit many biological activities and contain phytochemicals compounds, such as flavonoids, phenolics, tannin, and saponin. The plant can be investigated further to reveal the hidden compounds of potent nature.

REFERENCES

Abdel-Rahim, E., El-Saadany, S., & Wasif, M. (1986). Chemical and physical studies on *Balanites aegyptiaca* oil. *Grasas y Aceites, 37*(2), 81–85.

Abu-Al-Futuh, I. (1983). *Balanites aegyptiaca: an unutilized raw material potential ready for agro-industrial exploitation.* Wien, Austria: United Nations Industrial Development Organization.

Al Ashaal, H. A., Farghaly, A. A., El Aziz, M. A., & Ali, M. (2010). Phytochemical investigation and medicinal evaluation of fixed oil of *Balanites aegyptiaca* fruits (Balantiaceae). *Journal of Ethnopharmacology*, *127*(2), 495–501.

Anto, F., Aryeetey, M., Anyorigiya, T., Asoala, V., & Kpikpi, J. (2005). The relative susceptibilities of juvenile and adult *Bulinus globosus* and *Bulinus truncatus* to the molluscicidal activities in the fruit of Ghanaian *Blighia sapida*, *Blighia unijugata* and *Balanites aegyptiaca*. *Annals of Tropical Medicine & Parasitology*, *99*(2), 211–217.

Amalraj, V. A., & Shankanarayan, K. A. (1998). Ecological distribution of *Balanites roxburghii* plant in arid Rajasthan. *Journal of Tropical Forestry*, *2*(3), 183–187.

Badi, K.H., Ahmed, A.E.H., & Bayoumi, A. (1989). The forests of Sudan. Khartoum, Sudan: Ministry of Agriculture, Department of Forestry

Bagavan, A., Rahuman, A. A., Kamaraj, C., & Geetha, K. (2008). Larvicidal activity of saponin from *Achyranthes aspera* against *Aedes aegypti* and *Culex quinquefasciatus* (Diptera: Culicidae). *Parasitology Research*, *103*(1), 223–229.

Booth, F. E., & Wickens, G. E. (1988). Non-timber uses of selected arid zone trees and shrubs in Africa.Rome, Italy: Conservation Guide 19, Food & Agriculture Organization, FAO.

Chapagain, B., & Wiesman, Z. (2005). Variation in diosgenin level in seed kernels among different provenances of *Balanites aegyptiaca* Del. (Zygophyllaceae) and its correlation with oil content. *African Journal of Biotechnology*, *4*(11), 1209–1213.

Chapagain, B., & Wiesman, Z. (2011). Larvicidal effects of aqueous extracts of *Balanites aegyptiaca* (desert date) against the larvae of *Culex pipiens* mosquitoes. *African Journal of Biotechnology*, *4*(11), 1351–1354.

Chapagain, B. P., Saharan, V., & Wiesman, Z. (2008). Larvicidal activity of saponins from *Balanites aegyptiaca* callus against *Aedes aegypti* mosquito. *Bioresource Technology*, *99*(5), 1165–1168.

Chothani, D. L., & Vaghasiya, H. (2011). A review on *Balanites aegyptiaca* Del. (desert date): phytochemical constituents, traditional uses, and pharmacological activity. *Pharmacognosy Reviews*, *5*(9), 55.

Dubey, P., Yogi, M., Bharadwaj, A., Soni, M., Singh, A., & Sachan, A. K. (2011). *Balanites aegyptiaca* (L.) Del., a semi-arid forest tree: a review. *Academic Journals of Plant Science*, *4*(4), 12–18.

El Mastry, S., Ebeed, M., El Sayed, I., Nasr, M., & El Halafawy, K. (2010). Protective effect of *Balanites aegyptiaca* on antioxidant defense system against adriamycin-induced cardiac toxicity in experimental mice. *Egyptian Journal of Biochemistry and Molecular Biology*, *28*(1), 1502–1687.

Elfeel, A. A. (2010). Variability in *Balanites aegyptiaca* var. *aegyptiaca* seed kernel oil, protein and minerals contents between and within locations. *Agriculture and Biology Journal of North America*, *1*, 170–174.

Folarin, O. M., Eromosele, I. C., & Eromosele, C. O. (2011). Relative thermal stability of metal soaps of *Ximenia americana* and *Balanites aegyptiaca* seed oils. *Scientific Research and Essays*, *6*, 1922–1927.

Fregon, S. M. E. (2015). Physicochemical properties of *Balanites aegyptiaca (Laloub)* seed oil. Khartoum, Sudan: MSc. Thesis, Sudan University of Science and Technology.

Gad, M. Z., El-Sawalhi, M. M., Ismail, M. F., & El-Tanbouly, N. D. (2006). Biochemical study of the anti-diabetic action of the Egyptian plants fenugreek and *Balanites*. *Molecular and Cellular Biochemistry*, *281*(1–2), 173–183.

Gnoula, C., Mégalizzi, V., De Nève, N., Sauvage, S., Ribaucour, F., Guissou, P., & Lefranc, F. (2008). Balanitin-6 and-7: diosgenyl saponins isolated from *Balanites aegyptiaca* Del. display significant anti-tumor activity in vitro and in vivo. *International Journal of Oncology*, *32*(1), 5–15.

Hall, J., & Walker, D. (1991). *Balanites aegyptiaca: a monograph* (65 p.). Bangor, UK: School of Agricultural and Forest Sciences, University of Wales. (*904*(390), 75).

Karuppusamy, S., Rajasekaran, K., & Karmegham, N. (2002). Antimicrobial activity of *Balanites aegyptiaca. (L). Del. Journal of Ecotoxicology and Environmental Monitoring 12*, 2002, 67–8.

Levey, A., Gassman, J., Hall, P., & Walker, W. (1991). Assessing the progression of renal disease in clinical studies: effects of duration of follow-up and regression to the mean. Modification of Diet in Renal Disease (MDRD) Study Group. *Journal of the American Society of Nephrology, 1*(9), 1087–1094.

Mohamed, M. F. A. A. (2016). *Effect of aqueous extracts of Moringa oleifera leaves and Balanites aegyptiaca mesocarp on survival of snails and cercariae of Schistosoma haematobium.* Khartoum, Sudan: M.Sc. Thesis, Faculty of Veterinary, University of Khartoum.

Mohamed, A., Wolf, W., & Spiess, W. (2002). Physical, morphological and chemical characteristics, oil recovery and fatty acid composition of *Balanites aegyptiaca* Del. kernels. *Plant Foods for Human Nutrition, 57*(2), 179–189.

Nour, A. A. A., Ahmed, A. H. R., & Abdel-Gayoum, A. G. A. (1985). A chemical study of *Balanites aegyptiaca* L. (Lalob) fruits grown in Sudan. *Journal of the Science of Food and Agriculture, 36*(12), 1254–1258.

NRC, (National Research Council) (2008). *Lost Crops of Africa: Volume III: Fruits.* Washington, DC, USA: The National Academies Press.

Obidah, W., Nadro, M. S., Tiyafo, G. O., & Wurochekke, A. U. (2009). Toxicity of crude *Balanites aegyptiaca* seed oil in rats. *Journal of American Science, 5*(6), 13–16.

Okia, C. A., Agea, J. G., Kimondo, J. M., Abohassan, R. A. A., Okiror, P., Obua, J., & Teklehaimanot, Z. (2011). Use and management of *Balanites aegyptiaca* in drylands of Uganda. *Research Journal of Biological Sciences, 6*(1), 15–24.

Omer, S. (2015). *Variation in morphological, physical and chemical constituents of Balanites aegyptiaca Fruit between geographical sites.* Khartoum, Sudan: M. Sc. Thesis, Faculty of Forestry, University of Khartoum

Sands, B. E., Tremaine, W. J., Sandborn, W. J., Rutgeerts, P. J., Hanauer, S. B., Mayer, L., & Podolsky, D. K. (2001). Infliximab in the treatment of severe, steroid-refractory ulcerative colitis: a pilot study. *Inflammatory Bowel Diseases, 7*(2), 83–88.

Shanks, W., & Schwerdtner, W. (1991). Crude quantitative estimates of the original northwest-southeast dimension of the Sudbury Structure, south-central Canadian Shield. *Canadian Journal of Earth Sciences, 28*(10), 1677–1686.

Sidiyene, E. A. (1996). *Trees and spontaneous shrubs of the Adrar des Iforas.* Mali, Paris, France: Orstom.

Speroni, E., Cervellati, R., Innocenti, G., Costa, S., Guerra, M., Dall'Acqua, S., & Govoni, P. (2005). Anti-inflammatory, anti-nociceptive and antioxidant activities of *Balanites aegyptiaca* (L.) Delile. *Journal of Ethnopharmacology, 98*(1), 117–125.

Ufodike, E., & Omoregie, E. (1994). Acute toxicity of water extracts of barks of *Balanites aegyptiaca* and *Kigelia africana* to *Oreochromis niloticus* (L.). *Aquaculture Research, 25*(9), 873–879.

Usher, F. C. (1984). *Surgical mesh and method.* Google Patents.

Wiesman, Z., & Chapagain, B. P. (2006). Larvicidal activity of saponin containing extracts and fractions of fruit mesocarp of *Balanites aegyptiaca. Fitoterapia, 77*(6), 420–424.

Zaahkouk, S., Rashid, S. Z., & Mattar, A. (2003). Anti-diabetic properties of water and ethanolic extracts of *Balanites aegyptiaca* fruits flesh in senile diabetic rats. *The Egyptian Journal of Hospital Medicine, 10*, 90–108.

Chapter 28

Vangueria madagascariensis (Rubiaceae) as New Oil Source

DISTRIBUTION AND BOTANICAL DESCRIPTION

Vangueria madagascariensis belongs to the family Rubiaceae. Rubiaceae are mainly tropical woody plants and consist mostly of trees and shrubs, less often of perennial to annual herbs (Mongrand et al., 2005). It contains over 50 species distributed in Africa south of the Sahara with one species occurring in Madagascar (*V. madagascariensis*). The center of diversity is in East Africa (Kenya, Tanzania) and they are rare in West Africa (Lantz & Bremer, 2005). The tree is found near stream in water depressions in tall grass savanna in Kassala, Blue Nile (Ingesanna Hills), Kordofan (Nuhud-Abu-Zabad Road), Bahr El Ghazal, and Equatoria (Yirol, Yei), and is locally known as kirkir (El Ghazali, Abdalla, Khalid, Khalafalla, & Ahamad, 2003). Abdelmuti (1991) stated *V. madagascariensis* is widespread in Sudan in Darfur, and southern Kordofan states, it is a small tree, a shrub 2–3 m high or tree up to 10 m high branchlets brown, lenticellate, glabrous (El Ghazali et al., 2003). Leaves are clustered at end of branches, deciduous, opposite, elliptic, up to 20 × 8 cm, glabrous, paler beneath (Fig. 28.1). Inflorescences axillary branched panicles; flowers greenish, about 5-mm long (El Ghazali et al., 2003). Fruits subglobose, yellowish-brown when mature, edible, 4–5-seeded (Fig. 28.1); the size of the seed up to 3 cm across (Pino, Marbot, & Vazquez, 2004). Also Abdelmuti (1991) stated that fruits are about 8 mm in diameter and weighing 5.5–6.5 g. The reddish fruits, which are very sweet when ripe, are either eaten fresh or dried for consumption at a later date. The dried fruits are easily rehydrated by immersion in water. Ripe fruits are hand expressed and the juice is given to lactating mothers to relieve stomach, colic; also it is eaten as famine food (Abdelmuti, 1991). The soft flesh of the ripe fruit has a pleasant, slightly acid flavor. It is usually eaten fresh, in preserves and beverages, and it has a pleasant chocolate-like flavor. The fruit is also used to add flavor to beer (Orwa, Mutua, Kindt, Jamnadass, & Simons, 2009).

Unconventional Oilseeds and Oil Sources. http://dx.doi.org/10.1016/B978-0-12-809435-8.00028-7

FIGURE 28.1 *Vangueria madagascariensis* tree, fruits and kernels.

CHEMICAL COMPOSITION AND MAIN CONSTITUENTS

The volatile constituents of the fruits of *V. madagascariensis* growing in Cuba were isolated by simultaneous distillation/solvent extraction and analyzed by GC and GC/MS. Seventy-four compounds were identified in the aroma concentrate, of which 2-furfural and hexadecanoic acid were found to be the major constituents (Pino et al., 2004). Fentahun and Hager (2009) reported that *V. madagascariensis* wild fruits were rich in valuable nutrients and were accessible year-round in the Amhara region of Ethiopia with significant overlap at times of acute food and nutrient scarcity.

Vangueria fruits collected from Malawi have been found to contain at 26.5% dry matter; 78.1% total carbohydrate, 10.2%, 2.6% fat, 5.7% crude protein. Ascorbic acid content is 16.8 mg/100 g fresh weight (Saka, 1995). In Botswana, an ascorbic acid content of 4.7 mg/100 g has been reported for fruits with 64.4% moisture (Leakey, 1999). Abdelmuti (1991) reported that kirkir fruit contains 5.9% protein content, 1.2% fat content, 12.30% crude fiber, 4.90% ash content, and 75.70% carbohydrates content, while AbdelRahman (2011) reported that kirkir fruit contains 7.71% protein content, 2.35% fat content, 18.89% crude fiber content, 3.63% ash content, and 67.42% carbohydrate content. Also, AbdelRahman (2011) stated that the polyphenols content (tannins) of kirkir was 2.22%, ascorbic acid content (vitamin C) was 169.69 mg/100 g, β-carotene was 115.6 IU).

Stangeland, Remberg, and Lye (2009) revealed that the antioxidant activity, for *Vangueria apiculata* fruits decreased considerably as the fruit ripened. This could be due to different kinds of antioxidant compounds causing high antioxidant, for example, tannins, which are often more abundant in unripe fruits, while the opposite is often the case for anthocyanins.

A new biflavonoid, 5,7,3',5'',7'',4'''-hexahydroxy (4'-O-3''')-biflavone (**1**) and a polyketide derivative, methylcylohex-1-ene (**3**) were isolated from aerial parts of *Vangueria infausta*. In addition, other known flavonoids were also isolated from this plant (Mbukwa, Chacha, & Majinda, 2007).

The overall leaf fatty acid composition of *V. madagascariensis* was studied and results were found as linolenic, linoleic, and palmitic acids were the major fatty acids (Mongrand et al., 2005). Kotowaroo, Mahomoodally, Gurib-Fakim, and Subratty (2006) studied the *V. infausta* root and bark extracts which showed antimalarial activity against *Plasmodium berghei* in mice. It gave a parasite suppression of 73.5% in early infection and a repository effect of 88.7%. The plant extracts showed the presence of flavonoids, coumarins, tannins, terpenoids, anthraquinones, and saponins. Kotowaroo et al. (2006) studied the possible effects of *V. madagascariensis* on starch breakdown by alpha-amylase in vitro. The moisture content of *V. madagascariensis* seed was 6.4%.

MEDICINAL AND OTHER USES

Roots and bark of *V. madagascariensis* are used in traditional medicine; for example, in Tanzania, an extract from the roots is used to treat worm infections (Orwa et al., 2009). Jain and Srivastava (2005) stated that *V. madagascariensis* leaf bath is given for skin diseases and abscesses. This multibranched shrub or tree is popular as a source of both firewood and charcoal; also the tree wood is suitable for building construction, tool handles, and carving (Pino et al., 2004). The oil content of *V. madagascariensis* seed kernels was 40.0%. The crude protein was 22.2%, the ash content was 2.8%; while the crude fiber of *V. madagascariensis* seed kernel was 14.0%. Carbohydrates are the main energy reserves in the plant foods. Carbohydrates are of two types: available carbohydrate, which includes starches and sugars and nonavailable carbohydrate, which includes crude fiber and different dietary fiber constituents (Modgil, Modgil, & Kumar, 2004). The carbohydrate content of *V. madagascariensis* seed was 14.6%.

PHYSICOCHEMICAL PROPERTIES OF *Vangueria madagascariensis* KERNEL OIL

The oil obtained from *V. madagascariensis* kernels was odorless, of good color, and of good appearance. The oil showed higher values in refractive index, acid value, peroxide value (PV), saponification value, and specific gravity (Mustafa, 2011). The high content of oil, showing useful characteristics, such as odorless, and good color and appearance, make these seeds suitable for oil industrial applications. The oil content of *V. madagascariensis* seed kernel using cold and Soxhlet extraction methods was 40 and 41.2%, respectively.

Refractive index of the oil was 1.475. Compared with Codex standards 1999 for cottonseed, sunflower, sesame, and groundnut oils, the *V. madagascariensis* oil showed lower value of refractive index. The acid value of *V. madagascariensis* seed using cold and Soxhlet extraction methods was 0.6 and 0.22%, respectively. The low acid value indicates that the freshly extracted oil from the seed kernels has a negligible amount of free fatty acids. Acid value of *V. madagascariensis* oil compared with olive oil (0.2) showed that it was in range with the value in this work. Also Mohammed and Hamza (2008) reported different values of 0.45 and 0.5 for sesame oil. The PV of *V. madagascariensis* seed oil using cold and Soxhlet extraction were 0.8 and 1.0, respectively. The oil showed a saponification value of 182.6 and 181.4, respectively; it was comparable with the value of 186.0 of olive oil (Mustafa, 2011). The method of extraction has no significant effect on saponification value of the oil. Compared with Codex standards (1999) for cottonseed, sunflower, sesame, and groundnut oils, the *V. madagascariensis* oil showed lower saponification value. The specific gravity of *V. madagascariensis* oil extracted by two methods was 0.818 and 0.818, respectively.

FATTY ACID COMPOSITION

Fatty acid composition has been shown to influence the stability of oils, and polyunsaturated fatty acids have been found to contribute to the rancidification of several oils. Moreover, fatty acid composition has been found to be responsible for the odors and flavors associated with oil quality (León, Uceda, Jiménez, Martín, & Rallo, 2004). The major fatty acids of *V. madagascariensis* oil determined by GC/MS were palmitic (16:0), stearic (18:0), oleic (18:1n−9), and linoleic (18:2n−6) acids. The extraction method did not affect the fatty acid composition of the sample. The dominant fatty acid in the seed oil of *V. madagascariensis* was linoleic (C18:2); it represents 63.1 and 63.4% followed by oleic acid, which was 10.5 and 10.4% in VMCE and VMSE, respectively. Palmitic (C16:0) and stearic (C18:0) acids exhibited the third and fourth highest fatty acid contents in the two oils; palmitic acid was 9.7 and 9.8% in VMCE and VMSE, respectively; while stearic acid was 5.1 and 5.4% in VMCE and VMSE, respectively (Mustafa, 2011). The remaining fatty acids contributed only few percentages to the total fatty acid percent. Differences in the fatty acid composition of the seed oil may exist even within the same variety.

TOCOPHEROLS

As an important nutrient for human beings, vitamin E has been well known for its antioxidative properties (Guijun, Xiaoli, Zhongyue, Weidong, & Hongjie, 2007). Alpha-tocopherol was the most prominent and reported as 31.6 and 28.5 mg/100 g in VMCE and VMSE, respectively, and delta-tocopherol was 8.4 and 10.5 mg/100 g, respectively. The oils of VMCE and VMSE showed higher amounts of tocopherols, 110.5 and 107.9 mg/100 g, respectively, compared to other common oils, such as sesame (33–101), groundnut (17–130), and sunflower (44–152 mg/100 g) oils (CODEX-STAN 210-1999). The gamma-tocopherol content varied from 563 to 1095 mg/kg in oil and 293 to 569 mg/kg in sesame seed and was in broad agreement with the Codex range (Mustafa, 2011). Also in commercial sesame oils, the tocopherol content was 304–647 µg/g oil. In *V. madagascariensis* oils, β-tocopherol was predominant, constituting more than 59%. The other tocopherols in the oil of the two samples were below 1 mg/100 g each.

STEROL COMPOSITION

Recently phytosterols (plant sterols) have become a focus of interest due to their serum cholesterol lowering effect and consequently their protection against cardiovascular disease (Koschutnig et al., 2010). Moreover, sterols comprise the bulk of the unsaponifiables in many oils. They are of interest due to their impact on health (Gomaa & Ramadan, 2006). The total sterol composition of *V. madagascariensis* (kirkir) oil determined by GC–MS was found to be 41.9%. Seven sterols were identified, among these the three major ones are β-sitosterol (45.4%), campesterol (22.6%), and stigmasterol (20.0%) and with observation that in the two oils, β-sitosterol has the higher concentration, other sterols are present in small or minute amounts (Mustafa, 2011). Fig. 28.2 illustrates the chromatogram of sterols of kirkir oil by GC/MS.

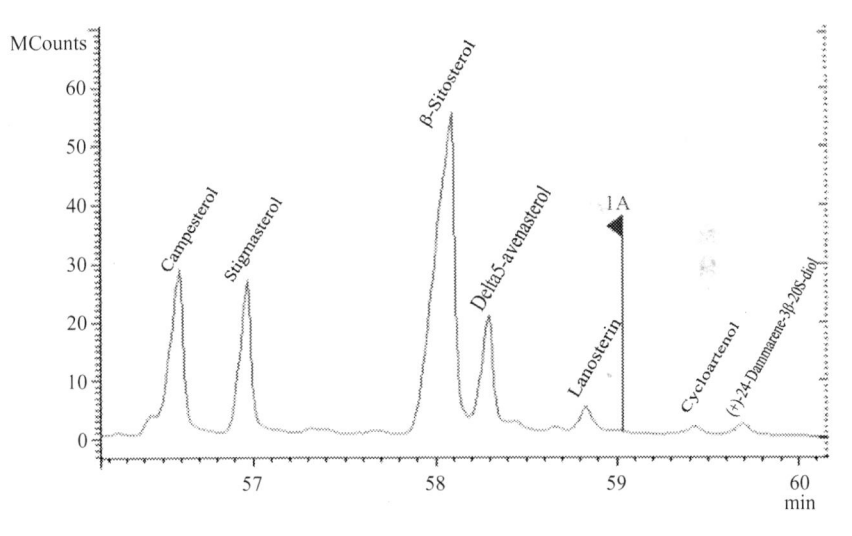

FIGURE 28.2 Chromatogram of sterols of *V. madagascariensis* (kirkir) oil by GCMS.

AMINO ACID PROFILE

An evaluation of the amino acid contents will reveal better the value of the protein in the seed. Presence of essential amino acids (EAAs) in the seed will enhance its value despite the low crude protein content (Amoo & Agunbiade, 2010). The amino acids profile of V. madagascariensis seed, analyzed by amino acid hydrolysis is presented in Table 28.1. Values are given for 16 different amino acids, and sums for EAAs and nonessential amino acids (NEAA). However, another EAA tryptophan, which acts as a precursor of the vitamin niacin and the neurotransmitter serotonin, cannot be determined because it is very labile in an acid environment and is completely destroyed by acid hydrolysis (HCI 6N)

TABLE 28.1 Amino Acid Composition of V. madagascariensis Seeds Compared With Egg, Sesame, and Bean (g/100 g protein)[a]

Amino acid	V. madagas-cariensis	Egg[b]	Sesame[b]	Broad bean[b]
Threonine	0.738 ± 0.4	0.634 ± 0.3	0.763 ± 0.3	0.159 ± 0.1
Methionine + cystine	0.206 ± 0.1	0.717 ± 0.5	0.988 ± 0.6	0.071 ± 0.1
Valine	1.031 ± 0.5	0.174 ± 0.1	0.985 ± 0.6	0.374 ± 0.2
Isoleucine	0.822 ± 0.5	0.778 ± 0.5	0.773 ± 0.5	0.222 ± 0.1
Leucine	1.608 ± 0.6	1.091 ± 0.4	1.433 ± 0.4	0.389 ± 0.1
Phenylalanine + tyrosine	1.311 ± 0.6	1.224 ± 0.5	1.614 ± 0.5	0.368 ± 0.1
Histidine	0.695 ± 0.6	0.301 ± 0.1	0.523 ± 0.2	0.126 ± 0.1
Lysine	0.823 ± 0.4	0.863 ± 0.5	0.585 ± 0.3	0.338 ± 0.1
Arginine	1.085 ± 0.6	0.754 ± 0.6	2.586 ± 1.0	0.760 ± 0.5
Total EAA	8.319	6.536	10.25	2.807
Aspartic acid	1.464 ± 0.7	0.892 ± 0.6	0.094 ± 0.1	0.659 ± 0.2
Glutamic acid	1.876 ± 0.6	0.121 ± 0.1	0.366 ± 0.1	0.781 ± 0.3
Serine	0.655 ± 0.4	0.672	0.072 ± 0.1	0.240 ± 0.1
Glycine	0.834 ± 0.1	0.302 ± 0.1	0.054 ± 0.1	0.201 ± 0.1
Alanine	1.058 ± 0.7	0.503 ± 0.2	0.140 ± 0.1	0.255 ± 0.1
Total NEAA	5.883	2.490	1.391	2.136
Total amino acids	14.202	9.026	11.641	4.943

EAA, Essential amino acid; NEAA, nonessential amino acids.
[a]All determinations were carried out in triplicate and mean value ± standard deviation (SD) reported.
[b]Source: Paul, A. A, & Southgate, D. A. T. (1979). McCance and Widdowson The composition of foods (4th ed.). Crown Copyright, London, UK.

(Friedman & Cuq, 1988). The total amount of the EAAs (phenylalanine, leucine, valine, threonine, isoleucine, methionine, tyrosine, histidine, arginine, cystine, and lysine) found in *V. madagascariensis* seeds were 8.3 g/100 proteins, and differed significantly. The percentage of sulfur-containing amino acids (methionine and cystine) in *V. madagascariensis* seed was 0.206 g/100 g, which was the lowest among the others. The percentage of aromatic amino acids (phenylalanine + tyrosine) was 1.311 g/100 g protein in *V. madagascariensis* (Mustafa, 2011). *V. madagascariensis* seeds when compared with egg, sesame, and broad bean showed a clear difference in their amino acids contents because those were known as a source of protein (Redead, 1989–90). All the EAAs with the exception of tryptophan, which was not analyzed were found to be present in high amount in *V. madagascariensis* seed, when compared to that of three different foods in Table 28.1.

The total amount of the EAAs of *V. madagascariensis* seeds was 58.6%. The individual EAAs of *V. madagascariensis* seeds were higher somewhat to that of egg and broad bean but egg was higher in methionine + cystine and isoleucine amino acids (Mustafa, 2011).

OXIDATIVE STABILITY OF KIRKIR OIL

The evaluation of oxidative stability is a key factor in developing the new oil for food applications. The oil stabilities during storage or upon heating are vital parameters for ensuring that the oil has good performance at elevated temperature (Rohman, Che Man, Ismail, & Hashim, 2011). Several methods have been used to evaluate the extent of oxidative deterioration, which are related to the measurement of the concentration of primary or secondary oxidation products. The most frequently used are PV. The PV is used for monitoring of peroxide formation in the early stages of oxidation. Changes in the PV as a measure of the primary oxidation in the two samples during storage at 70°C and measurement of the PV at certain time were reported by Mustafa (2011). She reported the initial PVs of *V. madagascariensis* Soxhlet extraction (VMSE) and *V. madagascariensis* cold press (VMCE) are 0.8 and 1.0 mEq. O_2/kg of oil, respectively. The initial PV of crude oils were less than the standard PV (10 mEq. O_2/kg) for vegetable oil deterioration (Anyasor, Ogunwenmo, Oyelana, Ajayi, & Dangana, 2009). The low PVs indicated slow oxidation of these oils. These values were considered satisfactory according to the international regulations on edible oils. At the end point of oxidation, the two oil samples reached PVs of 14.3 and 11.9, respectively. In VMSE and VMCE samples, the PV increased sharply and faster which means that the *V. madagascariensis* oil samples were less stable. In the VMSE and VMCE oils there was slight increase with storage time from 1.5 to 14.26 mEq. O_2/kg oil and from 1.7 to 11.9 mEq. O_2/kg oil up to the end point of oxidation. The VMSE and VMCE showed lower stability, which can be explained by the increase in PV (Mustafa, 2011).

TOTAL PHENOLIC CONTENT

Phenols are very important plant constituents because of their scavenging ability on free radicals due to their hydroxyl groups. Therefore, phenolic contents of plants may contribute directly to their antioxidant action (Ying et al., 2010). Total phenols were determined by the Folin–Ciocalteu method. The total phenolic compounds of methanolic extracts of V. *madagascariensis* bark, leaf, root, and seedcake, the results showed that the plant extracts contain extremely high contents of total phenols. The highest total phenol amounts were 169.5 and 167.9 g/kg plant extract as GAE for VMB, VML methanolic extracts, respectively. Overall, the lowest concentrations were found in the methanolic content of VMC extracts at 112.5 g/kg plant extract as GAE (Mustafa, 2011).

IDENTIFICATION OF PHENOLIC COMPOUNDS USING HPLC-DAD

HPLC-DAD was used to identify and know what is/are the responsible active ingredient(s) in the crude methanolic extracts of VML, VMB, and VMC. It was found that the crude methanolic extracts of VML, VMB, and VMC contain syringic, gallic acid, hydroxybenzoic, chlorogenic, vanillin, ferulic, and p-coumaric acids. These compounds have been identified according to their retention time and the spectral characteristics of their peaks compared to those of standards, as well as by spiking the sample with standards. Chlorogenic acid was detected to be the major phenolic component in all the methanolic extracts, and it is the higher in VML and VMC showing the levels of 1.027 and 1.226 mg/100 g dry weight contributing about 81.6 and 89.3% to the total amount, respectively (Mustafa, 2011) (Table 28.2).

TABLE 28.2 Phenolic Compound Content mg/100 g Dry Weight in V. *madagascariensis* Methanolic Extract

Compounds	VMB	VMC	VML
Syringic acid	0.007 ± 0.35	0.021 ± 0.22	0.018 ± 0.11
Hydroxybenzoic acid	0.00	0.052 ± 0.12	0.03 ± 0.21
Gallic acid	0.004 ± 0.18	0.061 ± 0.22	0.014 ± 0.11
Chlorogenic acid	0.00	1.226 ± 0.56	1.027 ± 0.67
Vanillin	0.015 ± 0.11	0.050 ± 0.21	0.026 ± 0.34
p-Coumaric	0.00	0.031 ± 0.23	0.005 ± 0.01
Ferulic acid	0.00	0.062 ± 0.18	0.030 ± 0.11

Values are means ± SD ($n = 3$), and they are given as mg/100 g dry weight of investigated kenaf seedcake extract/fractions. VMB, V. *madagascariensis* bark; VMC, V. *madagascariensis* seedcake; VML, V. *madagascariensis* leaves.

This result is predictable due to the weak selectivity of the Folin–Ciocalteu reagent, as it reacts positively with different antioxidant compounds (phenolic and nonphenolic substances). The previously mentioned HPLC-DAD results indicate that such phenolic compounds from the tree may show higher antioxidant activity. Isolation and characterization of such phenolic compounds may be useful in developing natural antioxidants (Mustafa, 2011).

The HPLC-DAD analysis of *V. madagascariensis* seedcake, leaves, and bark extracts revealed the presence of phenolic compounds. By this means, the crude methanolic extracts of VML, VMB, and VMC were analyzed. The chromatogram of the *V. madagascariensis* extracts, was monitored at 280 nm. Chlorogenic, hydroxybenzoic, *p*-coumaric, syringic acids, ferulic acid, vanillin, and gallic acid were detected. These compounds have been identified according to their retention time and the spectral characteristics of their peaks compared to those of standards, as well as by spiking the sample with standards (Mustafa, 2011).

TOTAL FLAVONOID

Plant flavonoids (mainly rutin) are partly responsible for the antioxidant characteristics of plant extracts (Guillén et al., 2008). The total flavonoid of methanolic extracts of *V. madagascariensis* bark, leaf, root, and seedcake was reported by Mustafa (2011). Her results showed that the leaves extracts of the plant had high flavonoid content. VML showed the highest amounts of 298.8. The extracts of the VMC recorded very low amount of flavonoid, 27.6 and 23.1 expressed as mg of rutin equivalents (RE)/100 g of extract. VMB showed medium concentrations as 123.6 mg/100 g of extract. Among the tested samples, VML methanolic extracts showed the most antioxidant activity (IC$_{50}$ 7.81 µg/mL), which may be attributed to the antioxidant effects of flavonoids (Mustafa, 2011).

1,1-DIPHENYL-2-PICRYLHYDRAZYL RADICAL SCAVENGING ACTIVITY

1,1-Diphenyl-2-picrylhydrazyl (DPPH) is a compound that possesses a nitrogen free radical and is readily destroyed by a free radical scavenger. This assay was used to test the ability of the antioxidative compounds functioning as proton radical scavengers or hydrogen donors (Chew, Lim, Omar, & Khoo, 2008). The extracts of *V. madagascariensis* were assayed over a range of dilutions to establish the concentration of each extract required to scavenge 50% of the DPPH radical present in the assay medium, referred to as the IC$_{50}$ defined as the concentration of the antioxidant needed to scavenge 50% of DPPH present in the test solution. Lower IC$_{50}$ value reflects better DPPH radical scavenging activity (Molyneux, 2004). The antioxidant activity of *V. madagascariensis* bark, leaf, root, and seedcake extracts was determined by Mustafa (2011). Free radical scavenging potentials of the two plants extracts at different concentrations were tested by the DPPH method and *V. madagascariensis* leaves, seedcake, and bark extracts presented IC_{50} of 7.81, 31.25, and 62.5 μg/mL, respectively.

ORAC ANTIOXIDANT ACTIVITY ASSAY

The oxygen radical absorbance capacity (ORAC) assay is the only method that takes free radical action to completion and uses an area-under-curve (AUC) technique for quantitation and thus, combines both inhibition percentage and the length of inhibition time of the free radical action by antioxidants into a single quantity (Taruscio, Barney, & Exon, 2004). The assay has been widely used in many recent studies related to plant (Atala, Vásquez, Speisky, Lissi, & López-Alarcón, 2009). Mustafa (2011) evaluated the antioxidant capacity of *V. madagascariensis* methanolic extract; they used seven samples of VML, VMB, and VMC, and their potencies were compared with the positive control: quercetin. The AUC was calculated for the sample, standard, and the positive control. ORAC results of VML displayed a higher level of antioxidant activity, 72.72 μM of Trolox than quercetin at 5 μg/mL (58.97 ± 0.02 μM of Trolox). VMB and VMC showed lower level of antioxidant (47.08 and 44.94 μM of Trolox, respectively) than quercetin at 5 μg/mL (58.97 ± 0.02 μM of Trolox) (Mustafa, 2011).

CYTOTOXIC ACTIVITY

The tetrazolium salt 3-[4,5-dimethylthiazol-2-yl]-2,5-diphenyl-tetrazolium bromide (MTT) is used to determine cell viability in assays of cell proliferation and cytotoxicity. MTT is reduced in metabolically active cells to yield an insoluble purple formazan product (Das, Das, Mazumder, Das, & Basu, 2009). MTT assay was used to evaluate the cytotoxic activity of extracts from different parts of *V. madagascariensis.* MTT is reduced to an insoluble purple formazan by mitochondrial dehydrogenase. Cell viability was measured by comparison of the purple color formation. Dead cells, on the other hand, did not form the purple formazan due to their lack of the enzyme. The effect of VML on the cell viability in vitro was studied by Mustafa (2011). Cells were cultured for 24 h with several different levels of VML extract (Table 28.3). The curves showed that the growth of melanoma cancer cells was inhibited in a dose-dependent manner.

Crude extracts of VML, VMB, and VMC were tested with a series of different doses on A549, PC-3, MCF-7, HepG2, HT-29, WRL-68, and WI-38, respectively, and after 24 h, cell viability was determined by the MTT assay; only VML was found effective. Test agents induced cell cytotoxicity in a concentration-dependent manner. These dose titration curves allowed determining of EC_{50} for the test agents toward different cell lines. These results indicated that cell lines differ in their sensitivity to the same test agent, which may be determined by multiple cell type-specific signaling cascades and transcription factor activities.

TABLE 28.3 Effect of *V. madagascariensis* Leaves Extract (VML) on Different Cells Type Expressed as EC_{50} Values in 24 h MTT Assay[a]

MTT assay (test agent)	$EC_{50} \pm SD$ (µg/mL)	
	VML	ASL
A549	42.54 ± 2.32	20.14 ± 1.18
PC-3	34.42 ± 2.97	36.91 ± 2.57
MCF-7	22.75 ± 1.98	19.42 ± 2.37
HepG2	28.40 ± 3.54	15.01 ± 1.05
HT-29	53.20 ± 4.24	17.84 ± 1.99
WRL-68	44.47 ± 1.27	17.64 ± 1.94
WI-38T	64.74 ± 2.92	41.63 ± 2.34

[a]*Results are mean ± SD (n = 3), results are given in mg/g extract.*

Thus, at the same time no significant cytotoxic effects toward normal hepatic cells and normal lung fibroblast compared with cancer cells (Mustafa, 2011).

STABILITY OF SUNFLOWER OIL AS AFFECTED BY THE ADDITION OF *Vangueria madagascariensis* EXTRACTS

PV is one of the most frequently determined quality parameters during oil production, storage, and marketing. The results of PV estimation give a clear indication of lipid autoxidation (Suja, Jayalekshmy, & Arumughan, 2005). The effects of individual plant extracts on the oxidative stability of sunflower oil, which is very rich in linoleic fatty acid, were evaluated by measuring the formation of PV at 70°C. The development of PV during the oxidation of sunflower oil was evaluated at 70°C. This temperature was ideal because at higher temperatures the peroxides will decompose very fast (Mariod, Matthäus, & Hussein, 2006). The PV of sunflower oil (control) with and without VML, VMC, VMB, and BHT showed a gradual increase. As demonstrated a maximum PV of 38.5 mEq. O_2/kg was reached after 56 h of storage in the control without addition of extract or BHT. The PVs of sunflower oil containing 100 mg of VML,VMC, and VMB or 20 mg BHT were found to be 12.5, 15.2, 17, and 8 mEq. O_2/kg, respectively. These samples showed an IO, after 56 h of storage, of 65, 64, 59 and 82%, respectively, in comparison to a control. From this, it can be assumed that all extracts added had a remarkable antioxidative activity to reduce oil degradation. VML showed the highest activity, followed by VMC, but BHT seem to be higher than the effect of these extracts (Mustafa, 2011).

STABILITY OF SUNFLOWER OIL MEASURED BY CONJUGATED DIENES AS AFFECTED BY THE ADDITION OF DIFFERENT EXTRACTS FROM *Vangueria madagascariensis* — VML, VMC, AND VMB

Several methods have been used to evaluate the extent of oxidative deterioration, which are related to the measurement of the concentration of primary or secondary oxidation products. One of the most frequently used is specific absorptivity in ultraviolet region at 232 and 270 nm. The specific absorptivity at 232 and 270 nm measures the contents of conjugated dienes (CDs) and conjugated trienes (CTs) (Rohman et al., 2011). CDs are formed as intermediates through a shift of a double bond of polyunsaturated acids (PUFA). These compounds can be measured by ultraviolet absorption at 232–234 nm (Farhoosh & Moosavi, 2009). Methanolic extracts from *V. madagascariensis* at 100, 250, 500 mg were tested as inhibitor of edible sunflower oil oxidation at 70°C at 234 nm (CDs were monitored). All the extracts at 500 mg/100 g oil appeared to be strong inhibitors of sunflower oil oxidation. The most active extract was *V. madagascariensis* leaves which exhibited an antioxidant activity, which seems to be similar to BHT at 200 mg/kg; the other two extracts (VMC and VMB) exhibited antioxidant activity lower than that of BHT (Mustafa, 2011). The extracts obtained from *V. madagascariensis* leaves, seedcakes, and bark seem to be good sources of natural antioxidants that can be used to inhibit edible sunflower oil oxidation and these results were in agreement with Shahidi, Desilva, and Amarowicz (2003) who studied different extracts from different sources to be used as inhibitor of edible oil oxidation.

The sunflower oil with antioxidants showed significantly less formation of CD compared to the oil without antioxidants (control). The CD values of sunflower oil containing *V. madagascariensis* extracts increased by two- to threefold at the end of the 56-h storage period, whereas the control samples showed a fivefold or more increase. The oxidation inhibitory activity of the extracts and controls used, decreased in the order: BHT < VML < VMC < VMB < control (Mustafa, 2011).

REFERENCES

Abdelmuti, O. M. (1991). Biochemical and nutritional evaluation of famine foods of the Sudan. PhD Thesis. Sudan: University of Khartoum.

AbdelRahman, N. A. (2011). Study of some Sudanese edible forest fruits and their nectars. PhD Thesis. Sudan: Sudan Academy of Science in Food Science and Technology.

Amoo, I. A., & Agunbiade, F. O. (2010). Some nutrients and anti-nutrients components of *Pterygota macrocarpa* seed flour. *Electronic Journal of Environmental, Agricultural and Food Chemistry*, 9(2), 293–300.

Anyasor, G. N., Ogunwenmo, K. O., Oyelana, O. A., Ajayi, D., & Dangana, J. (2009). Chemical analyses of groundnut (*Archis hypogaea*) oil. *Pakistan Journal of Nutrition*, 8(3), 269–272.

Atala, E., Vásquez, L., Speisky, H., Lissi, E., & López-Alarcón, C. (2009). Ascorbic acid contribution to ORAC values in berry extracts: an evaluation by the ORAC-pyrogallol red methodology. *Food Chemistry, 113*(1), 331–335.

Chew, Y. L., Lim, Y. Y., Omar, M., & Khoo, K. S. (2008). Antioxidant activity of three edible seaweeds from two areas in South East Asia. *LWT—Food Science and Technology, 41*, 1067–1072.

Codex Standard for named vegetable oils CODEX-STAN 210-1999.

Das, S., Das, M. K., Mazumder, P. M., Das, S., & Basu, S. P. (2009). Cytotoxic activity of methanolic extract of *Berberis aristata* DC on colon cancer. *Global Journal of Pharmacology, 3*(3), 137–140.

El Ghazali, G. E. B., Abdalla, W. E., Khalid, H. E., Khalafalla, M. M., & Ahamad, A. A. (2003). Medicinal plants of the Sudan Part V. Medicinal plants of the Ingassana Area. Khartoum: Ministry of Science & Technology Sudan Currency Printing Press.

Farhoosh, R., & Moosavi, S. M. R. (2009). Evaluating the performance of peroxide and conjugated diene values in monitoring quality of used frying oils. *Journal of Agricultural Science and Technology, 11*, 173–179.

Fentahun, M. T., & Hager, H. (2009). Exploiting locally available resources for food and nutritional security enhancement: wild fruits diversity, potential and state of exploitation in the Amhara region of Ethiopia. *Food Security, 1*, 207–219.

Friedman, M., & Cuq, J. L. (1988). Chemistry, analysis, nutritional value, and toxicology of tryptophan, a review. *Journal of Agricultural and Food Chemistry, 36*, 1079–1109.

Gomaa, A. M., & Ramadan, M. F. (2006). Characterization of bioactive lipid compounds of *Cistanche phlipaea*. *Electronic Journal of Environmental, Agricultural and Food Chemistry, 5*(2), 1306–1312.

Guijun, D., Xiaoli, L., Zhongyue, C., Weidong, P., & Hongjie, L. G. L. (2007). The dynamics of tocopherol and the effect of high temperature in developing sunflower (*Helianthus annuus* L.) embryo. *Food Chemistry, 102*(1), 138–145.

Guillén, R., Rodríguez, R., Jaramillo, S., Rodríguez, G., Espejo, J. A., & Fernandez-Bolaños, J. (2008). Antioxidant from asparagus spear: phenolics. *Acta Horticulturae, 776*, 247–254.

Jain, S. K., & Srivastava, S. (2005). Traditional uses of some Indian plants among islanders of The Indian Ocean. *Indian Journal of Traditional Knowledge, 4*, 345–357.

Koschutnig, K., Kemmo, S., Lampi, A. M., Piironen, M., Fritz-Ton, C., & Wagner, K. H. (2010). Separation and isolation of β-sitosterol oxides and their non-mutagenic potential in the *Salmonella* microsome assay. *Food Chemistry, 118*(1), 133–140.

Kotowaroo, M. I., Mahomoodally, M. F., Gurib-Fakim, A., & Subratty, A. H. (2006). Screening of traditional antidiabetic medicinal plants of Mauritius for possible alpha-amylase inhibitory effects in vitro. *Phytotherapy Research, 20*(3), 228–231.

Lantz, H., & Bremer, B. (2005). Phylogeny of the complex Vanguerieae (Rubiaceae) genera *Fadogia*, *Rytigynia*, and *Vangueria* with close relatives and a new circumscription of *Vangueria*. *Plant Systematics and Evolution, 253*, 159–183, http://en.wikipedia.org/wiki/Vangueria_ Categories:Vangueria.

Leakey, R. R. B. (1999). Potential for novel food products from agroforestry trees: a review. *Food Chemistry, 66*, 1–14.

León, L., Uceda, M., Jiménez, A., Martín, L. M., & Rallo, L. (2004). Variability of fatty acid composition in olive (*Olea europaea L.*) progenies. *Spanish Journal of Agricultural Research, 2*(3), 353–359.

Mariod, A. A., Matthäus, B., & Hussein, I. H. (2006). Antioxidant activities of extracts from *Combretum hartmannianum* and *Guiera senegalensis* on the oxidative stability of sunflower oil. *Emirates Journal of Agricultural Sciences, 18*(2), 20–28.

Mbukwa, E., Chacha, M., & Majinda, R. R. T. (2007). Phytochemical constituents of *Vangueria infausta*: their radical scavenging and antimicrobial activities. *ARKIVOC, ix*, 104–112.

Modgil, M., Modgil, R., & Kumar, R. (2004). Carbohydrate and mineral content of chyote (*Sechium edule*) and bottle gourd (*Lagenaria siceraria*). *Journal of Human Ecology, 15*(2), 157–159.

Mohammed, M. I., & Hamza, Z. U. (2008). Physicochemical properties of oil extracts from *Sesamum indicum* L. seeds grown in Jigawa State—Nigeria. *Journal of Applied Sciences and Environmental Management, 12*(2), 99–101.

Molyneux, P. (2004). The use of the stable free radical diphenylpicryl-hydrazyl (DPPH) for estimating antioxidant activity. *Songklanakarin Journal of Science and Technology, 26*, 211–219.

Mongrand, S., Badoc, A., Patouille, B., Lacomblez, C., Chavent, M., & Bessoule, J. (2005). Chemotaxonomy of the Rubiaceae family based on leaf fatty acid composition. *Phytochemistry, 66*, 549–559.

Mustafa, S. E. (2011). Evaluation of oil components of *Annona squamosa* (gishta) and *Vangueria madagascariensis* (kirkir) seeds and antioxidant activity of methanolic extracts from different parts of the two plants. PhD Thesis. Khartoum, Sudan: Department of Food Science and Technology, College of Agricultural Studies, Sudan University of Science and Technology.

Orwa, C., Mutua, A., Kindt, R., Jamnadass, R., & Simons, A. (2009). Agroforest tree database: a tree reference and selection guide version 4.0. Available from: http://www.worldagroforestry.org/af/treedb/

Pino, J., Marbot, R., & Vazquez, C. (2004). Volatile components of the fruits of *Vangueria madagascariensis* J. F. Gmel. F grown in Cuba. *Journal of Essential Oil Research, 16*(4), 318–320.

Redead, M. B., (1989–90). Utilization of tropical foods; compendium on technological and nutritional aspects of processing and utilization of tropical foods, both and plants, for purposes of training and field reference. Rome: Food & Agriculture Organization of the United Nations.

Rohman, A., Che Man, Y. B., Ismail, A., & Hashim, P. (2011). Monitoring the oxidative stability of virgin coconut oil during oven test using chemical indexes and FTIR spectroscopy. *International Food Research Journal, 18*, 303–310.

Saka, J. D. K. (1995). The nutritional value of edible indigenous fruits: present research status and future directions. In J. A. Maghembe, Y. Ntupanyama, & P. W. Chirwa (Eds.), *Improvement of indigenous fruit trees of the Miombo Woodlands of Southern Africa* (pp. 50–57). Nairobi, Kenya: ICRAF.

Shahidi, F., Desilva, C., & Amarowicz, R. (2003). Antioxidant activity of extracts of defatted seeds of Niger (*Guizotia abyssinica*). *Journal of the American Oil Chemists' Society, 80*, 443–450.

Stangeland, T., Remberg, S. F., & Lye, K. E. (2009). Total antioxidant activity in 35 Ugandan fruits and vegetables. *Food Chemistry, 113*, 85–91.

Suja, K. B., Jayalekshmy, A., & Arumughan, C. (2005). Antioxidant activity of sesame cake extract. *Food Chemistry, 91*(2), 213–219.

Taruscio, T. G., Barney, D. L., & Exon, J. (2004). Content and profile of flavonoids and phenolic acid compounds in conjunction with the antioxidant capacity for a variety of northwest *Vaccinium* berries. *Journal of Agricultural and Food Chemistry, 52*, 3169–3176.

Ying, Z., Lei, Y., Yuangang, Z., Xiaoqiang, C., Fuji, W., & Fang, L. (2010). Oxidative stability of sunflower oil supplemented with carnosic acid compared with synthetic antioxidants during accelerated storage. *Food Chemistry, 118*(3), 656–662.

Chapter 29

Phoenix dactylifera Date Palm Kernel Oil

INTRODUCTION

Date palm tree (*Phoenix dactylifera* L.) is grown extensively in arid and semi-arid regions of the world, like North Africa, the Arabian Peninsula, and Iran (Ahmed et al., 1995). Date palm constitutes the principal source of remuneration and the basis of economy for the people living in the Sahara. Date palm kernel (DPK) is a solid waste of date's industry and date consumption. The date palm kernel is about 12%–15% of the date palm fruit (wt./wt.) so if the world production is more than 7 million metric tons; the kernels (seeds, pits) shall be about one million metric tons that almost consider waste (Mirghani, 2012). Indeed, the production has tripled from 2,289,511 tons in 1974 to 6,772,068 tons in 2004 (von Braun, 2007). Date palm is one of the most important domesticated crops in the North African and the Near East countries. However, date palm plantations have been currently in danger to be completely destroyed by a vascular fusariosis (bayoud disease) caused by the *Fusarium oxysporum* sp. *albedinis* fungus. Up today, Tunisian date palm groves appear to be spared, but for several decades, they are seriously menaced by the plague. Hence, the search of early molecular markers associated with the disease has become imperative. In this scope, two mitochondrial plasmid-like DNAs have been identified as potential markers of resistance to the bayoud disease. Here, starting from a set of Tunisian date palm varieties, we report the availability of these molecules as reliable markers of the resistance to the plague and, how the use of PCR technology should allow a rapid and efficient diagnosis approach for identification of selected bayoud resistant individuals (Mokhtar, Abdelmajid et al.).

DATE PALM KERNELS

Date pits were examined for extractable high value-added components for including in functional foods. The objectives of this research were to conduct preliminary analysis of the pits from three leading varieties in the UAE and to identify potential uses in foods. Date pits were odorless and had a light to dark brown color and a bland taste with slight bitterness. They contained

Unconventional Oilseeds and Oil Sources. http://dx.doi.org/10.1016/B978-0-12-809435-8.00029-9

7.1%–10.3% moisture, 5.0%–6.3% protein, 9.9%–13.5% fat, 46%–51% acid detergent fiber, 65%–69% neutral detergent fiber, and 1.0%–1.8% ash. More than half of the pit protein could not be extracted consecutively with NaCl, ethanol, and acetic acid. Pits had a substantial amount of oil that needed to be characterized for constituent components, biological activities, and stability. Pits contained large quantities of fiber and, possibly, resistant starch that may have potential health benefits. Further research is needed to characterize isolated components and search for bioactive constituents with antimicrobial, antioxidant, and other health-promoting activities (Hamada, Hashim, & Sharif, 2002)

The seeds of four date palm *(P. dactylifera* L.) cultivars, Dekel Noor, Zahidi, Medjool, and Haiawy, grown in the Arava Valley of southern Israel were analyzed for their inorganic and some organic constituents. The following average values were obtained for the four cultivars on a dry-weight basis: protein 5.60%, oil 8.15%, fiber 16.13%, and ash 1.13%. Analysis of the mineral elements in the ash gave the following average values: Ca, 1.55%; Na, 0.97%; Mg, 8.07%; K, 27.60%; Fe, 0.73%; Cu, 0.13%; and Mn, 0.08%. The oil exhibited the following characteristics (average for the four cultivars): acid value (AV) ~1.04, iodine value (IV) 49.5, saponification value (SV) 221.0, and unsaponifiable matter 0.8%. Gas–liquid chromatography revealed that the major unsaturated fatty acid was oleic acid (42.3%), while the main saturated fatty acid was lauric (21.8%). Myristic, palmitic, and linoleic acids were also found, average values being 10.9, 9.6, and 13.7%, respectively. Potential uses for date seed oil is used in cosmetic, pharmaceutical and to a lesser degree for specialty products (Devshony, Eteshola, & Shani, 1992) (Table 29.1).

Date palm *(P. dactylifera)* is considered to be one of the oldest cultivatable crops. The DPK considered a waste product of many date processing plants producing pitted dates, date syrup, and date confectionery. Direct consumption of dates is also considered as a source of DPK. The extracted oil was then analyzed for color, refractive index, IV, SV, unsaponifiable matters, and total phenolic content as well as some other quality parameters, such as the AV and free fatty acid (FFA) content, and peroxide value (PV). Generally, the DPK of Deglect

TABLE 29.1 Characteristics of Seed (Kernel) of the Some Varieties of Date Palm

Variety	Moisture	Oil	Protein	Carbohydrates	Fiber
DPKDNO	07.34	09.17	06.58	76.38	16.06
DPKMO	08.52	07.30	07.25	74.92	15.48

DPK, Date palm kernel.
DPKDNO, Date Palm Kernel of Deglect Noor Oil; DPKMO, Date Palm Kernel of Moshkan Oil.

Noor (DPKDNO) found to have high oil content (9.67%) compared to DPK of Moshkan (DPKMO), which has 7.30% oil. The color of crude oil was found to be 5.6 R, 25 Y, and 0.2 blue (brownish) for DPKDNO while 2.3 R and 36 Y (yellowish) for DPKMO using Lovibond tentometer. IV and SV for DPKDNO and DPKMO were found to be 51.6 and 54.8, and 216.3 and 207.8, respectively. The unsaponifiable matters in both oils are almost the same which ranged between 0.8% and 1.4%. Total phenolic content in both oils was also in the same range of 0.96–0.98 mg/mL Gallic acid equivalent. The fatty acid composition using GC–MS showed that oleic acid is the main unsaturated fatty acid in both varieties (38.5 and 41.6%) while the main saturated fatty acid is lauric acid which was found to be 23.2 and 18.5% for DPKDNO and DPKMO, respectively. Other types of fatty acids, such as palmitic, myristic, stearic, and linoleic were also found in both varieties. Thus, the initiation study of this project may generate a new source of special oil which could be able to support the global demands of Halal source of specialty oil as cosmetic ingredient since it had been proven that it has antiwrinkle effect and is therefore of interest in antiaging skin care products (Mirghani et al., 2009).

In another study, (Besbes, Blecker, Deroanne, Drira, & Attia, 2004), the seeds of two date palm (*P. dactylifera* L.) cultivars, Deglet Nour and Allig, from the Degach region in Tunisia, were analyzed for their main chemical composition. Studies were also conducted on the properties of oil extracted from date pits. The following values (on a dry-weight basis) were obtained for Deglet Nour and Allig cultivars, respectively: protein 5.56 and 5.17%, oil 10.19 and 12.67%, ash 1.15 and 1.12%, and total carbohydrate 83.1 and 81.0%. Gas–liquid chromatography revealed that the major unsaturated fatty acid was oleic acid (41.3%–47.7%), while the main saturated fatty acid was lauric acid (17.8%) for the Deglet Nour cultivar and palmitic acid for the Allig cultivar (15.0%). Capric, myristic, myristoleic, palmitoleic, stearic, linoleic, and linolenic acids were also found. Thermal profiles of both date seed oils, determined by their DSC melting curves, revealed simple thermograms. Sensorial and physical profiles of Deglet Nour and Allig seed oil were based on studies of the CieLab (L*, a*, b*) color, oxidative stability, viscosity, and microstructure. Results showed that date seed oil could be used in cosmetic, pharmaceutical, and food products (Besbes et al., 2004) (Fig. 29.1; Tables 29.2 and 29.3).

ABOUT DATE PITS

The antimicrobial activity of plant oils and extracts has been recognized for many years. However, few investigations have compared large numbers of oils and extracts using methods that are directly comparable. In the present study, 52 plant oils and extracts were investigated for activity against *Acinetobacter baumannii*, *Aeromonas veronii* biogroup *sobria*, *Candida albicans*, *Enterococcus faecalis*, *Escherichia coli*, *Klebsiella pneumoniae*, *Pseudomonas aeruginosa*,

FIGURE 29.1 *Phoenix dactylifera* tree, fruits and kernels.

TABLE 29.2 Some Parameters of Date Palm Kernel Oils (DPKO) of the Two Cultivars Date Palm in This Study*

Variety	Melting point (°C)	FFA (%)	SV	IV	Unsaponifiable matter (%)	TPC (mg/mL)
DPCDNO	37.5	2.4	216.3	51.6	0.837	0.964
DPKMO	38.0	1.4	207.8	54.8	1.355	0.982

FFA, Free fatty acids; SV, Sapoinification value; IV, Iodine value; Unsab., Unsaponifiable matter; TPC, Total phenolic compounds.

TABLE 29.3 Fatty Acid Composition of Various Varieties of DPKO

Variety	Fatty acids (%)					
	C12:0	C16:0	C18:0	C18:1	C18:2	Others
DPCDNO	23.2	10.70	4.72	38.50	07.24	1–4
DPKMO	18.5	09.68	6.75	41.60	08.11	2–5

Salmonella enterica subsp. *enterica* serotype *typhimurium, Serratia marcescens,* and *Staphylococcus aureus,* using an agar dilution method. Many plant oils and extracts were investigated, using a broth microdilution method, for activity against *C. albicans, S. aureus,* and *E. coli.* The lowest minimum inhibitory concentrations were 0.03% (v/v) thyme oil against *C. albicans* and *E. coli* and 0.008% (v/v) vetiver oil against *S. aureus.* These results support the notion that plant essential oils and extracts may have a role as pharmaceuticals and preservatives (Hammer, Carson, & Riley, 1999).

REFERENCES

Ahmed, I. S. A., Al-Gharibi, K. N., Daar, A. S., & Kabir, S. (1995). The composition and properties of date proteins. *Food Chemistry, 53,* 441–446.

Besbes, S., Blecker, C., Deroanne, C., Drira, N. -E., & Attia, H. (2004). Date seeds: chemical composition and characteristic profiles of the lipid fraction. *Food Chemistry, 84*(4), 577–584.

Devshony, S., Eteshola, E., & Shani, A. (1992). Characteristics and some potential applications of date palm (*Phoenix dactylifera* L.) seeds and seed oil. *Journal of the American Oil Chemists Society, 69*(6), 595–597.

Hamada, J., Hashim, I., & Sharif, F. (2002). Preliminary analysis and potential uses of date pits in foods. *Food Chemistry, 76*(2), 135–137.

Hammer, K. A., Carson, C., & Riley, T. (1999). Antimicrobial activity of essential oils and other plant extracts. *Journal of Applied Microbiology, 86*(6), 985–990.

Mirghani, M. E. S. (2012). Processing of date palm kernel (DPK) for production of nutritious drink. *Advances in Natural and Applied Sciences, 6*(5), 575–583.

Mirghani, M. E. S., Kabbashi, N. A., Hussein, I. H., Yassin, A. A., Che Man, Y., Noor, M., & Ellyana, N. (2009). The potential of date palm kernel oil. http://irep.iium.edu.my/10618/1/

Mokhtar, T., Abdelmajid, R., & Mohamed, M. A molecular marker of date-palm (*Phoenix dactylifera* L.) resistance to bayoud disease. 604–612. Available from: http://www.pubhort.org/datepalm/datepalm2/datepalm2_75.pdf via the INTERNET

von Braun, J. (2007). *The world food situation: new driving forces and required actions.* Washington, DC: International Food Policy Research Institute.

Chapter 30

Durio zibethinus (Durian)

INTRODUCTION

There is a considerable interest in exotic fruits, especially those originating from tropical regions, because of their unique and exotic flavors. Both fruits and vegetables contain a variety of phytochemicals, including flavonoids, which have antioxidant and anticancer properties. Hence, it is very interesting and promising to investigate such fruits and vegetables and their waist and by-products, and extracts to make the best use of them in making a better life and growing the industry. A very popular tropical fruit, especially in Southeast Asia, is durian (*D. zibethinus* Murr.) (Fig. 30.1).

Durian (*D. zibethinus* Murray) is one of the most important seasonal fruits in tropical Asia. Durian cultivars are derived from *D. zibethinus* Murray, originating in the Malay Peninsula (Voon, Hamid, Rusul, Osman, & Quek, 2007). The importance of this fruit is mostly connected with its composition and antioxidant properties (Arancibia-Avila et al., 2008; Toledo et al., 2008). The name "Durian" comes from the Malay word "Duri," meaning "thorn." It is nutritionally rich in carbohydrates, protein, fat, phosphorous, iron, and vitamin A. In addition, it contains bioactive compounds with high antioxidant activity. Therefore, durian cv. Monthong is recommended as a nutritional supplement (around 5%–7%) to the normal diet (Poovarodom et al., 2010).

Moreover, this tree can produce oil that possesses a lot of benefit in many different fields and industries. This oil can be extracted from the seeds and arils of the durian fruit by different techniques. Hence, further investigations about durian oil are recently done by several scientists in order to evaluate and make a full profile on this oil and its uses.

The purpose of this research was to go through some previous work and reviews on the *D. zibethinus* (durian) tree and study the nature, parts, bioactive and nutrient compounds, fatty acids, antioxidant activity, and all benefits and usage of this tree, especially as unconventional source of oil.

Unconventional Oilseeds and Oil Sources. http://dx.doi.org/10.1016/B978-0-12-809435-8.00030-5

FIGURE 30.1 *Durio zibethinus* tree, fruits, and seeds.

Durio zibethinus TREE

D. zibethinus Murr. (Bombacaceae), a large tree native to Southeast Asia, is cultivated as an important fruit crop in Thailand, Malaysia, and Indonesia. The fruit is also found in the Philippines and some Indochina countries, such as Kampuchea, Vietnam, and Myanmar. Australia has started to plant durian in the Northern region to meet the increasing local demands for the fruit (Kwee & Sijam, 1990). However, Thailand is the leading country for durian production, with a yield of 686,500MTs in 2007 (Bumrungsri, Sripaoraya, Chongsiri, Sridith, & Racey, 2009).

Its fruit (durian) is large, weighing 1–3 kg, and has a green to brownish, thick and extremely tough skin covered with numerous, thick, sharp-pointed spikes. When ripe, durian is easily split into segments by cutting along the 4–5 faint lines, which run from the base to the apex of the fruit. Each of the segments usually contains one or more seeds embedded in a white or yellowish edible pulp (ad) which is soft and sweet and has an odor that is much liked by the people in the region. Durian odor has two distinct notes; one is a delicate, fruity note caused by esters, while the other is an onion-like note caused by thiols and thioesters (Baldry, Dougan, & Howard, 1972). According to (Stanton, 1966), The odor profile of the durian was essentially related to the occurrence of hydrogen sulfide, hydrodisulfides and dialkyl polysulfides, ethyl esters, and 1,1-diethoxyethane. But in spite of its unpleasant smell, the soft, yellow-white pulp is cherished as a luxury food and aphrodisiac by Southeast Asiatics.

Normally, the fruits are allowed to mature and develop their full flavor on the tree and are collected after they fall to the ground. Durian is eaten shortly after harvest, as it soon becomes sour and rancid due to chemical changes that follow (Irwandi & Che Man, 1996). Only one-third of the durian is edible, whereas the chestnut-shaped seeds (20%–25%) and the shell are usually thrown away although the seeds are edible. The aril is usually eaten fresh and sometimes is fermented or salted, or boiled with sugar (Wong & Tie, 1995). The aril of the fruit can also be frozen or processed into paste, processed into durian powder, or added into other products, such as ice cream, cakes, and confectionary (Nanthachai, 1994). The seeds are embedded in the arils. They are large, 5–7 cm long by c. 2–4 cm wide, hard but not stony and are covered with a thin, light-brown skin. A fresh seed may weigh as much as 40 g. The seeds are boiled or baked prior to consumption (Berry, 1980).

The growth pattern and element uptake during early years of production of durian (*D. zibethinus* Murr.) was studied by (Yaacob, 1983) using various ages of trees growing at Serdang, Malaysia. Growth was rapid until the 7th year when the dry matter production was nearly at its maximum. The proportion of underground to aboveground materials decreases as the tree ages and in a 7-year old tree, 30% is roots and 70% tops. The elements (N, P, K, Ca, and Mg) stored in the vegetative parts increases with the age of the tree in a similar manner to that of the total dry matter. However, the amount of elements stored in a 7-year old tree exceeds that which was applied as fertilizers suggesting that the once-jungle tree obtains additional elements from the soil.

The flavor of the fruit changes with age. As soon as the fruit falls, very rapid chemical changes take place in the aril. Hence, the aril is normally consumed immediately after opening the fruit otherwise it loses its flavor and firm, creamy texture, and becomes rancid and watery (Berry, 1980). These changes continue to a stage when the aril can no longer please the ordinary palate. It is said that the fruit cannot be kept for more than 5–6 days without losing its characteristic flavor and texture. These changes in flavor and texture may be attributed to oxidative and hydrolytic alterations in fatty acid composition of the fat. The chemical composition of the lipid invariably influences the textural properties of fat-containing foods (Forss, 1969). The products of oxidation, and hydrolysis of the unsaturated fatty acids in particular make the fruit taste rancid thus upsetting the delicate flavor balance. The fresh aril, however, provides a sufficient amount of essential fatty acids to the consumer (Berry, 1980). Experiments at prolonging the shelf life by cold storage have had only partial success. However, the aril along with seeds could be blanched, frozen, and stored below $-10°C$ for as long as 6 min without any easily detectable organoleptic changes.

General information exists on the contents of proteins, fats, carbohydrates, minerals, and vitamins in the pulp or seeds (Hegenauer, 1964). However, the durian has been widely thought to possess a tonic food value. This may be due to a psychological rather than a physiological effect. According to Kostermans, the aril is not easy to digest, and this indigestibility is perhaps why the fruit is used as an aphrodisiac Lumholz wrote that any very agreeable food can act favorably on the digestive organs and therefore induce a feeling of vigor (Lumholz 1920). Pratt and Rosario stated that the high percentage of a carbohydrate resembling erythrodextrin (unidentified substance?) adds to the food value (Pratt &Rosario, 1913). Griebel also found in the aril cells a dextrin which, with an iodine test, reacts like an erythrodextrin and a reducing sugar, such as mannite (Griebel, 1928).

The pulp contains:

H_2O	58%
Protein	2.8%
Fat	3.9%
Carbohydrate	34.1% (about 12% sugar; 12% starch)
Mineral	1.2%
Calcium	0.01%
Phosphorous	0.05%
Iron	1.0 g per 1000 g of pulp
Carotene	20 Int. Yit. A. unit
Vitamin C	25 g. per 1000 g pulp

Data from "Wealth of India," (1952). *3*, 119.

The Uses of Durian Tree

Due to their many applications, durians could make an important contribution to the beverage and food industry in the Western world. However, knowledge of the nature and content of nutrients present in the fruit and of the volatile components responsible for the characteristic odor and taste is still fragmentary and contradictory (Stanton, 1966).

It has been reported that durian has additional valuable health properties: polysaccharide gel, extracted from the fruit hulls, reacts on immune responses and is responsible for cholesterol reduction (Chansiripornchai, Chansiripornchai, & Pongsamart, 2008). The glycemic index of durian was the lowest in comparison with papaya and pineapple (Daniel, Aziz, than, & Thomas, 2008).

Although only a small number of studies has been conducted to explore the antidiabetic and antiobesity potential of durian, these studies show that durian can be further explored to detail its antidiabetic and antiobese potential. Durian exhibits potential effects on metabolic parameters in human and animal models (Roongpisuthipong, Banphotkasem, Komindr, & Tanphaichitr, 1991; Leontowicz et al., 2007; Toledo et al., 2008). When rats were fed durian in addition to a cholesterol enriched diet (1% cholesterol), it positively influenced the plasma lipid profile, plasma glucose, and antioxidant activity. These metabolically beneficial effects of durian might be due to the higher contents of bioactive compounds with various biological activities, such as metabolic enhancer and antioxidant. This suggests that durian consists of few critical bioactive components that can be further evaluated for hypoglycemic and antihyperlipidemic effects (Leontowicz et al., 2007).

Interestingly, in a small clinical trial, durian has been shown to improve glucose homeostasis by altering insulin secretion and its action (Roongpisuthipong et al., 1991). After ingestion of durian, the insulin response curve of 10 diabetic patients was significantly improved compared to the ingestion of other fruits (mango, pineapple, banana, rambutan) and control (no fruit) (Roongpisuthipong et al., 1991). Various durian cultivars have also been shown to possess antioxidant capacities due to the relatively high level of total polyphenols (Toledo et al., 2008). This antioxidant property of durian and its components can be useful for prevention of oxidative stress mediated induction of diabetic and obesity complications. Studies delineating the antiinflammatory properties of durian are yet to be reported.

The timber, although coarse and not durable, can be used for various purposes. In west Java, it is one of the commonest timbers in the market, and the villagers use it in building houses (Heyne, 1927). The impregnated timber is excellent for packing cases as it is strong and light. Various parts of the plant are used in folk medicine. Burkill (1935), Heyen (1927), and Ridley (1906) summarize the medical uses. In Malaya, the root is employed to treat fever, a decoction of root being smeared upon the pairi and an extract drunk (Ridley, 1906). In Java, the leaves are boiled, and the lukewarm water serves as a medicinal bath for jaundice. The leaves mixed with *Acorus* sp. are applied to epidermal swellings. (Heyne, 1927). A decoction of the valves of the fruit is also used for skin complaints (Heyne, 1927). In several places in Java, an extract obtained from the fruit valves is drunk and is believed to cure abortion. In Malaya, ashes of the valves are used by dyers for making silk white and as a lye with other dyes (Burkill, 1935).

Fruit leather, a product prepared by dehydration of fruit puree, is an established product, particularly in the North-American market (Raab & Oehler, 1976). Durian leather is an intermediate moisture food, for which the market is presently growing (Guilbert, 1987). The intermediate moisture products offer the potential advantage that can be eaten without preparation. They can be consumed directly as fruit snacks or combination with nuts and breakfast cereal. They can also be incorporated in various fabricated foods, such as biscuits, cookies and ice creams. Development of durian leather, as an alternative to diversify the uses of durian, has been reported by Irwandi and Che Man

(1996). Like other fruit leathers, durian leather was prepared by drying the aril after mixing with other ingredients into leathery sheets. Che Man, Jaswir, Yusof, Selamat, and Sugisawa (1997) reported that the additions of glucose syrup solid, sucrose, hydrogenated palm oil, and soy-lecithin into durian aril when mixing resulted in acceptable durian leather.

A small percentage of durian aril is processed to make lempuk and tempoyak, two products widely consumed in both Malaysia and Indonesia. Lempuk is a kind of cake prepared by mixing 10% sugar into durian aril before being cooked for about 6 h, while tempoyak is a fermented durian product prepared by adding durian aril with sugar and salt before being stored at room temperature for 3–7 days for fermentation (Irwandi & Che Man, 1996). However, the two products are not commercially produced in large quantities. Durian is also processed into a flavoring agent (Faridah, Sik, & Sharif, 1982)

Durian seed is an agricultural biomass waste of durian fruit. It can be a natural plant source of nonstarch polysaccharide gum with potential functional properties. The work investigating the effect of chemical extraction variables (i.e., the decoloring time, soaking temperature, and soaking time) on the physicochemical properties of durian seed gum was done by Amid & Mirhosseini, 2012), where the physicochemical and functional properties of chemically extracted durian seed gum were assessed by determining the particle size and distribution, solubility, and the water- and oil-holding capacity (WHC and OHC). And it revealed that the soaking time should be considered as the most critical extraction variable affecting the physicochemical properties of crude durian seed gum.

Recently, it has been shown that individuals who eat daily five servings or more of fruits and vegetables have approximately half the risk of developing a wide variety of cancer types, particularly those of the gastrointestinal tract (Gescher, Pastorino, Plummer, & Manson, 1998). Therefore, the antiproliferative activities of methanol extracts of Mon Thong durian at different stages of ripening on human cancer cell lines (Calu-6 for human pulmonary carcinoma and SNU-601 for human gastric carcinoma) were determined using MTT (3-(4,5-dimethylthiazol-2-yl)-2,5-diphenyltetrazolium bromide) assay. As far as we know there are no published results of such investigations.

Oil of the Durian Tree

Stanton (1966) gave a full description of the fruit and reviewed its proximate analyses. Recently, volatile flavoring constituents of the aril were reported by (Baldry et al., 1972). The seeds contain very little oil. The fresh aril, however, contains as much as 5% of oil with unknown fatty acid composition.

Since the fruit deteriorates rapidly and loses its edible value, it would be of practical and scientific interest to investigate the fatty acid composition of the aril. Additionally, the seeds may contain cyclopropene fatty acids (CPFA) that are commonly found in plants of the order Malvales to which durian belongs

(Carter & Frampton, 1964). More recently, CPFA were found to cause cancer in rainbow trout (Sinnhuber et al., 1976) and aortic atherosclerosis in rabbit (Ferguson, Wales, Sinnhuber, & Lee, 1976). These observations are of great concern because humans unknowingly consume seeds and nuts which may contain these fatty acids. Hence, a study was conducted by (Berry, 1980) to explore the fatty acid composition of both aril and seeds of the fruit durian using gas–liquid chromatography (GLC), and to investigate the influence of cooking temperature on the fate of CPFA in seeds.

In this study the oil was extracted from the freeze-dried aril pulp and oven-dried seeds powders separately with petroleum ether (bp 40–60°C) in Soxhlet apparatus for 16 h. The oil was recovered by evaporating the ether on a rotary evaporator under reduced pressure. GLC was used to analyze the mixture of fatty acid methyl esters derived from their respective oils. The dried aril yielded c. 15% of a yellow-colored oil. The fatty acid composition of the aril is given in Table 30.1. The oil from two durian strains, namely with cream- or yellow-colored aril, exhibited variation in fatty acid composition as evident from the data in Table 30.1. The oil, however, was primarily composed of palmitic and oleic acids which predominate in most fruit-coat fats (Hilditch & Williams, 1964). The unsaturated fatty acids constituted ca. 62%–70% of the total fatty acids. The linolenic acid content, especially in the yellow-colored aril, was relatively higher compared to most fruit-coat fats.

The composition of fatty acid in durian oil was unsaturated fatty acid: oleic acid (14.95%) and linoleic acid (12.46%) and saturated fatty acid which consists of stearic acid (45.84%) and palmitic acid (26.75%), relatively to pure stearic acid (Sumarno, 2001).

Both fresh and boiled seeds were examined for their oil content (1.8%). The fatty acid composition of the whole seeds, germ, and cotyledons is presented in Table 30.1. The whole seed oil contained c. 82% unsaturated fatty acids. The palmitic acid content of this oil was relatively lower compared to other seed oils of the family Bombacaceae (Hilditch & Williams, 1964). Of the unsaturated fatty acids, linoleic, and linolenic together constituted c. 22%. The cooking temperatures employed did not significantly affect the concentration of these fatty acids in seeds.

As a result of this study, the germ oil contained the highest amount of sterculic acid. The cooking temperatures employed reduced the malvalic and sterculic acid contents in seeds only by c. 22% and 19%, respectively. The oil samples from the whole seeds, germ, and cotyledons gave a positive Halphen test indicating the presence of CPFA, where these oils were found to contain a rather large amount of CPFA in which sterculic acid predominates; the highest amount is in germ oil.

In the investigation done by (Moser, Düvel, & Greve, 1980) to characterize the fatty acid composition of the lipids in the durian fruit, the Preliminary TLC investigations showed the durian fruit fatty acids to be present almost exclusively as triglycerides. The fatty acid composition was deter-mined, after

TABLE 30.1 Analytical Data on Durian Aril, Seed, and Their Oil

	Value					
	Aril		Whole		Seeds	
Property	Yellow	Creamy	Fresh	Cooked	Germ	Cotyledons
Composition (%)						
Moisture	66.00	ND[a]	77.00	61.60	77.00	77.00
Oil[b]	5.00	ND	0.50	0.50	1.54	0.20
Protein[b] (N × 6.25)	2.50	ND	1.57	ND	4.23	ND
Composition (area%)						
14:0	0.91	0.34	0.12	0.17	0.12	0.25
16:0	34.13	28.94	12.20	22.81	4.52	14.90
16:1	7.10	5.16	1.15	1.07	0.26	1.00
18:0	1.21	1.23	1.42	1.66	0.41	2.00
18:1	42.14	58.98	8.42	7.67	3.92	9.84
18:2	7.85	3.16	6.50	6.26	4.07	6.50
Dihydrosterculic[c]	—	—	2.52	3.09	1.66	4.05
18:3	5.69	2.21	11.30	13.61	5.62	16.64
20:0	Trace[d]	Trace				
Malvalic	—	—	15.72	12.28	1.76	20.65
Sterculic	—	—	38.53	31.34	77.68	20.89
22:0	—	—	1.21	Trace	—	2.82

[a]ND, Not determined.
[b]Wet basis.
[c]Tentative identification.
[d]Trace, <0.1%.

reesterification, separately for pericarp, arillus, and seed as well as according to harvest season and origin. Apart from distinct differences between harvests, the fatty acid concentrations also varied up to several per cent between the individual specimens. In Tables 30.2 and 30.3 the fatty acids identified and their average compositions are given. Oleic and palmitic acid were the main components in all parts of fruits from the spring harvest as well as in the arillus and seed of fruits from the summer harvest. Surprisingly arachidic acid predominated in the pericarp of specimens harvested in summer, and was present to a considerable extent in the seeds. Also palmitoleic, stearic, linoleic and linolenic acids were present in higher concentrations. 12:0, 14:0, 16:3, 20:1, 20:3, 22:0, and 22:1 fatty acids have been identified in individual fruits in amounts less than 3%.

TABLE 30.2 Fatty Acid Composition of Lipids (in % GLC Peak Area)[a]

Fatty acid	16:0	16:1	18:0	18:1	18:2	18:3	20:0	Others
Pericarp	30.9	4.2	1.8	52.8	4.2	4.0	<1	2.1
Arillus	33.9	4.9	1.0	51.2	3.0	5.1	<1	0.9
Seed	26.8	8.4	3.3	38.8	5.9	3.0	<1	13.8

GLC, Gas–liquid chromatography.
[a]*Mean values from seven fruits, relative variations up to 20%; origin, Chandburi/spring harvest.*

TABLE 30.3 Fatty Acid Composition of Lipids (in % GLC Peak Area)[a]

Fatty acid	16:0	16:1	18:0	18:1	18:2	18:3	20:0	Others
Pericarp	14.9	1.4	5.1	17.8	7.6	4.9	33.2	15.1
Arillus	31.9	6.9	1.0	51.0	3.0	5.4	<1	0.8
Seed	23.0	3.1	1.2	41.0	11.5	5.3	9.5	5.4

[a]*Mean values from ten fruits, relative variations up to 20%; origin, Prajeen Rayong/summer harvest.*

    The health properties of durian are based not only on the antioxidant properties, but also on its fatty acid composition. Cholesterol hypothesis implied that reducing the intake of saturated fats and cholesterol while increasing that of polyunsaturated oils is effective in lowering serum cholesterol, and thereby in reducing coronary heart disease. The protective activity is linked with a high supply of n–3 fatty acids coming from fish and seafood, and high consumption of wholegrain products, as well as fruits and vegetables (Siondalski & Lysiak-Szydlowska, 2007). Durian is rich in n–3 fatty acids, compared to some other fruits (Phutdhawong, Kaewkong, & Buddhasukh, 2005).

CONCLUSION

The durian is an important economic fruit crop in several Southeast Asian countries including Thailand. In recent years, it has gone from a relative curiosity into a horticultural crop of importance and interest beyond the regional market. All this interest for this tree was due to it being a multiplication tree that can be used in so many fields including the medical field and food industry.

    The aril of this tree is a good source of essential fatty acids, besides having aesthetic value. The seeds, on the other hand, contain a high amount of undesirable CPFA which are only partially destroyed during cooking. It would therefore seem extremely unwise to consume these seeds. However, this essential oil is extremely full of benefits for the human health.

REFERENCES

Amid, B. T., & Mirhosseini, H. (2012). Effect of different purification techniques on the characteristics of heteropolysaccharide-protein biopolymer from durian (*Durio zibethinus*) seed. *Molecules, 17*(9), 10875–10892.

Arancibia-Avila, P., Toledo, F., Park, Y. -S., Jung, S. -T., Kang, S. -G., Heo, B. G., et al. (2008). Antioxidant properties of durian fruit as influenced by ripening. *LWT—Food Science and Technology, 41*, 2118–2125.

Baldry, J., Dougan, J., & Howard, G. E. (1972). Volatile flavouring constituents of durian. *Phytochemistry, 11*(6), 2081–2084.

Berry, S. K. (1980). Cyclopropene fatty acids in some Malaysian edible seeds and nuts: I. Durian (*Durio zibethinus* Murr.). *Lipids, 15*(6), 452–455.

Bumrungsri, S., Sripaoraya, E., Chongsiri, T., Sridith, K., & Racey, P. A. (2009). The pollination ecology of durian (*Durio zibethinus*, Bombacaceae) in southern Thailand. *Journal of Tropical Ecology, 25*(01), 85–92.

Burkill, A. H. (1935). *A dictionary of the economic products of the Malay Peninsula.* London, England: Crown Agents of the Colonies.

Carter, F. L., & Frampton, V. L. (1964). Review of the chemistry of cyclopropene compounds. *Chemical Review, 64*, 497–525.

Chansiripornchai, N., Chansiripornchai, P., & Pongsamart, S. (2008). A preliminary study of polysaccharide gel extracted from the fruit hulls of durian (*Durio zibethinus*) on immune responses and cholesterol reduction in chicken. *Acta Horticulturae, 786*, 57–60.

Che Man, Y. B., Jaswir, I., Yusof, S., Selamat, J., & Sugisawa, H. (1997). Effect of different dryers and drying conditions on acceptability and physicochemical characteristics of durian leather. *Journal of Food Processing and Preservation, 21*(5), 425–441.

Daniel, R. S., Aziz, Al-s. I., Than, W., & Thomas, W. M. S. (2008). Glycemic index of common Malaysian fruits. *Asia Pacific Journal of Clinical Nutrition, 17*, 35–39.

Faridah, A. M., Sik, N. N., & Sharif, A. B. (1982). *Tanaman Durian. Risalah Pertanian Bilangan* 23 (p. 1). Malaysia, Kuala Lumpur: Jabatan Pertanian Semenanjung.

Ferguson, T. L., Wales, J. H., Sinnhuber, R. O., & Lee, D. J. (1976). Cholesterol levels, atherosclerosis and liver morphology in rabbits fed cyclopropenoid fatty acids. *Food and Cosmetology Toxicology, 14*, 15, through Chem. Abs. 84:135598v (1976).

Forss, D. A. J. (1969). Role of lipids in flavours. *Journal of Agricultural and Food Chemistry, 17*(4), 681.

Gescher, A., Pastorino, U., Plummer, S. M., & Manson, M. M. (1998). Suppression of tumour development by substances derived from the diet—mechanisms and clinical implications. *British Journal of Clinical Pharmacology, 45*, 1–12.

Griebel, C. (1928). *Zeitschrift für Untersuchung der Lebensmittel.* 55 89–93, 453–457.

Guilbert, S. (1987). Use of superðcial edible layer to protect intermediate moisture foods: applicationto the protection of tropical fruit dehydrated by osmosis. In Seow, C. C. (Ed.), *Proceedings of food preservation by moisture control.* Penang, Malaysia, 199È219.

Hegenauer, R. (1964). Chemoraxonomie der Ppanzen. *3*, 284.

Heyne, K. (1927). De nuttigeplanten van Neder-landsch Indi. *2*, 1057.

Hilditch, T. P., & Williams, P. N. (1964). *The chemical constitution of natural fats* (4th ed.). London: Chapman and Hall.

Irwandi, & Che Man, Y. B. (1996). Durian leather: development, properties and storage stability. *Journal of Food Quality, 19*, 479–489.

Kwee, L. T., & Sijam, K. (1990). Penyakit Tanaman Durian. 5. Dewan Bahasa Dan Pustaka, Kuala Lumpur.

Leontowicz, M., Leontowicz, H., Jastrzebski, Z., Jesion, I., Haruenkit, R., Poovarodom, S., et al. (2007). The nutritional and metabolic indices in rats fed cholesterol containing diets supplemented with durian at different stages of ripening. *Biofactors, 29*, 123–136.

Lumholz, C. (1920). Through central borneo. 267.

Moser, R., Düvel, D., & Greve, R. (1980). Volatile constituents and fatty acid composition of lipids in *Durio zibethinus*. *Phytochemistry, 19*(1), 79–81.

Nanthachai, N. (1994). *Durian: fruit development, postharvest physiology, handling and marketing in ASEAN (p. 1È6)*. Kuala Lumpur, Malaysia: Asean Food Handling Bureau.

Phutdhawong, W., Kaewkong, S., & Buddhasukh, D. (2005). GC-MS analysis of fatty acids in Thai durian aril. *Chiang Mai Journal of Science, 32*, 169–172.

Poovarodom, S., Haruenkit, R., Vearasilp, S., et al. (2010). Comparative characterisation of durian, mango and avocado. *International Journal of Food Science & Technology, 45*, 921–929.

Pratt, D. C., & del Rosario, J. I. (1913). The Philippine fruits and their composition. *Philippine Journal of Science, Section A, 8*, 74.

Raab, C., Oehler, N. (1976). *Making dried fruit leather* (p. 1È2). Oregon State University, Corvalis: Fact sheet 232.

Ridley, H. N. (1906). Malay drugs. *Agricultural Bulletin Straits & FMS, 5*(7), 245.

Roongpisuthipong, C., Banphotkasem, S., Komindr, S., & Tanphaichitr, V. (1991). Postprandial glucose and insulin responses to various tropical fruits of equivalent carbohydrate content in non-insulin-dependent diabetes mellitus. *Diabetes Research and Clinical Practices, 14*, 123–131.

Sinnhuber, R. O., Hendricks, J. D., Putnam, G. B., Wales, J. H., Pawlowski, N. E., Nixon, J. E., & Lee, D. J. (1976). Sterculic acid a naturally occurring cyclopropene fatty acid, a liver carcinogen to rainbow trout, *Salmo gairdenri*. *Federation Proceedings (Abs.), 35*, 1652.

Siondalski, P., & Lysiak-Szydlowska, W. (2007). Food components in the protection of the cardiovascular system. In Z. E. Sikorski (Ed.), *Chemical and functional properties of food components* (3rd ed., pp. 439–450). Boca Raton, FL: CRC Press LLC.

Stanton, W. R. (1966). *Tropical Science, 8*, 6; (1958) The genus Durio Adans, *Reinwardtia 3*(3), 357–460.

Sumarno, E. S. R. (2001). Fatty acid content in durian (*Durio zibethinus*, Murr) seed. *Indonesian Journal of Pharmacy, 12*(2), 65–71.

Toledo, F., Arancibia-Avila, P., Park, Y. -S., Jung, S. -T., Kang, S. -G., Heo, B. G., et al. (2008). Screening of the antioxidant and nutritional properties, phenolic contents and proteins of five durian cultivars. *International Journal of Food Sciences and Nutrition, 59*, 415–427.

Voon, Y. Y., Hamid, N. S. A., Rusul, G., Osman, A., & Quek, S. Y. (2007). Characterisation of Malaysian durian (*Durio zibethinus* Murr.) cultivars: relationship of physicochemical and flavour properties with sensory properties. *Food Chemistry, 103*(4), 1217–1227.

Wong, K. C., & Tie, D. Y. (1995). Volatile constituents of durian (*Durio zibethinus* Murr.). *Flavour and Fragrance Journal, 10*(2), 79–83.

Yaacob, O. (1983). The growth pattern and nutrient uptake of durian (*Durio zibethinus* Murr.) on an oxisol. *Communications in Soil Science & Plant Analysis, 14*(8), 689–698.

FURTHER READING

Joachim, A. W. R., & Pandittesekere, D. G. (1943). The analysis of Ceylon foodstuffs. XI-I The composition of some minor fruits, home-pounded grains, and vegetables. *Tropical Agriculturist Ceylon, 99*, 13–17.

Shaw, P. E., Chan, H. T., Jr., & Nagy, S. (Eds.). (1998). *Tropical and Subtropical Fruits*. Auburndale, FL, USA: AgScience.

Chapter 31

Jatropha curcas L. Seed Oil

INTRODUCTION

Jatropha curcas is an ornamental, medicinal and a multipurpose shrub belonging to the Euphorbiaceace family. The exact origin of *J. curcas* is still a subject of debate, but generally it is known to have come from Mexico and other states of Central America and there is a growing agreement to that (Achten et al., 2010). The leaves are used in traditional medicine against coughs or as antiseptics after birth, and the branches are chewing sticks (Gübitz, Mittelbach, & Trabi, 1999). The latex produced from the branches is useful for wound healing and other medical uses. It is widespread throughout arid and semiarid tropical region of the world. Each fruit contains 2–3 oblong seeds those are matured when the capsule changes from green to yellow. In addition to seeds, *Jatropha* contains other chemical compounds, such as saccharose, raffinose, staclyose, glucose, fructose, galactose, and protein. Furthermore, the plant also contains curcasin, arachidic, myristic, palmitic and stearic acid, and curcin (Janick & Paull, 2008). Table 31.1 shows the characteristics of Jatropha seeds. Chemicals are contained either in pulp or kernel that directly affected by the physical parameters. The average weight and volume of seed directly relate to the hardness of seeds that directly affects the process of analysis. Physical properties of one seed to another seed could distinct the product quality of *J. curcas* seed and its chemical. The Jatropha seed contains 46.27% oil. It has been reported that the toxicity and the disagreeable odor of seed is due to protein. The 4.56% w/w total ash content of seeds indicates presence of abrasive solids, soluble metallic soaps, and silica residue in the seed (Fig. 31.1).

Seeds of *J. curcas* contains some toxins compounds, such as those reported by King et al. (2009) a protein (curcin) and phorbol esters (diterpenoids), other researchers (Goel, Makkar, Francis, & Becker, 2007) suggested that the detoxification or complete removal of phorbol esters is essential before its use in industrial or medicinal applications. The major toxin phorbol ester is not vulnerable to heat, but can be hydrolyzed to less toxic substances extractable by either water or ethanol. Usman, Ameen, Lawal, & Awolola (2009) employed heat treatment, alkaline hydrolysis, and solvent extraction for the detoxification of the *J. curcas* seed cake (Aregheore, Becker, & Makkar, 2003). Despite the toxicity of *J. curcas*, edible varieties are known to exist in Mexico, which are

Unconventional Oilseeds and Oil Sources. http://dx.doi.org/10.1016/B978-0-12-809435-8.00031-7

TABLE 31.1 Characterization of *J. curcas* Seed

No.	Analytical parameter	Values
1	Weight of 1000 seeds	540.51 g
2	Volume of 1000 seeds	730.00 mL
3	Oil content (% v/w)	40–50
4	Moisture and volatilities (% w/w)	05.80
5	Ash content (% w/w)	4.56
6	Color	Light brownish
7	Odor	Disagreeable
8	Taste	Bitter
9	Protein (% w/w) (on dry basis)	22.50

not currently being exploited (King et al., 2009). The seed contains oil ranged between 40% and 50% (w/w) (Makkar, Becker, Sporer, & Wink, 1997).

The fact that the oil of *J. curcas* cannot be used for nutritional purposes without detoxification makes its use as an energy source for fuel production very attractive. The oil from the seeds has been found useful for the manufacture of candles and soap, in cosmetic industry and also as for medicinal purposes (Gübitz et al., 1999, Akbar, Yaakob, Kamarudin, Ismail, & Salimon, 2009).

OIL EXTRACTION

The amount of oil recovered was calculated as percentage of total oil present in *J. curcas* seed kernels. Three different solvents were used to evaluate their efficiency. The oil extraction capabilities of petroleum ether, hexane, and isopropanol are shown in Table 31.2. The extraction yield with petroleum ether was found in range of 10%–53%. Similarly, the extraction yield with hexane was found to be in the range of 20%–58% and with isopropanol was found to be near 15%–59%. The best result is obtained by using hexane and Soxhlet extraction method. Use of aqueous oil extraction (AOE), as control was found to yield 21.4% oil. The same data that is AOE yield in the range of 17%–21% was reported by other authors (Shah, Sharma, & Gupta, 2005). In AEOE method, the enzymes used, such as cellulase and pectinase failed to extract large amount of oil. Their yield is very less (in the range of 2%) when compared to that described by other authors (Shah et al., 2005).

Referred to Table 31.2 reveal that extraction Soxhlet extraction is better compared to other methods, because the process is continuous and there is a complete oil recovery. Among the solvents, hexane is considered as the best solvent because its yield is maximum when compared with other solvents (Sayyar, Abidin, Yunus, & Muhammad, 2009). But, the oil recovered by hexane and isopropanol is slightly

FIGURE 31.1 *Jatropha curcas* fruits and seeds.

TABLE 31.2 Extraction of Oil by Use of Various Solvents

Extraction method (% yield)

Organic solvent	Separating funnel	Centrifugation	Filtration	Soxhlet extraction
Petroleum ether	28.60 ± 4.16	48.30 ± 0.57	10.30 ± 4.0	53.46 ± 1.0
Isopropanol	19.54 ± 2.3	53.33 ± 1.52	15.66 ± 4.0	55.06 ± 2.5
Hexane	36.00 ± 9.53	52.66 ± 4.04	20.00 ± 6.2	58.36 ± 0.9

yellow in color. This might cause problem in using oil for further biodiesel production. Thus, it is better to use petroleum ether for efficient biodiesel production.

PHYSICOCHEMICAL PROPERTY OF OIL

The physical property, that is, color of the oil extracted by different organic solvents was also studied. All solvents except hexane yielded colorless oil and without impurities. Hexane yield faint yellow color oil. The chemical property, that is, %FFA content of the oil was found to be 2.24% (mg KOH/g). The composition of *J. curcas* oil from Nigeria consists of main fatty acid, such as palmitic acid (13%), stearic acid (2.53%), oleic acid (48.8%), and linoleic acid (34.6%) (Martinez-Herrera, Siddhuraju, Francis, Davila-Ortiz, & Becker, 2006). *J. Curcas* oil contains a high percentage of unsaturated fatty acid, which is about 78%–84%.

Experimental results showed that Jatropha oilseeds have FFA content 2.24%. The FFA and moisture contents have significant effects on the transesterification of glycerides with alcohol using catalyst in the production of biodiesel from Jatropha oil. The high FFA content (>1% w/w) will result in soap formation and the separation of products will be exceedingly difficult and as a result, it has low yield of biodiesel product. A two-step process was investigated for feedstock having the high FFA content (Akbar et al., 2009).

The iodine value was found to be 51.27 g 100/g oil as shown in Table 31.3. The value is below 100 and as such the oil can be classified as nondry

TABLE 31.3 Characteristics of the Prepared *J. curcas* Seed Oil

Property	Value
Saponification value	155 mg KOH/g oil
Peroxide value	7.20 mEq/g oil
Iodine value	51.27 g 100/g oil
Free fatty acids	0.0718%
Acid value	0.1428 mg KOH/g oil
Viscosity	8.2 cSt
Boiling point	124°C
Specific gravity	0.8480
Flash point	150°C
Cloud point	14°C
PH	5.2
Dielectric strength	22 kV
Pour point	4°C
Density at 27°C	0.725 g/cm^3

oil. This value also represents the decrease in unsaturation of oil, which is beneficial in the sense that the lower the unsaturation of oils and fats, the greater will be its oxidation stability; it was noticed that oil with high iodine, cloud, and pour point values exhibits poor cold performance (Muhammad et al., 2011). Oxidation of lipids is a major cause of their deterioration, and hydroperoxide formed by the reaction between oxygen and the unsaturated fatty acids is the primary product of these reactions. Hydroperoxide has no flavor or odor, but breaks down rapidly to form aldehydes, which have the disagreeable flavor and odor (McDonnell, Ward, & Timoney, 1995). SV is a measure of the alkali reactive groups in fats and oils and is useful in predicting the type of glycerides in an oil sample. SV is obtained by determining potassium hydroxide (in mg) required to saponify 1 g of fat. *J. curcas* oil with low SV of 155 mg KOH/g oil as shown in Table 31.3 was found better because oil with a high SV causes exhaust emissions during the burning of the engine (Kumar & Sharma, 2008). SV indicates the average molecular weight of the oil as by measuring SV means molecular mass can be obtained because it is, SV, inversely related to mean molecular mass. The oil of *J. curcas* seed was found to have a SV of 155 mg KOH/g as shown in Table 31.3. A higher SV indicates that there is a greater portion of low molecular weight fatty acids. Peroxide value (PV) is a measure of the extent of glycerides constituent decomposition by lipase action, which is added by light, air, and moisture. It is also an indication of the level of rancidity of the oil. The PV of *J. curcas* oil was found to be 7.20 mEq/g oil, (Table 31.3), indicating that the oil is still good and it has a low level of rancidity. Refining of oils can result in a considerable decrease of PV of various oils. This low value of 7.20 mEq/g shows the efficiency of the refining process of this study and this makes the oil ideal for consumption and storage (Kumar & Sharma, 2008).

SUITABILITY FOR BIODIESEL PRODUCTION

Viscosity is the most important property of transformer oil since it affects the operation of fuel injection equipment, particularly at low temperature when an increase in viscosity affects the fluidity of the biodiesel as fuel. Jatropha seed oil with viscosity of 8.2 cSt as shown in Table 31.3 is close to that of diesel fuel. As the oil temperature increases, its viscosity decreases. High viscosity leads to poorer atomization of biodiesel spray and less accurate operation of the fuel injection. The lower the oil viscosity, the easier it is to pump and atomize, and achieve finer results. A low viscosity of 8.2 cSt for Jatropha oil is good in this aspect. The operating temperature greatly affects the viscosity of a fluid though the viscosity of the Jatropha seed oil was taken at 27°C. Usually at a higher temperature, the viscosity becomes lower, which shows that there is an inverse relation between viscosity and temperature. For a smooth oil operation in electrical equipment, the temperature needs to remain around the mild range. (Bashi, Abdullahi, Yunus, & Nordin, 2006). Contaminants, such as water (Mirghani, Kabbashi, Alam, Qudsieh, & Ma'an, 2011), sediments, and conducting particles reduce the dielectric strength of insulating oil. Clean, dry oil has an inherently high dielectric strength but this does not necessarily indicates the absence of all contaminates; it merely indicates that the amount of contaminant present between the electrodes is not that large enough to affects the average breakdown voltage of the oil (Makkar, Becker, & Schmook, 1998). Further purification of vegetable oils can be carried out which will reduce the turbidity and hence raises the breakdown voltage (Bashi et al., 2006).

A higher density means more mass of fuel per unit volume. In this case, the vegetable oil has a higher density compared to crude oil. Jatropha seed oil has a density of 0.725 g/cm^3, which is high. The high mass of oils would give a higher energy available for work output per wit volume (Raja, Smart, & Lee, 2011).

The flashpoint is an important specification for safety during transport, storage, and handling (Raja et al., 2011). The flash point of *J. curcas* seed oil was found to be 150°C (Table 31.3). This is good value in preventing accidental ignition. The flash point has shown that the oil can safely be used even where the temperature is expected to be very high (Raja et al., 2011). The limit expected for most applications is 100°C. The appearance and the relative density are in agreement with the standard limit. With a flash point of 150°C, *J. curcas* oil can prevent auto ignition and fire hazard at high temperatures during transportation and storage (Makkar et al., 1998).

Cloud point is the temperature where wax begins to appear visible when the fuel is cooled, while the pour point is the temperature where the amount of wax from a solution is sufficient to gel the oil. The cloud point for *J. curcas* oil is 14°C (Table 31.3), which means that the oil can perform satisfactorily even in cold climatic conditions. The higher cloud point can affect the engine performance and emission adversely under cold climatic conditions.

The pour point of 4°C for *J. curcas* oil is low. In general, a higher pour point often limits the area of oils as fuels for transformer in cold climatic conditions. Cloud and pour point are the criteria used for low temperature performance of oil.

When the ambient temperature is below the pour point, wax precipitates in the vegetable oil and it loses its flow characteristics. Wax can block the filters and fuel supply line. Under these conditions, fuel cannot be pumped through the machine. Therefore, the pour point should be low so that oil can remain flow even at low temperature (Table 31.4).

The fatty acid composition of *J. curcas* oil was analyzed by gas chromatography. Table 31.5 shows major long-chain fatty acids present in the *J. Curcas* oil, which is palmitic acid (16.69%), stearic acid (7.67%), oleic acid (40.39%), linoleic acid (33.09%), and linolenic acid (0.28%). *J. Curcas* oil contains a high percentage of unsaturated fatty acid, which is about 75.64%.

TABLE 31.4 Fatty Acid Composition of *J. curcas*

Composition (fatty acid)	(%)
Palmitic acid (C16:0)	11.3
Stearic acid (C18:0)	17.0
Oleic acid (C18:1)	12.8
Linoleic acid (C18:2)	47.3
Arachidic acid (C20:0)	04.7
Arachidonic acid (C20:1)	01.8
Behenic acid (C22:0)	00.6
C24:0	04.4

Source: Data taken from Adebowale, K., & Adedire, C. (2006). Chemical composition and insecticidal properties of the underutilized Jatropha curcas seed oil. *African Journal of Biotechnology, 5*(10), 901.

TABLE 31.5 Fatty Acid Composition of Jatropha Seed Oil by Gas Chromatography

Composition (fatty acid) (%)	(%)
Palmitic acid (C16:0)	15.3974
Stearic acid (C18:0)	6.2653
Oleic acid (C18:1)	44.9305
Linoleic acid (C18:2)	33.4068
Unsaturated fatty acid	75.64
Saturated fatty acid	24.36

OTHER USES OF JATROPHA OIL

Oil for soap making is the most profitable use; the fruit of Jatropha contains an viscous oil that can be used for soap making (Openshaw, 2000). It is rich in palmitic acid, with high levels of hydrophobicity, and makes a soft, durable soap under even the simplest of manufacturing processes. It is also used in West Africa, Zambia, Tanzania, and Zimbabwe as a soap stock, including the manufacture of soft laundry soap (Pratt, Henry, Mbeza, Mlaka, & Satali, 2002). Jatropha oil gives a very good foaming, white soap with positive effects on the skin, partly due to the glycerine content of the soap (Henning, 2000). Jatropha soap is said to have medicinal characteristics and is therefore used by people with various skin diseases and sensitivity to regular soap (Messemaker, 2008). The 36% linoleic acid (C18:2) content in Jatropha kernel oil is of possible interest for skin care (Pratt et al., 2002; Benge, 2006). The oil has a strong purgative action and is also widely used for skin diseases and to soothe pain, such as that caused by rheumatism (Henning, 2003). The seed oil can be applied to treat eczema (Heller, 1996). The oil is also an ingredient in hair conditioners (Brittaine & Lutaladio, 2010). The use of Jatropha for soap industry (alternative karitee butter) and cosmetics is regarded as one of its nonenergy use (Ouwens et al., 2007).

Additionally, the oil is used as fuels for jet, transformer, and high hydrocarbon. It has also been used for stimulating hair growth and making candles and soap (Aderibigbe, Johnson, Makkar, Becker, & Foidl, 1997). The oil and aqueous extract from oil has been used in the control of insect pests of cotton including bollworm and on pests of pulses potato and corn. The seed oil of the plant has been used as an ingredient in the treatment of rheumatic conditions itch and parasitic skin diseases, and in the treatment of fever, jaundice and gonorrhea, as well as a diuretic agent and mouth wash (Dijkstra & Van Opstal, 1987, El-refaie, Salem, & Ahmed, 2009).

REFERENCES

Achten, W. M., Nielsen, L. R., Aerts, R., Lengkeek, A. G., Kjær, E. D., Trabucco, A., & Akinnifesi, F. K. (2010). Towards domestication of *Jatropha curcas*. *Biofuels*, *1*(1), 91–107.

Aderibigbe, A., Johnson, C., Makkar, H., Becker, K., & Foidl, N. (1997). Chemical composition and effect of heat on organic matter-and nitrogen-degradability and some antinutritional components of Jatropha meal. *Animal Feed Science and Technology*, *67*(2), 223–243.

Akbar, E., Yaakob, Z., Kamarudin, S. K., Ismail, M., & Salimon, J. (2009). Characteristic and composition of *Jatropha curcas* oil seed from Malaysia and its potential as biodiesel feedstock feedstock. *European Journal of Scientific Research*, *29*(3), 396–403.

Aregheore, E., Becker, K., & Makkar, H. (2003). Detoxification of a toxic variety of *Jatropha curcas* using heat and chemical treatments, and preliminary nutritional evaluation with rats. *The South Pacific Journal of Natural and Applied Sciences*, *21*(1), 51–56.

Bashi, S. M., Abdullahi, U. U., Yunus, R., & Nordin, A. (2006). Use of natural vegetable oils as alternative dielectric transformer coolants. *The Institution of Engineers, Malaysia*, *67*(2), 4–9.

Benge, M. (2006). *Assessment of the potential of Jatropha curcas, (biodiesel tree,) for energy production and other uses in developing countries*. USAID report.

Brittaine, R., & Lutaladio, N. (2010). *Jatropha: a smallholder bioenergy crop: the potential for pro-poor development* (Vol. 8). Rome, Italy: Food and Agriculture Organization of the United Nations (FAO).

Dijkstra, A. J., & Van Opstal, M. (1987). *Process for producing degummed vegetable oils and gums of high phosphatidic acid content*. Google Patents.

El-refaie, E. -s. M., Salem, M. R., & Ahmed, W. A. (2009). Prediction of the characteristics of transformer oil under different operation conditions. *World Academy of Science, Engineering and Technology*, *53*, 764–768.

Goel, G., Makkar, H. P., Francis, G., & Becker, K. (2007). Phorbol esters: structure, biological activity, and toxicity in animals. *International Journal of Toxicology*, *26*(4), 279–288.

Gübitz, G. M., Mittelbach, M., & Trabi, M. (1999). Exploitation of the tropical oil seed plant *Jatropha curcas* L.. *Bioresource Technology*, *67*(1), 73–82.

Heller, J. (1996). *Physic nut. Jatropha curcas L. Promoting the conservation and use of underutilized and neglected crops. 1*. Roma: IBPGR.

Henning, R. (2000). *The Jatropha manual. A guide to the integrated exploitation of the Jatropha plant in Zambia.* Germany: Deutsche Gesellschaft für Technische Zusammenarbeit GTZ.

Henning, R. (2003). *The Jatropha booklet, a guide to the Jatropha system and its dissemination in Africa.* Weissensberg, Germany: Baganí: GbR.

Janick, J., & Paull, R. E. (2008). *The encyclopedia of fruit and nuts.* Egham, Surrey, UK: CABI.

King, A. J., He, W., Cuevas, J. A., Freudenberger, M., Ramiaramanana, D., & Graham, I. A. (2009). Potential of *Jatropha curcas* as a source of renewable oil and animal feed. *Journal of Experimental Botany, 60*(10), 2897–2905.

Kumar, A., & Sharma, S. (2008). An evaluation of multipurpose oil seed crop for industrial uses (*Jatropha curcas* L.): a review. *Industrial Crops and Products, 28*(1), 1–10.

Makkar, H., Becker, K., Sporer, F., & Wink, M. (1997). Studies on nutritive potential and toxic constituents of different provenances of *Jatropha curcas. Journal of Agricultural and Food Chemistry, 45*(8), 3152–3157.

Makkar, H., Becker, K., & Schmook, B. (1998). Edible provenances of *Jatropha curcas* from Quintana Roo state of Mexico and effect of roasting on antinutrient and toxic factors in seeds. *Plant Foods for Human Nutrition, 52*(1), 31–36.

Martinez-Herrera, J., Siddhuraju, P., Francis, G., Davila-Ortiz, G., & Becker, K. (2006). Chemical composition, toxic/antimetabolic constituents, and effects of different treatments on their levels, in four provenances of *Jatropha curcas* L. from Mexico. *Food Chemistry, 96*(1), 80–89.

McDonnell, K., Ward, S., & Timoney, D. (1995). Hot water degummed rapeseed oil as a fuel for diesel engines. *Journal of Agricultural Engineering Research, 60*(1), 7–14.

Messemaker, L. (2008). The Green Myth? Assessment of the Jatropha value chain and its potential for pro-poor biofuel development in Northern Tanzania, The Green Myth. MSc Thesis, the Faculty of Geosciences, Utrecht University, The Netherlands.

Mirghani, M. E., Kabbashi, N. A., Alam, M. Z., Qudsieh, I. Y., & Ma'an, F. (2011). Rapid method for the determination of moisture content in biodiesel using FTIR spectroscopy. *Journal of the American Oil Chemists' Society, 88*(12), 1897–1904.

Muhammad, N., Bamishaiye, E., Bamishaiye, O., Usman, L., Salawu, M., Nafiu, M., & Oloyede, O. (2011). Physicochemical properties and fatty acid composition of *Cyperus esculentus* (Tiger Nut) tuber oil. *Bioresearch Bulletin, 5*, 51–54.

Openshaw, K. (2000). A review of *Jatropha curcas*: an oil plant of unfulfilled promise. *Biomass and Bioenergy, 19*(1), 1–15.

Ouwens, K. D., Francis, G., Franken, Y. J., Rijssenbeek, W., Riedacker, A., Foidl, N., & Bindraban, P. (2007). Position paper on *Jatropha curcas*. State of the art, small and large scale project development. *Agronomy and Genetics*, 26–28.

Pratt, J., Henry, E., Mbeza, H., Mlaka, E., & Satali, L. (2002). Malawi agroforestry extension project marketing & enterprise program, main report. *Malawi Agroforestry, 47*, 139.

Raja, S. A., Smart, D. R., & Lee, C. L. R. (2011). Biodiesel production from Jatropha oil and its characterization. *Research Journal of Chemical Sciences, 1*(1), 81–87.

Sayyar, S., Abidin, Z. Z., Yunus, R., & Muhammad, A. (2009). Extraction of oil from Jatropha seeds-optimization and kinetics. *American Journal of Applied Sciences, 6*(7), 1390.

Shah, S., Sharma, A., & Gupta, M. (2005). Extraction of oil from *Jatropha curcas* L. seed kernels by combination of ultrasonication and aqueous enzymatic oil extraction. *Bioresource technology, 96*(1), 121–123.

Usman, L., Ameen, O., Lawal, A., & Awolola, G. (2009). Effect of alkaline hydrolysis on the quantity of extractable protein fractions (prolamin, albumin, globulin and glutelin) in *Jatropha curcas* seed cake. *African Journal of Biotechnology, 8*(22), 6374–6378.

Chapter 32

Mesua ferrea Nahar Tree Seed Oil

INTRODUCTION

Nahar tree is commonly known as Cobra's saffron (English) and its scientific name is *Mesua ferrea* Lin, belong to the family Guttiferae. *M. ferrea* is called other names in different location for instance, Nagakeshara (Hindi), Nagasampige (Kannada), Nageshwar (Assam), Nagachampakam (Tamil) (Chahar, Sanjaya, Geetha, Lokesh, & Manohara, 2013).It is native to Sri Lanka (a state tree of Tripura) and can be found in Burma, Thailand, New guinea, Indochina, and India (Chahar et al., 2013; Keawsa-ard, Liawruangrath, & Kongtaweelert, 2015). *M. ferrea* that the following features (1) medium size glabrous tree (18–30 m height), (2) ash straight trunk, (3) grayish or reddish-brown bark, (4) oblong-lanceolate or acute and leathery leaves, (5) large, white and fragrant flower, (6) ovoid fruits, (7) angular, smooth, and chestnut brown seeds (2.5–5 cm long), and (8) extremely hard wood (Chahar et al., 2013; Chanda, Rakholiya, & Parekh, 2013).

Different parts of *M. ferrea* have been utilized in preparation of traditional medicine for treatment of ailments, such as antiseptic, purgative, blood purifier, worm control, tonic properties, treat fever, cold, asthma, and as carminative, expectorant, cardiotonic, diuretic, antipyretic agent, wound healing, snake antivenom, stomachic, expectorant, astringent, etc. (Chahar et al., 2013). Scientists have equally investigated potentials of *M. ferrea* bioactive compounds as antioxidant, antiinflammatory, immunomodulatory, anticancer, antimicrobial, antivenom, hepatoprotective, antineoplastic, etc. Efforts to isolate pure bioactive compounds of *M. ferrea* have also yield identification of compounds like xanthrones, phenolics, coumarins (Jayanthi, Kamalraj, Karthikeyan, & Muthumary, 2011), etc. Nonmedicinal use of *M. ferrea* oils has also been stated to include diesel fuel-blends, production of coatings, PVC plasticizers, cosmetics, etc. Detail of current findings on characteristics, potentials use, and biological properties of *M. ferrea* are provided in the subsequent sections (Fig. 32.1).

Unconventional Oilseeds and Oil Sources. http://dx.doi.org/10.1016/B978-0-12-809435-8.00032-9

FIGURE 32.1 *Mesua ferrea* **tree, seeds and kernels.**

APPLICATIONS OF *M. ferrea*

As Bioactive Compounds for Medicinal Drugs

M. ferrea has found useful as active ingredients in different types of Ayurvedic formulations commonly used for healing purposes. According to Chahar et al. (2013), *M. ferrea* is present in Nagakeshara-adi-churna used in bacillary dysentery; Nagakeshara yoga used in piles; Vyaghrihareetaki avaleha in Shwasa, kasa, and pinasa; Eladi churna in Carminative, vomiting, and indigestion anorexia; and Lavangadi churna used in cough, diarrhea, dysentery, mouth diseases, dental caries, anemia, and fever.

Production Epoxidized Oil

Epoxidized oils have been identified as the most commercially viable alternatives to traditional plasticizers for improvement of PVC properties. Epoxidized *M. ferrea* seed oil was used to produced thermostable PVC and PVC-clay nanocomposites (Das & Karak, 2012a). *M. ferrea* seed oil was also used to synthesis nanoclay/epoxy nanocomposites which displayed improved in tensile strength, scratch hardness, gloss, and thermal stability (Das & Karak, 2012b).

Polyester Resin Preparation

Polyesters are used in industrial construction and installation, molding compounds, coatings, adhesive, and laminates. Although, production of petroleum-based polyesters have trend to use renewable resources due to environmental reasons. Fats and oil from either plant or animal have been identified as alternative resources for polyesters production (Bakare, Pavithran, Okieimen, & Pillai, 2006). Oil of *M. ferrea* seed has equally been utilized for polyester resin production. Das and Karak (2009), reported on utilization of monoacylglycerols of *M. ferrea* L., seed oil for synthesis of epoxy resins that are suitable as coating materials. Reaction of phthalic and/or maleic anhydride with monoacylglycerol of *M. ferrea* seed oil produced by alcoholysis method was also reportedly used for synthesis of polyester production (Dutta, Karak, & Dolui, 2004). Low-cost stoving paint was also synthesis using Nahar seed oil polyester resin produced through initial conventional alcoholysis with glycerol followed by polycondensation with phthalic anhydride using the azeotropic distillation technique (Dutta, Karak, & Dolui, 2007).

Thermally treated vegetable oil (including *M. ferrea* seed oil) is being admired as alternative for production of coating materials against petroleum products. Vegetable oils are advantageous because they are reasonably inexpensive, nontoxic, renewable, biodegradable, readily available, easily processed, and environmentally benign. Thermal treatment of vegetable oil at high temperature led to viscose product through polymerization reactions including Diels–Alder reactions, ester interchange, cyclisation, oxidative polymerization, etc. Thermally treated *M. ferrea* seed oil exhibited better coating performance and suitable for production film with short drying time (Dutta, Karak, & Dolui, 2005).

Fuel Blends

Direct use of Nahar oil as a suitable fuel in engines is hindered by its higher viscosity, density, and low volatility. Studies have shown the possibilities of blending Nahar oil with diesel up to 15% (by volume) which can be used in a diesel engine (CI) without any major engine modification (Kushwah, Mahanta, & Mishra, 2008). In another study, blend of waste cooking methyl ester and Nahar oil methyl ester (60:40 or 80:20) is the optimum blend for engine application (Papu, Dash, & Lingfa, 2015).

CHARACTERISTICS OF *M. ferrea* AND ITS OILS

Previous works on *M. ferrea* have elucidated the characteristics of *M. ferrea* and its oil. According to Table 32.1, the lipid content is highest in seed of *M. ferrea* while the least was found in the leave. This is possible because the composition analysis was carried out on fresh leave with high moisture content.

TABLE 32.1 Proximate Composition of Different Parts of *M. ferrea*

Part	Total lipid (%)	Moisture (%)	Ash (%)	Total protein (%)	Water soluble protein (%)	Starch (%)	Crude fiber (%)	Carbohydrate (%)
Seed	66.91–70.23	4.02–5.05	1.46–1.50	6.99–7.19	2.98–3.11	5.51–5.85	1.22–1.98	15.88–18.68
Leaves	2.32–2.44	65.12–72.19	2.60–2.71	4.23–4.85	1.47–2.01	3.06–3.27	3.12–3.29	14.82–22.30
Bark	—	—	—	—	—	—	—	—
Flower	—	—	—	—	—	—	—	—

Adapted from Sayeed, M. A., Ali, M. A., Sohel, F. I., Khan, G. R. M. A. M., & Yeasmin, M. S. (2004). Physico-chemical characteristics of *Mesua ferrea* seed oil and nutritional composition of its seed and leaves. *Bulletin of the Chemical Society of Ethiopia, 18*(2), 157–166.

TABLE 32.2 Physicochemical Properties of *M. ferrea* Seed Oil

Properties	Amount
Specific gravity	0.9287–09312
Refractive index	1.4690–1.4739
Solidification point	−4.0–(−4.3)
Pour point	−1.0–(−1.3)
Cloud point	5.5–6.0
Flash point	90–98
Fire point	110–116
Smoke point	44–47
Iodine value	89.17–93.01
Saponification value	199.03–206.40
Saponification equivalent	271.80–281.86
Acid value	9.64–11.87
Free fatty acid	4.85–5.96
Ester value	188.95–1.95.44
Unsaponifiable matter	1.44–1.50
Acetyl value	2.70–2.84
Peroxide value	3.58–3.64
Reichert-Meissl value	5.852–6.031
Polenske number	0.7891–0.8401

Adapted from Sayeed, M. A., Ali, M. A., Sohel, F. I., Khan, G. R. M. A. M., & Yeasmin, M. S. (2004). Physico-chemical characteristics of *Mesua ferrea* seed oil and nutritional composition of its seed and leaves. *Bulletin of the Chemical Society of Ethiopia, 18*(2), 157–166.

Table 32.2 shows physicochemical properties of samples of seed oil from *M. ferrea* obtained from different location in Bangladesh (Sayeed, Ali, Sohel, Khan, & Yeasmin, 2004). In another study, the physical properties of seed oil of *M. ferrea* obtain from India has the following physical properties: acid value (34 mg KOH/g), iodine value (89.26 gI$_2$ per 100 g), saponification value (241 mg KOH/g), specific gravity (0.89 at 28°C), and cloud point (8°C) (Dutta et al., 2004). Estimation of seed oil of *M. ferrea* was stated to make up of mostly triglyceride (87.65%–89.50%) with little content of monoglycerides (1.05%–1.35%) and diglycerides (2.12%–2.32%). Fractions of total lipid extracts were neutral lipid (89.83%–92.18%), glycolipid (3.65%–4.15%) and phospholipid (1.98%–2.68%). The saturated and unsaturated fatty acids in the oils were ranged from 27.40% to 29.11% and 65.85% to 68.31%, respectively (Sayeed et al., 2004).

Nahar (*M. ferrea* L.) seeds contain about 70%–75% of nondrying oil made up of oleic acid (52.3%) and linoleic acid (22.3%) as unsaturated fatty acids and stearic acid (9.5%) and palmitic acid (15.9%) as saturated fatty acids (Das & Karak, 2009). Composition analysis of other parts of *M. ferrea* Linn carried out by high resolution GC and HRGCMS revealed present of between 32 and 50 different components. The composition of the oil varied depending on the part of the plant. The bark oil was rich in (E)-α-bisabolene (31.3%) and α-selinene (12.2%), the tender and mature leaves were rich in α-copaene (19.3 and 99%) and β-caryophyllene (18.8 and 26.0%), the bud and flower were rich in α-copaene (28.7 and 20.2%), and germacrene D (190 and 16.1%) (Choudhury, Ahmed, Barthel, & Leclercq, 1998). The following compounds have been reportedly isolated from *M. ferrea*: mesuaxanthone A (1,5-dihydroxy-3-methoxy-xanthone), mesuaxanthone B (1,5,6-trihydroxyxanthone), 1,5-dihydroxyxanthone (II), euxanthone 7-methyl ether (IV) and β-sitosterol, Ferruol A, $C_{23}H_{30}O_5$ (a 4-alkylcoumarin), mesuaferrone-b (a biflavonone), β-amyrin, β-sitosterol, mesuaferrol (a cyclohexadione), Ferrxanthone (a 1,3,5,6-tetraoxygenated xanthone), mammeigin, and mesuol (Govindachari, Pai, Subramaniam, Ramdas, & Muthukumaraswamy, 1967; Chow & Quon, 1968; Bala & Seshadri, 1971; Raju, Srimannarayana, Rao, Bala, & Seshadri, 1976; Walia & Mukerjee, 1984; Dennis, Kumar, & Srimannarayana, 1988). More recently, α-amyrin and β-amyrin, lupeol, ellipticine, doxorubicin, and β-sitosterol have been successfully isolated from *M. ferrea* stems (Keawsa-ard et al., 2015).

BIOLOGICAL PROPERTIES

Antibacterial

Disc diffusions and broth dilution methods were used to determine antimicrobial activities of crude oil obtained from Nahar seed and leaves. Crude Nahar seed oil extracted using different organic solvents (*n*-hexane, ethanol, methanol, petroleum ether, and chloroform) exhibited antimicrobial activities against *Escherichia coli, Pseudomonas aeruginosa, Bacillus subtilis,* and *Staphylococcus aureus.* The minimum inhibition concentration of the various organic solvents crude extracts of Nahar seed oil ranged from 240 to 980 μg/mL while the minimum bactericidal concentration was 2 mg/mL (Adewale, Mirghani, Muyibi, Daoud, & Abimbola, 2010). Further study revealed that Nahar seed oil exhibited disinfection potentials and that its kinetic studies at 1 mg/mL dosage followed first-order models with a *k* value of −0.04 (Adewale, Mirghani, Muyibi, Daoud, & Abimbola, 2013). Also, lipophilic extracts (nonpolar) of Nahar seed extracted using seven different solvents exhibited more antibacterial activities against Gram-positive bacteria compared to Gram-negative ones (Chanda et al., 2013). In another report, aqueous extract of Nahar seed caused about 4.5 and 3.5 mm while more active methanolic counterpart caused 19.5 and 22.5 mm inhibition zone, respectively against *Klebsiella pneumonia* and *Proteus mirabilis* (Parekh & Chanda, 2007).

Methanolic extract of leaves of Nahar was found to be a potent bactericide against several bacteria strains (Mazumder, Dastidar, Basu, Mazumder, & Kumar, 2003). Ethanol and methanol extracts of Nahar leaves also exhibited antibacterial properties against *E. coli*, *P. aeruginosa*, *B. subtilis,* and *S. aureus*. The zone of inhibition of extracts against the microbes ranged from 16.0 to 18.05 mm and the MIC and MBC ranged from 0.625 to 2.5 mg/mL, respectively. However, cytotoxicity test showed that Nahar leave extracts were lethal against brine shrimps with LC_{50} of 500 ppm (μg/mL) (Adewale, Mirghani, Muyibi, Daoud, & Abimbola, 2011, 2012). Also, methanol extract of whole flowers of Nahar at concentration between the ranges of 100–150 μg/mL inhibited both Gram-positive and Gram-negative. In vivo test showed that 100–200 μg/g body weight protects Swiss strain of albino mice against *Salmonella typhimurium* ATC 6539 (Mazumder, Dastidar, Basu, Mazumder, & Singh, 2004). Study of different solvents extract of parts (seed, stem bark, and leaves) of Nahar revealed antimicrobial potential against bacterial with little effect against fungi (Sohel & Yeasmin, 2004).

Antiinflammatory

Ethanolic extract of Nahar was reported to possess antiinflammatory potential by ameliorated carrageenan-induced paw edema in rat model (Nandy & Tiwari 2012). The antiinflammatory properties of oral administration of Nahar flower ethanolic extract was dose dependent. Mesuol isolated from *M. ferrea* seed oil was investigated for immunomodulatory potential in cellular immune response model. Mesuol not only restored the hematological profile in cyclophosphate-amide induced myelosuppression model but also potentiated adhesion of neutrophile in test rats and carbon phagocytosis (Chahar, Kumar, Lokesh, & Manohara, 2012). Another study also shows that hind paw edema, cotton pellet implantation, and granuloma pouch tests in carrageenan-induced albino rats were reportedly reduced upon oral administration of Mesuaxanthone A and B from *M. ferrea* (Chahar et al., 2013).

Anticonvulsant

Anticonvulsant potential of oral administration of ethanolic extract (500–2000 mg/kg) of Nahar flower was investigated using albino mice by maximum electroshock seizure induction test. The extract did not exhibit any cytotoxicity effect in the animal but only show sign of depression as a result of decrease in spontaneous activity. Anticonvulsant properties of the ethanolic extract was revealed through the increased in onset time and decreased duration of seizures by eletroconvulsive shock. The anticonvulsant effect was dose dependent and might be due to inhibition of the activity of gamma amino butyric acid at its receptors (Tiwari, Irchhaiya, & Jain, 2012).

Antioxidant

Methanolic extract of Nahar flower exhibited in vivo antioxidant and hepatoprotective properties in male Wister rate. After daily administration of Nahar methanolic extract for 1 week, the following biochemical parameters of the rat were determined: alanine aminotransferase (ALT), aminotransferase (AST), phosphorkinase (CPK), catalase (CAT), super oxide dismutase (SOD), glutathione peroxidase (GPx) and glutathione reductase (GR), creatinine, and urea. Antioxidant and hepatoprotective effect was evident in rat administered with 100 mg/kg of body weight from the increase in biochemical parameters while that of AST was decreased (Garg, Sharma, Ranjan, Attri, & Mishra, 2009). Phenolic extracts of Nahar seeds and pericarp were reported to have shown antioxidant activities in terms of DPPH, ABTS, and FRAP (Surveswaran, Cai, Corke, & Sun, 2007). Antioxidant activity of isolated endophyte from *M. ferrea* (*Phomopsis* sp. GJJM07) was investigated using DPPH radical scavenging assay. The result showed that ethyl acetate extract of fungus showed potent antioxidant activity with IC50 value of 31.25 μg/mL compared to the IC50 value of standard ascorbic acid, 11.11 μg/mL (Jayanthi et al., 2011). Although, chloroform and ethanol extracts of *M. ferrea* exhibited significant antioxidant activities, there were variations in their specificity. Ethanol extracts showed >90% protection to erythrocytes, Hb, and DNA while chloroform extract showed <90% but >70% antioxidant protective activity. This variation was probably due to difference in their phenolics and total flavonoid content (Rajesh, Manjunatha, Krishna, & Kumara Swamy, 2013).

Anticancer

Dichloromethane extract of *M. ferrea* displayed anticancer activities against KG-oral, MCF-7, and NCI-H187 cell lines. Furthermore, Friedelin was isolated from dichloromethane extract and chemical analysis showed that it was a mixture of α-amyrin and β-amyrin, lupeol, and β-sitosterol. The mixture of α-amyrin and β-amyrin showed anticancer potential against MCF-7 cell line with the IC50 value 28.45 μg/mL and lupeol demonstrated anticancer activities against KB, MCF-7, and NCI-H187 cell lines with the IC50 values of 30.12, 34.25, and 21.56 μg/mL, respectively. All the extracts and the isolated compounds were noncytotoxic to Vero cells (Keawsa-ard et al., 2015).

REFERENCES

Adewale, A. I., Mirghani, M. E. S., Muyibi, S. A., Daoud, J. I., & Abimbola, M. M. (2010). Extraction and antibacterial activity of Nahar (*Mesua ferrea*) seed kernels' oil. *ACT-Biotechnology Research Communications 1:1 (2011)*, *28*, 32.

Adewale, A. I., Mirghani, M. E. S., Muyibi, S. A., Daoud, J. I., & Abimbola, M. M. (2011). In-vitro anti-microbial and brine-shrimp lethality potential of the leaves extract of Nahar (*Mesua ferrea*) plant. In *2nd International Conference on Biotechnology Engineering (ICBioE 2011)*. Kuala Lumpur: The Legend Hotel.

Adewale, A. I., Mirghani, M. E. S., Muyibi, S. A., Daoud, J. I., & Abimbola, M. M. (2012). Antibacterial and cytotoxicity properties of the leaves extract of Nahar (*Mesua ferrea*) plant. *Advances in Natural and Applied Sciences*, *6*(5), 583–588.

Adewale, A. I., Mirghani, M. E. S., Muyibi, S. A., Daoud, J. I., & Abimbola, M. M. (2013). Disinfection studies of Nahar (*Mesua ferrea*) seed kernel oil using pour plate method. *African Journal of Biotechnology, 10*(81), 18749–18754.

Bakare, I. O., Pavithran, C., Okieimen, F. E., & Pillai, C. K. S. (2006). Polyesters from renewable resources: preparation and characterization. *Journal of Applied Polymer Science, 100*(5), 3748–3755.

Bala, K. R., & Seshadri, T. R. (1971). Isolation and synthesis of some coumarin components of *Mesua ferrea* seed oil. *Phytochemistry, 10*(5), 1131–1134.

Chahar, M. K., Kumar, D. S. S., Lokesh, T., & Manohara, K. P. (2012). In-vivo antioxidant and immunomodulatory activity of mesuol isolated from *Mesua ferrea* L. seed oil. *International Immunopharmacology, 13*(4), 386–391.

Chahar, M. K., Sanjaya, K. D. S., Geetha, L., Lokesh, T., & Manohara, K. P. (2013). *Mesua ferrea* L.: a review of the medical evidence for its phytochemistry and pharmacological actions. *African Journal of Pharmacy and Pharmacology, 7*(6), 211–219.

Chanda, S., Rakholiya, K., & Parekh, J. (2013). Indian medicinal herb: Antimicrobial efficacy of *Mesua ferrea* L. seed extracted in different solvents against infection causing pathogenic strains. *Journal of Acute Disease, 2*(4), 277–281.

Choudhury, S., Ahmed, R., Barthel, A., & Leclercq, P. A. (1998). Volatile oils of *Mesua ferrea* (L.) from Assam, India. *Journal of Essential Oil Research, 10*(5), 497–501.

Chow, Y. L., & Quon, H. H. (1968). Chemical constituents of the heartwood of *Mesua ferrea*. *Phytochemistry, 7*(10), 1871–1874.

Das, G., & Karak, N. (2009). Epoxidized *Mesua ferrea* L. seed oil-based reactive diluent for BPA epoxy resin and their green nanocomposites. *Progress in Organic Coatings, 66*(1), 59–64.

Das, G., & Karak, N. (2012a). Epoxidized *Mesua ferrea* L. seed oil-plasticized thermostable PVC and PVC-clay nanocomposites. *Journal of Vinyl and Additive Technology, 18*(3), 168–177.

Das, G., & Karak, N. (2012b). *Mesua ferrea* L. seed oil based amido-amine modified nanoclay/ epoxy nanocomposites. *Journal of Applied Polymer Science, 124*(3), 2403–2414.

Dennis, T. J., Kumar, K. A., & Srimannarayana, G. (1988). A new cyclo hexadione from *Mesua ferrea*. *Phytochemistry, 27*(7), 2325–2327.

Dutta, N., Karak, N., & Dolui, S. (2004). Synthesis and characterization of polyester resins based on Nahar seed oil. *Progress in Organic Coatings, 49*(2), 146–152.

Dutta, N., Karak, N., & Dolui, S. K. (2005). Structural analysis, rheological behaviour and the performance of films for surface coatings' application of heated and unheated Nahar seed oil. *Polymer Degradation and Stability, 88*(2), 317–323.

Dutta, N., Karak, N., & Dolui, S. K. (2007). Stoving paint from *Mesua ferrea* L. seed oil based short oil polyester and MF resins blend. *Progress in Organic Coatings, 58*(1), 40–45.

Garg, S., Sharma, K., Ranjan, R., Attri, P., & Mishra, P. (2009). In vivo antioxidant activity and hepatoprotective effects of methanolic extract of *Mesua ferrea* Linn. *International Journal of PharmTech Research, 1*(4), 1692–1696.

Govindachari, T. R., Pai, B. R., Subramaniam, P. S., Ramdas Rao, U., & Muthukumaraswamy, N. (1967). Constituents of *Mesua ferrea* L.—II: : Ferruol A, a new 4-alkylcoumarin. *Tetrahedron, 23*(10), 4161–4165.

Jayanthi, G., Kamalraj, S., Karthikeyan, K., & Muthumary, J. (2011). Antimicrobial and antioxidant activity of the endophytic fungus *Phomopsis* sp. GJJM07 isolated from *Mesua ferrea*. *International Journal of Current Science, 1*, 85–90.

Keawsa-ard, S., Liawruangrath, B., & Kongtaweelert, S. (2015). Bioactive compounds from *Mesua ferrea* stems. *Chiang Mai Journal of Science, 42*(1), 185–195.

Kushwah, Y. S., Mahanta, P., & Mishra, S. C. (2008). Some studies on fuel characteristics of *Mesua Ferrea*. *Heat Transfer Engineering*, *29*(4), 405–409.

Mazumder, R., Dastidar, S. G., Basu, S. P., Mazumder, A., & Kumar, S. (2003). Emergence of *Mesua ferrea* Linn. leaf extract as a potent bactericide. *Ancient Science of Life*, *22*(4), 160.

Mazumder, R., Dastidar, S. G., Basu, S., Mazumder, A., & Singh, S. (2004). Antibacterial potentiality of *Mesua ferrea* Linn. flowers. *Phytotherapy Research*, *18*(10), 824–826.

Nandy, S., & Tiwari, P. (2012). Screening of anti-inflammatory activity of *Mesua ferrea* Linn flower. *International Journal of Biomedical Research*, *3*(5), 245–252.

Papu, N. H., Dash, S. K., & Lingfa, P. (2015). Performance and emission characteristics of a diesel engine run by blend of Nahar and waste cooking oil methyl ester. *International Journal for Innovative Research in Science and Technology*, *1*(11), 532–538.

Parekh, J., & Chanda, S. (2007). In vitro screening of antibacterial activity of aqueous and alcoholic extracts of various Indian plant species against selected pathogens from Enterobacteriaceae. *African Journal of Microbiology Research*, *1*(6), 92–99.

Rajesh, K. P., Manjunatha, H., Krishna, V., & Kumara Swamy, B. E. (2013). Potential in vitro antioxidant and protective effects of *Mesua ferrea* Linn. bark extracts on induced oxidative damage. *Industrial Crops and Products*, *47*, 186–198.

Raju, M. S., Srimannarayana, G., Rao, N. V. S., Bala, K. R., & Seshadri, T. R. (1976). Structure of mesuaferrone-b a new biflavanone from the stamens of *Mesua ferrea* Linn. *Tetrahedron Letters*, *17*(49), 4509–4512.

Sayeed, M. A., Ali, M. A., Sohel, F. I., Khan, G. R. M. A. M., & Yeasmin, M. S. (2004). Physicochemical characteristics of *Mesua ferrea* seed oil and nutritional composition of its seed and leaves. *Bulletin of the Chemical Society of Ethiopia*, *18*(2), 157–166.

Sohel, F. I., & Yeasmin, M. S. (2004). Antimicrobial screening of *Cassia fistula* and *Mesua ferrea*. *Journal of Medical Science*, *4*(1), 24–29.

Surveswaran, S., Cai, Y. -Z., Corke, H., & Sun, M. (2007). Systematic evaluation of natural phenolic antioxidants from 133 Indian medicinal plants. *Food Chemistry*, *102*(3), 938–953.

Tiwari, P. K., Irchhaiya, R., & Jain, S. K. (2012). Evaluation of anticonvulsant activity of *Mesua ferrea* Linn. ethanolic flower extract. *International Journal of Pharmacy & Life Sciences*, *3*(3).

Walia, S., & Mukerjee, S. K. (1984). Ferrxanthone, a 1,3,5,6-tetraoxygenated xanthone from *Mesua ferrea*. *Phytochemistry*, *23*(8), 1816–1817.

Chapter 33

Nephelium lappaceum L. Rambutan Kernel Oil

INTRODUCTION

Name "rambutan" originates from the Malayan word "rambut" which means "hair," due to numerous hairlike spikes on the surface of the fruit. The rambutan tree *Nephelium lappaceum* is an evergreen, medium-sized bushy tree, growing to a height of 20 m, with a dense, low, round, and spreading crown (Augustin & Chua, 1988). Rambutan has pinnate leaves composed of 3–11 oval leaflets with smooth edges and pointed tips. The leaves are dark green in color and are alternately arranged on branches. The leaflets are elliptic, 7.5–20 cm long, and 3.5–8 cm wide. Flowers are greenish white, fragrant (emit sweet aroma), very small, petal less, disk-shaped, arranged in the multibranched, erect clusters at the end of the branches. Rambutan produces round-shaped, oblong berries, 4–5 cm long arranged in dense clusters. Fruit has thin reddish or orange-yellow rind covered with thick, coarse hairs or long, soft spines on the surface. Pulp or the aril is edible, white, opaque, translucent, juicy and sweet, surrounding a glossy brown seed that cultivated in most parts of the Philippines. Also reported in India to Indochina and Malaysia, and extensively cultivated in Java and Malaysia (Augustin & Chua, 1988).

Rambutan produces fruit two times per year. It starts to produce fruit 5–6 years after planting. Each tree can produce 5000–6000 fruit per season. The utilized parts of the tree are the roots, leaves, and bark. Its fruit is considered to be astringent, stomachic, vermifuge, and febrifuge. The seeds of the fruit are somewhat bitter and of narcotic property (Harahap, Ramli, Vafaei, & Said, 2012).

There are folkloric uses of all products of the rambutan (*N. lappaceum*) tree in the different parts where it is cultivated. The Malays have used a decoction of its roots for fever, its bark for diseases of the tongue, and its fruit's pulp for diarrhea, dysentery, and as a refrigerant in fever. Elsewhere the seeds have been used to extract oil and fascinatingly they are also roasted and eaten. Depending on the cultivar, rambutan produces functionally male, functionally female, or hermaphroditic flowers (with both types of reproductive organs). Flowers are rich source of nectar which attracts bees, ants, and flies, responsible for the pollination of this plant (Fig. 33.1).

Unconventional Oilseeds and Oil Sources. http://dx.doi.org/10.1016/B978-0-12-809435-8.00033-0

FIGURE 33.1 *Nephelium lappaceum* fruity tree and seeds.

The rind of rambutan (*N. lappaceum*) was found to contain extremely high antioxidant activity when assessed using several methods. Although having a yield of only 18%, the ethanolic rambutan rind extract had a total phenolic content of about 762 mg GAE/g extract, which is comparable to that of a commercial preparation of grape seed extract. Comparing the extract's prooxidant capabilities with vitamin C, a-tocopherol, grape seed, and green tea, the rind had the lowest prooxidant capacity. In addition, the extract at 100 µg/mL was seen to limit oxidant-induced cell death (DPPH at 50 µM) by apoptosis to an extent similar to that of grape seed. This extract, either alone or in combination with other active principles, can be used in cosmetic, nutraceutical and pharmaceutical applications (Palanisamy et al., 2008).

CONSTITUENTS OF THE SEED

The rambutan canning industry is well established in Thailand and canners in Malaysia are also producing canned rambutans in syrup (Kheiri & bin Mohammad Som, 1979; Augustin & Chua, 1988). The rambutan fruits are de-seeded during processing and the seeds remain as a wasted by-product of the canning industry. There is potential for exploitation of this by-product but this can only be realized when more information on the composition of the kernel and its components are known (Augustin & Chua, 1988).

The weight of the seed constitutes about 5.6%–7.4% of the whole fruit. When the kernel is combusted, the following has been observed to make up the seeds contents: 11.9%–14.1% protein, 37.1%–38.9% crude fat, 2.8%–6.6%

crude fiber, 2.6–2.9 ash on dry weight basis and it also had 34.1%–34.6% moisture.

Solid Fat Content of Rambutan Seed

Solid fat content (SFC) is a significantly indicator of hardness. The lowest of SFC in fats almost used in the chocolate industry resulting in the softer texture of the products, because of chocolate made with softer fat containing low crystals less than with a hard fat. The SFC profile was affected to relative tendency of chocolate hardness, which it also resulting in the pure fat system is value consideration. The solids profile of rambutan seed fat was affected in the amount of SFC by temperature. Rambutan seed fat is softer than cocoa butter at low temperatures and has a harder consistency at higher temperatures. This behavior is probably due to the composition difference (Solís-Fuentes, & Durán-de-Bazúa, 2004). So, rambutan fat would be useful in filled chocolate manufacture as a softer filling fat compatible with cocoa butter (Lannes, Medeiros, & Gioielli, 2003).

Rambutan seed fat native in Southeast Asia, rambutan (*N. lappaceum* Linn.) belongs to the same family (Sapindaceae) as the subtropical fruits lychee and longan (Marisa, 2006). Rambutan is a seasonal fruit native of west Malaysia and Sumatra. It is cultivated widely in Southeast Asian countries. For commercial crop in Asia, this fruit is important. Normally, this fruit is consumed fresh, canned, or processed, and appreciated for its refreshing flavor and exotic. The rambutan fruits are deseeded during processing and these seeds (~4–9 g/100 g) are waste by-product of the canning industry (Tindall, 1994). Some studies had reported that rambutan seed possesses a relatively high amount of fat with values between 14 and 41 g/100 g. (Sirisompong, Jirapakkul, & Klinkesorn, 2011). And other information on the seed had showed that rambutan possesses a relatively high amount of fat between 17% and 39% (Morton, 1987; Zee, 1993). Furthermore, the demand of rambutan consumption for the industry was increased. Therefore, the extracted fat from the rambutan seed not only could be used for manufacturing candles, soaps, and fuels, but it also has a possibility to be a source of natural edible fat with feasible manufacturing use.

EXTRACTION OF SEED OIL

Properties of rambutan seed oil and fatty acid composition of the oil were studied by (Issara, Zzaman, & Yang, 2014). The physicochemical characterization of rambutan kernel oil was studied by Kheiri & Som (1979). Rambutan fat and oil extracts in the present study contained mainly palmitic (7.39%–10.33%), stearic (12.21%–16.58%), arachidic (12.34%–16.22%), and behenic (6.53%–8.91%) as SFA, oleic (50.17%–52.18%) as MUFA, and linolenic (2.02%–3.04%) as PUFA. The oral administration of rambutan seed fat and oil extracts at dose levels up to 5000 mg/kg body weight did not affect body weight. Table 33.1 showed the average of fatty acid composition form various studies.

TABLE 33.1 Main Fatty Acids in Rambutan (*N. lappaecum* Linn.) Seed Fat

Fatty acid		Average (%)
Palmitic acid	(C16:0)	8.1
Palmitoleic acid	(C16:1)	1.5
Stearic acid	(C18:0)	14.5
Oleic acid	(C18:1)	51.1
Arachidic acid	(C20:0)	15.4
Gondoic acid	(C20:1)	4.3
Behenic acid	(C22:0)	2.7
Nonidentify	—	1.2
SFA[a]		40.7
USFA[a]		56.9

[a]SFA, Saturated fatty acids; USFA, unsaturated fatty acids.

BIOCHEMICAL

Fatty acid composition of N. lappaceum seed fat and oil extracts: The major components in fat and oil extracts from *N. lappaceum* seeds by SC-CO_2 at 35 MPa and 45°C were palmitic acid (C16:0), stearic acid (C18:0), arachidic acid (C20:0) and behenic acid (C22:0) as SFA, oleic acid (C18:1 *cis*-9) as MUFA, and linolenic acid (C18:3) as PUFA. Oleic acid constituted over 50% of total fatty acids in fat and oil extracts. There were statistically significant ($P < 0.05$) differences in some fatty acids between fat and oil extracts. The differences may be due to the different solubility of individual fatty acids in SC-CO_2. These values were only in partial agreement with previous studies (Solís-Fuentes, Camey-Ortíz, del Rosario Hernández-Medel, Pérez-Mendoza, & Durán-de-Bazúa, 2010; Sirisompong et al., 2011; Manaf, Marikkar, Long, & Ghazali, 2013; Azhari, 2015), who reported higher ($P < 0.05$) proportions of C20:0 and C18:3 fatty acids, and lower ($P < 0.05$) proportions of C16:0, C18:0, C22:0, and C18:1 *cis*-9 fatty acids. This was attributed to different extraction methods using different solvents that may affect the proportion of individual fatty acids when the results of different trials were compared.

Acute Oral Toxicity

In the 15-day observation period during the acute oral toxicity study, the animals did not show any signs of mortality or toxicity. There were no significant differences ($P > 0.05$) in mean body weights following administration of rambutan seed fat and oil extracts at doses up to 5000 mg/kg body weight at the 15-day study between the control and treatment groups in male and female rats.

Macroscopic examination of all the animals at the end of the study did not find any gross pathological abnormalities. From this study, the oral LD50 of rambutan seed fat and oil extracts in Wistar rats was greater than 5000 mg/kg body weight. The fat and oil extracts were considered to be nontoxic.

Dermal Irritation Study

The application of *N. lappaceum* seed fat extract to rabbit skin caused very slight erythema, barely perceptible (score 1) in one of three total and no edema (score 0) in all animals in the initial 24 h. The occurrence of erythema responses following the application of fat and oil extracts was resolved by 48 h. After completion of the 72 h observation period, there were no signs of skin irritation or dermal abnormalities observed in all animals.

Dermal Toxicity Study

There were no significant differences ($P > 0.05$) in the body weights in either sex when the control and treatment groups were compared (Eiamwat et al., 2014). In male rats treated with fat and oil extracts, mean values of body weights were generally similar to control group animals in the 15-day study. Similarly, in the female group, the mean body weights during the course of study were comparable to the respective controls. These results suggest that the application of the fat and oil extracts at dose level of 2000 mg/kg body weight to rats did not affect body weights. Hence, the acute dermal LD50 in Wistar rats was greater than 2000 mg/kg body weight.

Fat and oil extracts of rambutan (*N. lappaceum*) seeds are potential material inputs for industrial food and cosmetic applications. In the study of dermal irritation, there were no irritant effects on skin with rambutan seed fat and oil extracts. Examining dermal toxicity with the application of fat and oil extracts at dose level of 2000 mg/kg body weight to rats did not result in a significant change in body weight. There were no treatment-related deaths or any toxic effects in all animals until the terminal necropsy. Findings of the present study are not possible to compare with data of other investigators because there are no reports on toxicity studies on rambutan seed fat and oil extracts. In summary, these results suggest that rambutan seed fat and oil extracts are practically nontoxic. This data provides a basis of toxicity information for the use of rambutan seed fat and oil extracts.

Antioxidant and Antibacterial Activities

Plants contain a large variety of substances possessing antioxidant activity including natural antioxidant compound, such as polyphenols, carotene, tocopherol, vitamin C, vitamin E, xanthophylls and tannins (Thitilertdecha, Teerawutgulrag, & Rakariyatham, 2008; Issara et al., 2014; Zzaman, Issara, Febrianto,

& Yang, 2014), and fruit/vegetable that have specific bioactive compounds had concern much attention due to health benefit effect (Febrianto et al., 2012). Moreover, these compounds are able to protect the oxidative damage in human body's cell and tissue. The phenolics compound can be found in all parts of the plant for sources of natural antioxidants (Chanwitheesuk, Teerawutgulrag, & Rakariyatham, 2005). Thitilertdecha et al. (2008) noted that rambutan (*N. lappaceum* Linn.) peel and seed parts showed a high antioxidant and antibacterial activities, and more potential activities were found in the peel extracts more than the seed extracts by using methanol solvent for extraction of antioxidant and antibacterial substances which is the best solvent for extraction when compare with other solvents. It is as a result to providing high extraction yields and also strong antioxidant and antibacterial activities. The natural antioxidant in lipid-containing product and lipid-based product, such as oil, fat, margarine, butter, etc. is considered insufficient and/or had been removed on the purification process because it is considered as impurity which would adversely affect in subsequent use (Febrianto et al., 2012). In addition, Febrianto et al. (2012) studied the effect of fermentation time and roasting process in the rambutan seed fat and found that they could improve the antioxidant activity and total phenolics compound of rambutan seed fat. Higher antioxidant activity, which is resulted from fermentation process could be enhanced further by applying roasting process also. However, the appropriate fermentation process should not longer than 6 days which resulting to efficiently increase the total phenolic compounds of rambutan seed fat.

PHYSIOCHEMICAL

Fats containing highly saturated or long-chain fatty acids are commonly have a higher melting point than unsaturated or short-chain fatty acids. Unsaturated fatty acids have different isomeric forms that have different melting points. They naturally expose in the *cis*-form, but can be converted to the *trans*-form during partial hydrogenation processing (Dziezak, 1989). Crystalline forms in which fats may exist categorized as alpha, beta, and beta-prime. Weiss, Tabor, and Carnevale (1983) classified a number of fats according to their crystallizing nature due to have a reported that rambutan seed fat have some physical properties, such as characteristic of melting in the room temperature like a cocoa butter, but it was found that cocoa butter have a temperature range of melting point and crystallization occur narrow than rambutan seed fat. Beside, cocoa butter does not contain many triglycerides and majority composed as plamito oleosterin (Pérez Martínez & Lorenzo Parra, 2007). Ghotra, Dyal, and Narine (2002) and McClements, Decker, and Weiss (2007) reported that rambutan seed have crystalline forms β and β' in the amounts of 84.70 and 15.30%, respectively. It was shown that the rambutan seed fat had crystal stability. In general the crystallization of rambutan seed fat is usually analyzed by using differential scanning calorimetry (DSC). Solís-Fuentes et al. (2010) described the crystallization curve and melting cure of rambutan seed fat. The melting point of rambutan seed fat also observed by the last peak of heating curve ($\sim 45^\circ$C) showed higher than the cocoa butter which normally useful in the chocolate manufacture (Thitilertdecha, Teerawutgulrag, Kilburn, & Rakariyatham, 2010).

SUGGESTED USAGES

The physical properties and sensory, and consumer perception are important factors influencing in the confectionary products. Although rambutan seed fat is similar to those of cocoa butter and can be used to substitute cocoa butter in chocolate; the use of rambutan seed fat in food and other industry branches will need to be approved by regulatory authorities in each country. However, for the effort the alternative to fine other fat to substitute cocoa butter in chocolate product is highly consider, which resulting to the final product quality. Further studies require integration of texture and aroma profile in rambutan seed fat and do the sensory evaluation to study consumer acceptance in the chocolate product that produces by rambutan seed fat.

This study shows that the extract of the peels of rambutan can be used as an alternative source of antioxidants for the stabilization of sunflower oil. It can be observed that the subfraction of the rambutan extract works more effectively than the crude extract. For the crude extract at 300 ppm, comparable antioxidant activities with tocopherol were shown for only up to 1 year of storage. In order to improve the protective effect of the crude extract, higher concentrations of the sample can be incorporated into sunflower oil.

REFERENCES

Augustin, M., & Chua, B. (1988). Composition of rambutan seeds. *Pertanika, 11*(2), 211–215.

Azhari, F. (2015). Hubungan Kadar Timbal Pada Urin dan Karakteristik Individu Dengan Kejadian Anemia Pada Pedagang Wanita di Terminal Bus Kampung Rambutan Jakarta Timur Tahun 2014.

Chanwitheesuk, A., Teerawutgulrag, A., & Rakariyatham, N. (2005). Screening of antioxidant activity and antioxidant compounds of some edible plants of Thailand. *Food Chemistry, 92*(3), 491–497.

Dziezak, J. (1989). Spices. *Journal of Food Technology, 43*, 102–116.

Eiamwat, J., Pengprecha, P., Tiensong, B., Sematong, T., Reungpatthanaphong, S., Natpinit, P., ..., Hankhunthod, N. (2014). Fatty acid composition and acute oral toxicity of rambutan (*Nephelium lappaceum*) seed fat and oil extracted with SC-CO$_2$. *110 Determination of D-saccharic acid-1,4-lactone (DSL) in fermentation tea (Kombucha) by capillary electrophoresis*, 312.

Febrianto, F., Hidayat, W., Bakar, E. S., Kwon, G. -J., Kwon, J. -H., Hong, S. -I., & Kim, N. -H. (2012). Properties of oriented strand board made from Betung bamboo (*Dendrocalamus asper* (Schultes. f) Backer ex Heyne). *Wood science and technology, 46*(1–3), 53–62.

Ghotra, B. S., Dyal, S. D., & Narine, S. S. (2002). Lipid shortenings: a review. *Food Research International, 35*(10), 1015–1048.

Harahap, S. N., Ramli, N., Vafaei, N., & Said, M. (2012). Physicochemical and nutritional composition of rambutan anak sekolah (*Nephelium lappaceum* L.) seed and seed oil. *Pakistan Journal of Nutrition, 11*(11), 1073.

Issara, U., Zzaman, W., & Yang, T. (2014). Rambutan seed fat as a potential source of cocoa butter substitute in confectionary product. *International Food Research Journal, 21*(1), 25–31.

Kheiri, M., & bin Mohammad Som, M.N. (1979). Physico-chemical characteristics of rambutan [*Nephelium lappaceum*] kernel fat [Peninsular Malaysia]. *Agricultural Product Utilisation Report (Malaysia). 186.*

Kheiri, M., & Som, M. N. M. (1979). *Physico-chemical characteristics of rambutan kernel fat.* Agricultural Product Utilization, Report No. 186, MARDI, Serdang, Selangor, Malaysia.

Lannes, S. C. S., Medeiros, M. L., & Gioielli, L. A. (2003). Physical interactions between cupuassu and cocoa fats. *Grasas y Aceites, 54,* 253–258.

Manaf, Y. N. A., Marikkar, J. M. N., Long, K., & Ghazali, H. M. (2013). Physico-chemical characterisation of the fat from red-skin rambutan (*Nephellium lappaceum* L.) seed. *Journal of Oleo Science, 62*(6), 335–343.

Marisa, M. W. (2006). Ascorbic acid and mineral composition of longan (Dimocarpus longan), lychee (Litchi chinensis) and rambutan (Nephelium lappaceum) cultivars grown in Hawaii. *Journal of Food Composition and Analysis, 19,* 655–663.

McClements, D. J., Decker, E. A., & Weiss, J. (2007). Emulsion-based delivery systems for lipophilic bioactive components. *Journal of Food Science, 72*(8), R109–R124.

Morton, J. F. (1987). Rambutan. In Julia F. Morton (Ed.), *Fruits of warm climates* (pp. 262–265). Miami, FL: Julia F. Morton.

Palanisamy, U., Cheng, H. M., Masilamani, T., Subramaniam, T., Ling, L. T., & Radhakrishnan, A. K. (2008). Rind of the rambutan, *Nephelium lappaceum,* a potential source of natural antioxidants. *Food Chemistry, 109*(1), 54–63.

Pérez Martínez, V. T., & Lorenzo Parra, Z. (2007). El impacto del déficit mental en el ámbito familiar. *Revista Cubana de Medicina General Integral, 23*(3), 1–8.

Sirisompong, W., Jirapakkul, W., & Klinkesorn, U. (2011). Response surface optimization and characteristics of rambutan (*Nephelium lappaceum* L.) kernel fat by hexane extraction. *LWT—Food Science and Technology, 44*(9), 1946–1951.

Solís-Fuentes, J. A., & Durán-de-Bazúa, M. C. (2004). Mango seed uses: thermal behaviour of mango seed almond fat and its mixtures with cocoa butter. *Bioresource Technology, 92,* 71–78.

Solís-Fuentes, J. A., Camey-Ortíz, G., del Rosario Hernández-Medel, M., Pérez-Mendoza, F., & Durán-de-Bazúa, C. (2010). Composition, phase behavior and thermal stability of natural edible fat from rambutan (*Nephelium lappaceum* L.) seed. *Bioresource Technology, 101*(2), 799–803.

Thitilertdecha, N., Teerawutgulrag, A., & Rakariyatham, N. (2008). Antioxidant and antibacterial activities of *Nephelium lappaceum* L. extracts. *LWT—Food Science and Technology, 41*(10), 2029–2035.

Thitilertdecha, N., Teerawutgulrag, A., Kilburn, J. D., & Rakariyatham, N. (2010). Identification of major phenolic compounds from *Nephelium lappaceum* L. and their antioxidant activities. *Molecules, 15*(3), 1453–1465.

Tindall, C. (1994). Issues of evaluation. *Qualitative methods in psychology: a research guide.* Rome, Italy: Food and Agricultural Organization (pp. 142–159).

Weiss, J., Tabor, M., & Carnevale, G. (1983). The Painlevé property for partial differential equations. *Journal of Mathematical Physics, 24*(3), 522–526.

Zee, F. (1993). Rambutan and pili nuts: potential crops for Hawai. In J. Janick, & J. E. Simon (Eds.), *New Crops* (pp. 461–465). New York: Wiley and Sons Inc.

Zzaman, W., Issara, U., Febrianto, N., & Yang, T. (2014). Fatty acid composition, rheological properties and crystal formation of rambutan fat and cocoa butter. *International Food Research Journal, 21*(3), 983–987.

Chapter 34

Mangifera indica Mango Seed Kernel Oil

INTRODUCTION

Mango (*Mangifera indica* L.) seed kernel oil could be extracted using Soxhlet extraction with petroleum ether, ethanol, and hexane. Oil extracted with hexane has better overall quality. Its acid, peroxide, iodine saponification values, and phenolic content were 0.10 mg KOH/g oil, 8.72 mg/g oil, 38.50 mg/100 g oil, 207.5 mg KOH/g oil, and 98.7 mg/g, respectively (Kittiphoom & Sutasinee, 2013). The main fatty acids found in the mango seed kernel oil were steric acid and oleic acid. The results of studies suggested that mango seed kernel oil is a good source of the unsaturated fatty acid, phenolic compounds, and has the potential to be used as nutrient rich food oil or as ingredients for functional or enriched foods (Dhingra & Kapoor, 1985; Nzikou et al., 2010). The acidity, iodine, peroxide, and saponification values are the major characterization parameters for oil quality. The mango kernel oils were very light yellow in color (Fig. 34.1). The analysis showed that different extraction methods influenced the physicochemical properties of the oils. The physicochemical properties of mango kernel oil (MKO) are shown in Table 34.1.

The main fatty acids found in MKO are about 45% oleic acid and 38% steric acid (Nzikou et al., 2010). Oleic acid is an 18-carbon monounsaturated fatty acid, essential in human nutrition, and helps reducing triglycerides, LDL-cholesterol, total cholesterol, and glycemic index. Also, the increase in stability over oxidation of vegetable oil is attributed to oleic acid (Abdulkarim, Long, Lai, Muhammad, & Ghazali, 2007). Stearic acid, a long C18 straight-chain saturated fatty acid, has been found to bind and plasticize composites (Netravali, 2003), human serum albumin (Bhattacharya, Grüne, & Curry, 2000) and helical sites in biomolecules (Villa, Lopez-Forment, & Prescott, 1998). The fatty acid composition and phenolic contents, of mango seed kernel oil were examined in many studies (Kittiphoom & Sutasinee, 2013) (Table 34.2).

Many studies showed that mango kernels contain phenolic compounds and stable fat rich in saturated fatty acids that makes it suitable source of natural antioxidants (Parmar & Sharma, 1984; Dinesh, Boghra, & Sharma, 2000; Puravankara, Boghra, & Sharma, 2000; Núñez-Sellés, 2005). Both phenolic compounds and mango kernel crude oil were recommended as good natural

Unconventional Oilseeds and Oil Sources. http://dx.doi.org/10.1016/B978-0-12-809435-8.00034-2

FIGURE 34.1 *Mangifera indica* tree, fruits, seed, and kernels.

antioxidants due to delaying oxidation in different edible oils (Abdalla, Darwish, Ayad, & El-Hamahmy, 2007).

In a study, Youssef (1999) indicated that adding 1% of crude oil extracted from mango kernel exhibited antioxidant potency similar to that of 200 ppm of BHT against oxidation of sunflower oil. Besides the phospholipids and phenolic compounds, other factors, such as tocopherols and carotenoids may also be involved in the effectiveness of mango seed kernel powder in extending the shelf life of cheese and ghee (Parmar & Sharma, 1986; Dinesh et al., 2000; Puravankara et al., 2000).

APPLICATIONS AND USES OF MANGO KERNEL OIL

MKO has been used in the cosmetics industry as an ingredient in soaps, shampoos, and lotions because it is a good source of phenolic compounds (Soong and Barlow, 2004) including microelements like selenium, copper, and zinc (Schieber & Carle, 2005). In addition, the extract of mango seed kernel exhibited the highest degree of free radical scavenging and tyrosinase-inhibition activities compared with methyl gallate and phenolic compounds from the mango seed kernel and methyl gallate in emulsion affected the stability of the cosmetic emulsion systems. Fatty acids are important as nutritional substances and metabolites in living organisms. Many kinds of fatty acids play an important role in the regulation of a variety of physiological and biological functions (Zhao & Zhou, 2007).

TABLE 34.1 Physicochemical Property of Mango Seed Kernel Oils Extracted With Different Solvents (Means ± SD)

Physicochemical property	Ethanol extract	Hexane extract	Petroleum ether extract	Nzikou et al. (2010)	Khammuang and Sarnthima (2011)*
Total oil yield (g/g)	6.96 ± 0.13[a]	8.46 ± 0.01[b]	8.04 ± 0.07[b]	13.0	2.53
Acid value (mg KOH/g oil)	27.55 ± 0.55[b]	0.10 ± 0.012[a]	0.15 ± 0.083[a]	5.35	NR
Peroxide value (mg/g oil)	26.35 ± 2.1[b]	8.72 ± 3.4[a]	8.82 ± 2.8[a]	NR	NR
Iodine value (mg/100 g oil)	59.17 ± 2.3[b]	38.50 ± 3.9[a]	37.25 ± 3.0[a]	39.5	NR
Saponification value (mg KOH/g oil)	206.0 ± 13.8[b]	207.5 ± 14.2[b]	190.2 ± 12.8[a]	207.5	NR
Total phenolic content (mg GAE/g)	53.5 ± 7.5[a]	98.7 ± 8.8[c]	77.0 ± 7.4[b]	117	118.1 ± 0.2

NR, Not reported.
*The Keaw variety.
[a,b,c]Means within the same row not followed by the same letter differ significantly (P < 0.05). Each experiment was performed in triplicates.

TABLE 34.2 Fatty Acid Composition (% w/w) of Oils Extracted With Different Solvents (Means SD)

Fatty acid	Ethanol extract	Hexane extract	Petroleum ether extract	Nzikou et al. (2010)	Dhingra and Kapoor (1985)
Palmitic acid (C16:0)	8.50 ± 0.26[a]	8.97 ± 0.15[b]	8.73 ± 0.15[ab]	6.48	7.18
Steric acid (C18:0)	38.50 ± 0.45[b]	37.37 ± 0.23[a]	37.7 ± 0.16[a]	37.94	38.90
Oleic acid (C18:1)	43.45 ± 0.46[a]	43.77 ± 0.37[a]	44.75 ± 0.40[b]	45.76	42.60
Linoleic acid (C18:2)	6.48 ± 0.53[b]	6.78 ± 0.25[b]	5.67 ± 0.31[a]	7.45	5.70
Linolenic acid (C18:3)	0.49 ± 0.009[a]	0.79 ± 0.18[b]	0.59 ± 0.19[a]	2.37	5.3
Eicosanoic acid (C20:1)	2.51 ± 0.18[b]	2.19 ± 0.19[a]	2.48 ± 0.16[b]	—	—

[a,b]Means within the same row not followed by the same letter differ significantly ($P < 0.05$). Each experiment was performed in triplicates.

Three types of nonconventional oils were extracted, analyzed, and tested for toxicity. Date palm kernel oil (DPKO), MKO, and rambutan seed oil (RSO) were extracted and tested for toxicity by Mirghani, Jaswir, M Salih, Hashim, and Che Man (2009). Three varieties of mango were found to contain about 10% oil in average. The phenolic compounds in DPKO, MKO, and RSO were 0.98, 0.88, and 0.78 mg/mL gallic acid equivalent, respectively. Oils were analyzed for their fatty acid composition and they are rich in oleic acid C18:1 and showed the presence of (dodecanoic acid) lauric acid C12:0, which reported to appear some antimicrobial activities. All extracted oils, DPKO, MKO, and RSO showed no toxic effect using prime shrimp bioassay. Since these oils are stable, melt at skin temperature, have good lubricity and are great source of essential fatty acids; they could be used as highly moisturizing, cleansing, and nourishing oils because of high oleic acid content. They are ideal for use in such halal cosmetics, such as skin care and massage, haircare, soap, and shampoo products

Mango seed can help you to get rid of dandruff. Take mango seed butter and apply it on your hair for luster and strength. You can also mix it with mustard oil and leave it out in the sun for few days. Application of this mixture can control alopecia, hair loss, early greying, and dandruff.

REFERENCES

Abdalla, A. E., Darwish, S. M., Ayad, E. H., & El-Hamahmy, R. M. (2007). Egyptian mango by-product 2: antioxidant and antimicrobial activities of extract and oil from mango seed kernel. *Food Chemistry, 103*(4), 1141–1152.

Abdulkarim, S., Long, K., Lai, O., Muhammad, S., & Ghazali, H. (2007). Frying quality and stability of high-oleic *Moringa oleifera* seed oil in comparison with other vegetable oils. *Food Chemistry, 105*(4), 1382–1389.

Bhattacharya, A. A., Grüne, T., & Curry, S. (2000). Crystallographic analysis reveals common modes of binding of medium and long-chain fatty acids to human serum albumin. *Journal of Molecular Biology, 303*(5), 721–732.

Dhingra, S., & Kapoor, A. C. (1985). Nutritive value of mango seed kernel. *Journal of the Science of Food and Agriculture, 36*(8), 752–756.

Dinesh, P., Boghra, V., & Sharma, R. (2000). Effect of antioxidant principles isolated from mango (*Mangifera indica* L.) seed kernels on oxidative stability of ghee (butter fat). *Journal of Food Science and Technology, 37*(1), 6–10.

Khammuang, S., & Sarnthima, R. (2011). Antioxidant and antibacterial activities of selected varieties of thai mango seed extract. *Pakistan Journal of Pharmaceutical Sciences, 24*(1), 37–42.

Kittiphoom, S., & Sutasinee, S. (2013). Mango seed kernel oil and its physicochemical properties. *International Food Research Journal, 20*(3), 1145–1149.

Mirghani, M. E. S., Jaswir, I., M., Salih, H., Hashim, Y. Z. H.-Y., & Che Man, Y. (2009). Special oils for halal cosmetics. *3rd IMT-GT international symposium on halal science and management KLIA*. Sepang, Selangor, Malaysia: Universiti Putra.

Netravali, A. (2003). Green Composites from Cellulose Fabrics & Soy Protein Resins. National Textile Center Project F01-CR01 Annual Report. Cornell University. http://www.human.cornell.edu/units/txa/research/ntc/F01-CR01-A3.pdf

Núñez-Sellés, A. J. (2005). Antioxidant therapy: myth or reality? *Journal of the Brazilian Chemical society, 16*(4), 699–710.

Nzikou, J., Kimbonguila, A., Matos, L., Loumouamou, B., Pambou-Tobi, N., Ndangui, C., & Desobry, S. (2010). Extraction and characteristics of seed kernel oil from mango (*Mangifera indica*). *Research Journal of Environmental and Earth Sciences, 2*(1), 31–35.

Parmar, S., & Sharma, R. (1984). Mango seed kernels: a review on chemical composition and edible uses. *Indian Food Packer, 38*(5), 40.

Parmar, S., & Sharma, R. (1986). Use of mango seed kernels in enhancing the oxidative stability of ghee. *Asian Journal Dairy Research, 5*, 91.

Puravankara, D., Boghra, V., & Sharma, R. S. (2000). Effect of antioxidant principles isolated from mango (*Mangifera indica* L.) seed kernels on oxidative stability of buffalo ghee (butter-fat). *Journal of the Science of Food and Agriculture, 80*(4), 522–526.

Schieber, A., & Carle, R. (2005). Occurrence of carotenoid *cis*-isomers in food: technological, analytical, and nutritional implications. *Trends in Food Science & Technology, 16*(9), 416–422.

Soong, Y. Y., & Barlow, P. J. (2006). Quantification of gallic acid and ellagic acid from longan (Dimocarpus longan Lour.) seed and mango (Mangifera indica L.) kernel and their effects on antioxidant activity. *Food Chemistry, 97*(3), 524–530.

Villa, C., Lopez-Forment, W., & Prescott, C. V. (1998). Not all sigmodontine rodents in the sugarcane fields in coastal Veracruz, Mexico, are pests.

Youssef, A. (1999). Utilization of the seeds of mango processing wastes as a secondary source of oil and protein. *Alexandria Journal of Agricultural Research, 44*(3), 149–166.

Zhao, J. H., & Zhou, M. F. (2007). Neoproterozoic adakitic plutons and arc magmatism along the western margin of the Yangtze block, South China. *The Journal of Geology, 115*(6), 675–689.

Chapter 35

Moringa oleifera Seed Oil

INTRODUCTION

Moringa oleifera Lam belongs to a single genus family Moringaceae, which has fourteen species (Morton, 1991; Anwar & Bhanger, 2003). It is commonly called Ben or Behen oil tree, and drumstick tree in English language. The three grows rapidly in most regions and climatic conditions tropical areas. It is known as Rawag tree in Sudan, as Zogeli, Gawara, or Habiwal among the Hausa speaking people of Nigeria, "Okwe oyibo" in Igbo, "Ewe Igbale" in Yoruba. *M. oleifera* is grown and widely cultivated in many countries in tropical Africa (Anjorin, Ikokoh, & Okolo, 2010) as well as in many Asian countries (Ramachandran, Peter, & Gopalakrishnan, 1980). The *M. oleifera* is rapidly growing tree even in poor soil and is little affected by drought (Sengupta & Gupta, 1970) and can be easily grown in poor third world countries. The compositional and nutritional attributes of seeds were described by (Oliveira, Silveira, Vasconcelos, Cavada, & Moreira, 1999). The production of useful oil from its seeds could be of economic benefit to the native population of the areas where the tree is cultivated (Morton, 1991). The tree is rather slender, with drooping branches that grow to approximately 10 m in height. In cultivation, it is often cut back annually to 1–2 m and allowed to regrow so the pods and leaves remain within arm's reach. Almost all parts of the tree can be utilized for several purposes that help sustain living conditions. The leaves, fruits, flowers, and immature pods of this tree are edible and they form a part of traditional diets in many countries of the tropics and subtropics (Siddhuraju & Becker, 2003) that makes *M. oleifera* is esteemed as a versatile plant due to its multiple substantial nutritional benefits, it also has a great potential as a medicinal plant. The leaves are a good source of protein, vitamin A, B and C and minerals, such as calcium and iron. The flowers, leaves and roots are used for the treatment of ascites, rheumatism and venomous bites and as cardiac and circulatory stimulants in folk remedies (Dahot, 1988; Anhwange, Ajibola, & Oniye, 2004).

The seeds extracts also have antimicrobial activity (Idris et al., 2013) and are utilized for waste-water treatment (Ali, Muyibi, Salleh, Alam, & Salleh, 2010). In some developing countries, the powdered seeds of *M. oleifera* are traditionally utilized as a natural coagulant for water purification because of their strong coagulating properties for sedimentation of suspended undesired particles (Kalogo, Rosillon, Hammes, & Verstraete, 2000; Anwar & Rashid, 2007) (Fig. 35.1).

Unconventional Oilseeds and Oil Sources. http://dx.doi.org/10.1016/B978-0-12-809435-8.00035-4

FIGURE 35.1 *Moringa oleifera* **fruits and seeds.**

Various parts of the plant have been shown to be useful, such as the roots have been experimentally shown to have antiinflammatory action (Ndiaye et al., 2001; Jamil, Shahid, Khan, & Ashraf, 2007; Talreja, 2010; Mishra et al., 2011), the leaves, stem bark, and seeds have been reported to have therapeutic properties (Anwar & Rashid, 2007). The seed powder of *M. oleifera* works as a natural coagulant, which clarifies very turbid water (Mangale, Chonde, Jadhav, & Raut, 2012).

MORINGA SEED OIL

Furthermore, the seeds of *M. oleifera* had been found to be a potential new source of vegetable oil, especially with the advent of the need for edible oils, oleochemicals, lipids, and lipid-derived fuels (biodiesel) all over the world (Anwar & Rashid, 2007). The oil yield of *M. oleifera* seeds varied from 27 up to 42% as reported by various reports from various parts of the globe (Fuglie, 1999; Tsaknis, Lalas, Gergis, Dourtoglou, & Spiliotis, 1999; Anwar & Bhanger, 2003; Abdulkarim, Long, Lai, Muhammad, & Ghazali, 2005). The oil is light golden yellow in color, liquid at room temperature with palatable flavor. The plant has been under discovered plants as a source of edible oil and there is lack information on physicochemical properties of the seed oil; however, the results of the quantity characterization of the Moringa seed oil are recorded in Table 35.1.

TABLE 35.1 Characterization and Physicochemical Properties of Moringa Seed Oil

Parameter	*M. oleifera* oil	Comments
Oil yield	25%–44%	Varies due to many variables
Specific gravity	0.91–1.1827	
Density at 24°C	0.9037	Cold press
Viscosity (mPa·s)	103	Cold press
Refractive index (25°C)	1.4713–1.4725	
Melting point (°C)	28.0	
Saponification value (mg KOH/g)	171.9–191	
Peroxide value (mmol/kg) (mEq/kg)	8.1–15	Fresh crude oil
Iodine value (I_2/100 g oil)	66–85.3	
Acid value (mg/KOH/g)	3.8–5.04	Fresh crude oil
Free fatty acid (% as oleic acid)	0.5–2.51	Differs for fresh, de-gummed, refined, etc.
Smoke point (°C)	190–201	Cold press

This percentage yield is considerable compared to the other plants known as sources of edible oil. The saponification value (SV) ranged between 172 and 180 indicating that it has long carbon chain and can be used for soap and cosmetics production. The free fatty acid is 2.51, which is an indication that its rancidity is low and it should have a long shelf life (Salaheldeen, Aroua, Mariod, Cheng, & Abdelrahman, 2014).

However, the acid value was higher than the Codex standard value for virgin vegetable oils. The peroxide value (PV) was 15.96 mEq/kg. The value was higher than the Codex standard value (10 mEq/kg) for refined vegetable oil and lower than the maximum value (20 mEq/kg) allowed for unrefined olive oil (Abiodun, Adegbite, & Omolola, 2012). The iodine value (IV) was low compared to other seed oils that signify low degree of unsaturation and the lesser the liability of the oil to become rancid by oxidation.

The SV (172–191 mg/g) is ranged between low and high low compared to the values recorded (Abiodun et al., 2012; Ogunsina et al., 2011), soybean, peanut, and cotton seed oil (Codex, 1993). The lower the SVs of oil the lower the lauric acid content of that oil. The lauric acid content and the SV of an oil serve as important parameters in determining the suitability of an oil in soap making (Asuquo, Anusiem, & Etim, 2010; Abiodun et al., 2012). The refractive index at 25°C (1.47) and specific gravity (0.91) were slightly lower than the values obtained for shea butter oil. The refractive index of Moringa oil was slightly higher than the value reported by Anwar and Rashid (2007) at 40°C. Both the refractive index and specific gravity of Moringa seed oil were similar to that of palm oil and groundnut oil reported by (Abiodun et al., 2012). The smoke point was 138.00°C. This value was low when compared with canola oil (220–230°C) (Copp et al., 2005; Przybylski, Mag, Eskin, & McDonald, 2005). This provides a useful characterization of its suitability for frying.

Table 35.2 showed that the total unsaturated fatty acids in Moringa seed oil extracted by various techniques were found to be higher than 80%; the major fatty acid was oleic (C18:1) at concentrations of 75.39% cold press (CP), 73.60% solvent extraction using n-hexane (H), and oil produced with solvent extraction with chloroform/methanol (50:50) was 73.91% (CM), followed by gadoleic (C20:1) [2.54% (CP), 2.40% (H), and 2.46% (CM)]. Behenic acid (C22:0) with n-hexane; CM, oil produced with solvent extraction with chloroform/methanol (50:50).

There was no statistical difference (at the level of 95%) in the fatty acid compositions of the oils produced from the three different ways of extraction. On the basis of the results obtained, the fatty acid composition of *M. oleifera* seed oil showed that it falls in the oleic acid oil category (Sonntag, 1979; Tsaknis et al., 1999). The *M. oleifera* seed oil (from Kenya) had about the same content of C18:1 but much less C18:2 than olive oil. Moringa oil was less unsaturated than the olive oil. Results are shown in Table 35.2.

The PV of *M. oleifera* seed oil fell in the range selected as satisfactory. The oil extracted using the CP had lower PV followed by the one extracted using a solvent mixture of chloroform: methanol (1:1) and n-hexane. This result was not expected, because cold pressure extracted oil remained in contact with air for a longer time and high PV explanation can be given on that product (Lalas and Tsaknis 2002b).

The induction period of Moringa oil was more than 3 times longer than that of olive oil before degumming and up to 2 times longer after degumming. The unstable oxidation behavior of the three oils could not be related to the ratio of tocopherol/C18:2 (tocopherol/C18:2 ratios of Moringa oil were 44.16, 54.40, and 31.23 for the cold pressure, n-hexane and chloroform: methanol, respectively). The oxidative stability of olive oil is related to some extent to the presence of α-tocopherol (Kiritsakis, 1988).

M. oleifera seed oils could be partly attributed to the resistance to oxidation (Tsaknis et al., 1999) due to the content of α-, β- and especially δ-tocopherol, which are significantly higher quantity than in virgin olive oil, (Table 35.3) (Lalas and Tsaknis 2002a). In addition, olive oil contained linoleic and linolenic acid in higher amounts compared to their quantity in *M. oleifera* seed oil and more easily underwent oxidation and degradation than C18:1. Furthermore, the higher oxidative stability of *M. oleifera* seed oil than olive oil should be attributed to other constituents of the nonglyceride fraction of the oil, which possess antioxidant properties (Tsaknis, Lalas, Gergis, & Spiliotis, 1998).

TABLE 35.2 Fatty Acid Composition of the Degummed Oils[a]

Determination	GLC (%)			
Fatty acid	CP	H	CM	Olive oil
C8:0	0.03 ± 0.01	0.03 ± 0.01	0.02 ± 0.01	ND
C14:0	0.11 ± 0.07	0.11 ± 0.07	0.11 ± 0.05	<0.01
C16:0	5.73 ± 0.311	6.04 ± 0.406	5.81 ± 0.356	11.2 ± 0.663
C16:1 *cis* ω9	0.10 ± 0.05	0.11 ± 0.06	0.10 ± 0.06	0.60 ± 0.09
C16:1 *cis* ω7	1.32 ± 0.89	1.46 ± 0.88	1.44 ± 0.91	ND
C17:0	0.09 ± 0.02	0.09 ± 0.03	0.09 ± 0.03	0.1 ± 0.01
C18:0	3.83 ± 0.16	4.14 ± 0.19	4.00 ± 0.20	2.80 ± 0.11
C18:1 *cis* ω9	75.39 ± 0.75	73.60 ± 0.77	73.91 ± 0.79	72.21 ± 0.78
C18:2 *cis* (9,12)	0.72 ± 0.36	0.73 ± 0.41	0.71 ± 0.39	4.2 ± 0.49
C18:3 *cis* (9,12,15)	0.20 ± 0.07	0.22 ± 0.08	0.20 ± 0.04	0.5 ± 0.10
C20:0	2.52 ± 0.65	2.76 ± 0.55	2.70 ± 0.49	0.6 ± 0.29
C20:1	2.54 ± 0.44	2.40 ± 0.38	2.46 ± 0.39	0.2 ± 0.06
C22:0	5.83 ± 0.56	6.73 ± 0.29	6.38 ± 0.36	<0.01
C22:1 *cis*	0.15 ± 0.06	0.14 ± 0.07	0.14 ± 0.08	ND
C26:0	0.96 ± 0.11	1.08 ± 0.12	1.06 ± 0.13	ND
Total saturated	19.1	20.98	20.17	14.72

[a]*Results are the men of three replicates standard deviation (SD); CP, cold press; H, n-hexane; CM, chloroform/methanol; ND, not detected.*
Source: Data taken from Tsaknis, J., Lalas, S., Gergis, V., Dourtoglou, V., & Spiliotis, V. (1999). Characterization of *Moringa oleifera* variety Mbololo seed oil of Kenya. *Journal of Agricultural and Food Chemistry, 47*(11), 4495–4499; Lalas and Tsaknis (2002b).

TOTAL PHENOLIC, FLAVONOID, AND ANTIOXIDANT CAPACITY

The high oil yield of the seed of *M. oleifera* from total phenolic, flavonoid, and antioxidant capacity of *M. oleifera* oil and palm oil (Table 35.4) showed that there was significant difference ($P < 0.05$) in these characteristics of range of both *M. oleifera* oil and palm oil with palm oil having higher values. Siddhuraju and Becker (2003) had studied the antioxidant properties of various solvent extracts of total phenolic constituents from three different agroclimatic origins of *M. oleifera* leaves.

Total phenol (mg gallic acid Eq/g), total flavonoids (mg Rutin Eq/g), and total antioxidant capacity (mg ascorbic acid Eq/g) were >40, ≥18, >37 for *M. oleifera* oil, respectively. *M. oleifera* oil showed a concentration dependent DPPH free radical scavenging and reducing power capabilities. The study has

TABLE 35.3 Tocopherol Composition of Nondegummed Oils[a]

M. oleifera variety Periyakulam					M. oleifera variety Mbololo		
(mg/kg)	CP	H	CM	V. olive oil	CP	H	CM
α-Tocopherol	5.06 ± 0.67	15.38 ± 0.68	2.42 ± 0.37	88.5 ± 6.33	101.46 ± 10.23	98.82 ± 6.32	105.02 ± 9.09
γ-Tocopherol	25.40 ± 1.16	4.47 ± 0.87	5.52 ± 0.69	9.9 ± 0.65	39.54 ± 4.51	27.90 ± 1.23	33.45 ± 3.33
δ-Tocopherol	3.55 ± 0.45	15.51 ± 0.99	12.67 ± 0.55	1.6 ± 0.10	75.67 ± 6.33	71.16 ± 6.96	77.60 ± 4.52

[a]Results are the men of three replicates standard deviation (SD); CP, cold press; H, n-hexane; CM, chloroform/methanol; ND, not detected.
Source: Lalas and Tsaknis (2002b)

TABLE 35.4 Total Phenol, Flavonoid, and Antioxidant Capacity of *M. oleifera* Seed Oil and Palm Oil

Parameter	*M. oleifera* oil[a]	Palm oil[a]
Total phenol (mg GAE g)	40.17 ± 0.01	62.32 ± 0.04
Total flavonoid (mg RE g)	18.24 ± 0.01	33.13 ± 0.03
Total antioxidant capacity (mg AAE g)	37.94 ± 0.02	68.27 ± 0.02

[a]Values are ± SEM

shown that *M. Oleifera* gave high oil yield, which has good antioxidant capacity with potential for industrial, nutritional, and health applications (Ogbunugafor et al., 2011).

USES OF MORINGA SEED OIL

M. oleifera appears to be a potentially valuable crop, yielding useful oil that might be an acceptable substitute for high-oleic oils, such as olive and high-oleic sunflower oils in the dietary fats. Many investigations and nutritional evaluation revealed that *M. oleifea* seed oil has a very good potential for edible and industrial purposes as well as for developing nutritionally balanced, high-stability blended formulations with other high-linoleic oils (Lalas and Tsaknis 2002a,b; Anwar & Bhanger, 2003). The investigated *M. oleifera* oil, naturally high in oleic acid and coincident with genetically modified oilseeds, might be well positioned in the international oil trade, provided it can be cultivated in wide scale production. The frying quality and stability of high-oleic *M. oleifera* seed oil had been investigated and compared to other vegetable oils (Abdulkarim, Long, Lai, Muhammad, & Ghazali, 2007). The characterization of the oil from the seeds of *M. oleifera* showed that this oil could be utilized successfully as a source of edible oil for human consumption (Tsaknis et al., 1999; Asuquo et al., 2010). It contains high monounsaturated to saturated fatty acids ratio and might be an acceptable substitute for highly monounsaturated oils, such as olive oil in diets.

(Rashid, Anwar, Moser, & Knothe, 2008) studied successfully the *M. oleifera* seed oil as a possible source of biodiesel production. Another study was done by (da Silva et al., 2010) on characterization of *M. oleifera* seed oil for biodiesel production. There are many studies had shown that *M. Oleifera* gave high oil yield, which has good antioxidant capacity with potential for industrial, nutritional and health applications, therefore large scale cultivation of this economic plant could be used as poverty alleviation strategy in many countries (Lalas and Tsaknis 2002a; Ogbunugafor et al., 2011).

REFERENCES

Abdulkarim, S., Long, K., Lai, O., Muhammad, S., & Ghazali, H. (2005). Some physico-chemical properties of *Moringa oleifera* seed oil extracted using solvent and aqueous enzymatic methods. *Food Chemistry, 93*(2), 253–263.

Abdulkarim, S., Long, K., Lai, O., Muhammad, S., & Ghazali, H. (2007). Frying quality and stability of high-oleic *Moringa oleifera* seed oil in comparison with other vegetable oils. *Food Chemistry, 105*(4), 1382–1389.

Abiodun, O., Adegbite, J., & Omolola, A. (2012). Chemical and physicochemical properties of Moringa flours and oil. *Global Journal of Science Frontier Research, 12*(5–C), 12–18.

Ali, E. N., Muyibi, S. A., Salleh, H. M., Alam, M. Z., & Salleh, M. R. M. (2010). Production of natural coagulant from *Moringa oleifera* seed for application in treatment of low turbidity water. *Journal of Water Resource and Protection, 2*(3), 259.

Anhwange, B., Ajibola, V., & Oniye, S. (2004). Chemical studies of the seeds of *Moringa oleifera* (Lam) and *Detarium microcarpum* (Guill and Sperr). *Journal of Biological Sciences, 4*(6), 711–715.

Anjorin, T. S., Ikokoh, P., & Okolo, S. (2010). Mineral composition of *Moringa oleifera* leaves, pods and seeds from two regions in Abuja, Nigeria. *International Journal of Agriculture and Biology, 12*(3), 431–434.

Anwar, F., & Bhanger, M. (2003). Analytical characterization of *Moringa oleifera* seed oil grown in temperate regions of Pakistan. *Journal of Agricultural and Food Chemistry, 51*(22), 6558–6563.

Anwar, F., & Rashid, U. (2007). Physico-chemical characteristics of *Moringa oleifera* seeds and seed oil from a wild provenance of Pakistan. *Pakistan Journal of Botany, 39*(5), 1443–1453.

Asuquo, J., Anusiem, A., & Etim, E. (2010). Extraction and characterization of sheabutter oil. *World Journal of Applied Science and Technology, 2*(2), 282–288.

Codex (1993). Alimentairus Commission. Graisses et huiles vegetales, Division 11, version abregee FAO/WHO Codex Stan 20-1981, 23 -1981.

Copp, G. H., Bianco, P., Bogutskaya, N., Erős, T., Falka, I., Ferreira, M., & Grabowska, J. (2005). To be, or not to be, a non-native freshwater fish? *Journal of Applied Ichthyology, 21*(4), 242–262.

da Silva, J. P., Serra, T. M., Gossmann, M., Wolf, C. R., Meneghetti, M. R., & Meneghetti, S. M. (2010). *Moringa oleifera* oil: studies of characterization and biodiesel production. *Biomass and Bioenergy, 34*(10), 1527–1530.

Dahot, M. U. (1988). Vitamin contents of the flowers and seeds of *Moringa oleifera. Pakistan Journal of Biochemistry, 21*(1–2), 21–24.

Idris, M. A., Muyibi, S. A., Karim, A., Ismail, M., Jamal, P., & Jami, M. S. (2013). Investigating antibacterial activity of defatted *Moringa oleifera* seed extract on selected gram-negative and gram-positive microbes. *International Journal of Academic Research, 5*(1), .

Jamil, A., Shahid, M., Khan, M., & Ashraf, M. (2007). Screening of some medicinal plants for isolation of antifungal proteins and peptides. *Pakistan Journal of Botany (Pakistan), 39*(1), 211–221.

Kalogo, Y., Rosillon, F., Hammes, F., & Verstraete, W. (2000). Effect of a water extract of *Moringa oleifera* seeds on the hydrolytic microbial species diversity of a UASB reactor treating domestic wastewater. *Letters in Applied Microbiology, 31*(3), 259–264.

Kiritsakis A. (ed.) (1988). *The Olive Oil* (pp. 52, pp. 61-64, pp. 234). Thessaloniki, Greece: Agricultural Co-operative Editions.

Lalas, S., & Tsaknis, J. (2002a). Characterization of *Moringa oleifera* seed oil variety "Periyakulam 1". *Journal of Food Composition and Analysis, 15*(1), 65–77.

Lalas, S., & Tsaknis, J. (2002b). Extraction and identification of natural antioxidant from the seeds of the *Moringa oleifera* tree variety of Malawi. *Journal of the American Oil Chemists' Society, 79*(7), 677–683.

Fuglie L. J. (1999). *The Miracle Tree: Moringa oleifera: Natural Nutrition for the Tropics* (68 pp.). Dakar: Church World Service.(revised in 2001 and published as The Miracle Tree: The Multiple Attributes of Moringa, pp. 172).

Mangale, S., Chonde, S., Jadhav, A., & Raut, P. (2012). Study of *Moringa oleifera* (drumstick) seed as natural absorbent and antimicrobial agent for river water treatment. *Journal of Natural Product and Plant Resources*, 2(1), 89–100.

Mishra, G., Singh, P., Verma, R., Kumar, S., Srivastav, S., Jha, K., & Khosa, R. (2011). Traditional uses, phytochemistry and pharmacological properties of *Moringa oleifera* plant: an overview. *Der Pharmacia Lettre*, 3(2), 141–164.

Morton, J. F. (1991). The horseradish tree, *Moringa pterygosperma* (Moringaceae)—a boon to arid lands? *Economic Botany*, 45(3), 318–333.

Ndiaye, M., Dieye, A., Mariko, F., Tall, A., Sall, D. A., & Faye, B. (2001). [Contribution to the study of the anti-inflammatory activity of *Moringa oleifera* (Moringaceae)]. *Dakar Medical*, 47(2), 210–212.

Ogbunugafor, H., Eneh, F., Ozumba, A., Igwo-Ezikpe, M., Okpuzor, J., Igwilo, I., & Onyekwelu, O. (2011). Physico-chemical and antioxidant properties of *Moringa oleifera* seed oil. *Pakistan Journal of Nutrition*, 10(5), 409–414.

Ogunsina, B. S., Indira, T. N., Bhatnagar, A. S., Radha, C., Debnath, S., & Gopala Krishna, A. G. (2011). Quality characteristics and stability of Moringa oleifera seed oil of Indian origin. *Journal of Food Science and Technology*, 51(3), 503–510.

Oliveira, J. T. A., Silveira, S. B., Vasconcelos, I. M., Cavada, B. S., & Moreira, R. A. (1999). Compositional and nutritional attributes of seeds from the multiple purpose tree *Moringa oleifera* Lamarck. *Journal of the Science of Food and Agriculture*, 79(6), 815–820.

Przybylski, R., Mag, T., Eskin, N., & McDonald, B. (2005). Canola oil. *Bailey's Industrial Oil and Fat Products*, 2, 61–121.

Ramachandran, C., Peter, K., & Gopalakrishnan, P. (1980). Drumstick (*Moringa oleifera*): a multipurpose Indian vegetable. *Economic Botany*, 34(3), 276–283.

Rashid, U., Anwar, F., Moser, B. R., & Knothe, G. (2008). Moringa oleifera oil: a possible source of biodiesel. *Bioresource technology*, 99(17), 8175–8179.

Salaheldeen, M., Aroua, M., Mariod, A., Cheng, S. F., & Abdelrahman, M. A. (2014). An evaluation of *Moringa peregrina* seeds as a source for bio-fuel. *Industrial Crops and Products*, 61, 49–61.

Sengupta, A., & Gupta, M. (1970). Studies on the seed fat composition of Moringaceae family. *Fette, Seifen, Anstrichmittel*, 72(1), 6–10.

Siddhuraju, P., & Becker, K. (2003). Antioxidant properties of various solvent extracts of total phenolic constituents from three different agroclimatic origins of drumstick tree (*Moringa oleifera* Lam.) leaves. *Journal of Agricultural and Food Chemistry*, 51(8), 2144–2155.

Sonntag, N. (1979). Composition and characteristics of individual fats and oils. *Bailey's Industrial Oil and Fat Products*, 1, 289–477.

Talreja, T. (2010). Screening of crude extract of flavonoids of *Moringa oleifera* against bacteria and fungal pathogen. *Journal of Phytology*, 2(11), 31–35.

Tsaknis, J., Lalas, S., Gergis, V., & Spiliotis, V. (1998). A total characterisation of Moringa oleifera Malawi seed oil. *Rivista Italiana delle Sostanze Grasse*, 75, 21–28.

Tsaknis, J., Lalas, S., Gergis, V., Dourtoglou, V., & Spiliotis, V. (1999). Characterization of Moringa oleifera variety Mbololo seed oil of Kenya. *Journal of Agricultural and Food Chemistry*, 47(11), 4495–4499.

Chapter 36

Ziziphus spina-christi (Christ's Thorn Jujube)

INTRODUCTION

Out of the 7000 cultivated plant species worldwide (Hammer and Khoshbakht 2005; Raman 2006), food security has become increasingly dependent on a small number of crops (Altieri 1999; Borlaug 2002). Nevertheless, an extrapolation by FAO (1999) indicates that a number of 1,8000–2,5000 wild-collected species are used as food. Among those indigenous fruit trees play a very important role in the livelihoods of rural people, especially for those living in the dry areas (Maydell 1989), where crop failure often results in the poor nutrition of the local population (Maxwell 1991). These fruit trees are part of local and regional agriculture and food procurement systems and are important genetic resources in global efforts to maintain biodiversity (Grivetti and Ogle 2000).To meet the needs of an increasing world population the development of alternative crops to improve the range of commodities available is needed (El-Siddig et al. 1999). This is particularly important for crops which can withstand harsh conditions, such as drought, heat, and salinity stress. An example of such a species is *Ziziphus spina-christi* (L.) Willd. Commonly known as Christ's thorn.

Most of the commentators and translators of the Qur'an in English have considered *Sidrah* to be the *Lote-Tree* which in Arabic is known as either *Nabq* or *'Unnab,* and in European languages as *jujube* or *Ziziphus.* Before trying to identify true *Sidrah,* let us consider the opinion of those who claim it to the *Lote-Tree* of the genus *Ziziphus.* According to Paxton, the name *Ziziphus* has been derived from the Arabic word *Ziziofun,* which means a spiny shrub. In fact all the species of *Ziziphus* are armed with spines or thorns and are not deep-rooted plants. The three species of *Ziziphus* occurring in Arabia are *Ziziphus mauritiana,* which is the source of jujube fruits, *Ziziphus lotus,* which produces the edible fruits known as *'Unnab* (in Arabic), and *Z. spina-christi,* thorns of which are said to have been used to make the crown for Holy Christ. These species are generally spiny shrubs but sometimes attain the size of a small tree. All of them do produce edible fruits although in taste none can be rated as good as the ones described in the Qur'an, namely date, grape, figs, pomegranate, and olive. *Ziziphus* trees are good source of fuel but by no measures any of the three

Unconventional Oilseeds and Oil Sources. http://dx.doi.org/10.1016/B978-0-12-809435-8.00036-6

is held suitable for timber. All these plants are found in tropical regions of Africa and Asia and are not found in hilly regions having a cool climate. Orchards of pure *Ziziphus* are seldom planted. These are either found wild or cultivated in patches, or around gardens as protective barriers or hedges (Farooqi, 2016).

Ziziphus possess multiple medicinal properties, such as analgesic, antiinflammatory, antibacterial, antiaging, antitumor, and antioxidant properties. *Ziziphus* have a significant role in the treatment of cancer, diabetes, and have also been used in folk medicine to treat jaundice, diarrhea, ulcers, and fevers. It also possesses the capability to use as an immunomodulator. This is a remedy for hypertonias, anemia, nephritis, and nervous diseases, Fruits and roots of *Ziziphus* are used to cure sunstroke, healing cuts, inhibiting gastrointestinal infections, and removing stones. Other applications in cosmetic products, lac culture, wine production, food preservation due to the presence of antimicrobial compounds and in tonics to enhance the immune system and blood nourishment. Its oil also possesses hair growth promoting activity. *Ziziphus* species are promising plants for arid regions due to their drought resistance characteristics.

This paper aims to review the experiments and studies made on Christ's thorn tree as a multiapplication fruit tree, especially as an unconventional source of oil.

Botanical Description and Chemical Composition of the Tree

The genus *Ziziphus* belongs to the Rhamnaceae family which consists of about 100 species of deciduous or evergreen trees and shrubs distributed throughout the tropical and subtropical regions of the world, from which 12 species are cultivated (Hammer 2001). *Z. spina-christi* is a spiny shrub or small tree (Fig. 36.1) that strongly resists heat and drought. Normally, the species grows into a tree form, but it often acquires a bush form due to intensive grazing during the later part of the dry seasons (Obeid and Mahmoud 1971) and heavy destructive cutting for fencing material and fuel (Miehe 1986).

Usually, in Arabic, the fruits have the name of the tree, but in the case of *Z. spina-christi*, the tree is called siddir and the fruit nabag indicating the specific importance of this plant to local people. Gebauer (2005) reported that *Z. spina-christi* is the most abundant wild fruit tree used in home gardens in El Obeid, the capital of North Kordofan state in central Sudan.

Z. spina-christi (Christ's Thorn Jujube) is a wild tree today found in Jordan, Palestine, Egypt, and some parts of Africa; this tree is divided into four parts: leaves, branches with leaves, branches, and stems. And it has a fruit which is a globose drupe about 1–1.5 cm in diameter, red-brown, with a hard stone surrounded by a sweet edible pulp. The flesh of *Z. spina-christi* fruits is rich in carbohydrates (80.6% in dry matter) notably starch (21.8%), sucrose (21.8%), glucose (9.6%) and fructose (16%), and in iron (3 mg 100/g dried fruit; Nour et al. 1987; Abdelmuti 1991). One hundred gram dried fruit pulp contains 314 calories, 4.8 g protein, 0.9 g fat, 140 mg calcium, 0.04 mg of thiamine, 0.13 mg

FIGURE 36.1   *Ziziphus spina-christi* tree and fruits.

riboflavin, 3.7 mg niacin, and 30 mg ascorbic acid (Duke 1985; Berry-Koch et al. 1990). In another study, Eromosele et al. (1991) found 98 mg of ascorbic acid in the mesocarp of *Z. spina-christi*. The leaves are rich in calcium (1270 mg 100/g dry weight), iron (7.2 mg 100/g), and magnesium (169 mg 100/g; Anonymous 1992).

SEED OIL AND FATTY ACIDS

The seed contains 28.5% lipid and 18.6% protein. Its proteins contain high levels of sulfur-rich amino acids. Analysis of fatty acids revealed the presence of 13 fatty acids, linoleic acid predominates in its lipids (45%) followed by linolenic acid (20%), which showed high activity against *Escherichia coli* and *Bacillus subtilis*. Cholesterol and β-sitosterol represent the major constituents of the unsaponifiable fraction (c. 49%) (Nazif 2002).

USES OF THE CHRIST'S THORN JUJUBE TREE

The Christ's Thorn Jujube tree was already in use as a medicinal plant in ancient Egypt. In ancient Egyptian prescriptions, it was used in remedies against swellings, pain, and heat, and thus should have antiinflammatory effects. Nowadays, *Z. spina-christi* is used in Egypt (by Bedouins, and Nubians), the Arabian Peninsula, Jordan, Iraq, and Morocco against a wide range of illnesses, most of them associated with inflammation. However, little research has been conducted on the rich nutritional content of its fruits and leaves (Nour et al. 1987; Berry-Koch et al. 1990; Eromosele et al. 1991) as well as on its well-known antimicrobial effects (Nazif 2002; Harami et al. 2006). On the other hand, pharmacological research undertaken to date suggests that it possesses a lot of protective effects.

The genus *Ziziphus* is known for its medicinal properties as a hypoglycemic, hypertensive, antiinflammatory, antimicrobial, antioxidant, antitumor, and liver protective agent and as an immune system stimulant (Said et al. 2006). According to Dr. Nidal A. Jaradat, *Z. spina-christi* is used as food, diuretic, for teeth decay, increase hair growth, washing the dead body with them make it longer standing, tonic, antiinflammatory, antidiarrhea, decrease thirst, (2005). Its fruits are eaten to treat diarrhea and malaria and as an antispasmodic. The powder of the twigs is used externally to treat rheumatism and scorpion sting (El-Kamali and El-Khalifa 1999).

In the United Arab Emirates, leaves are boiled in water and used as a shampoo or mixed with lemon and applied to the face and hair to soften or to soothe it. Ash of wood mixed with vinegar is applied to heal snake bites and a tea made of fruit is used to treat measles. Fruits and crashed kernels are eaten to treat chest pains, respiratory problems, and as a tonic (Jongbloed 2003). Moreover, in Oman fruits are regarded as having purifying properties, such as cleansing the stomach, removing impurities from blood as well as being a restorative for the whole system. The water in which crushed leaves had been boiled is given to women in prolonged labor or with a retained placenta for its oxytocic properties (Miller and Morris 1988).

In India, a preparation from the bark is used to clean wounds and sores, while the gum that exudes from the tree in hot weather is used in eye remedies. The chewed leaves are said to numb the taste buds, which provides a method of suppressing the unpleasant flavor of some oral medicines. Table 36.1 shows the different parts of *Ziziphus* plants and their role in human health.

Aside from the medical field benefits, *Z. spina-christi* has edible fruits and a number of other beneficial applications that include the use of leaves as fodder, branches for fencing. The wood is used as a source of fuel and it produces an excellent charcoal. The timber is used for tool handles, fence posts, bedstead legs, walking sticks, furniture, bentwood chairs, roofing beams, doors, windows, and turned items. It is hard and heavy and is known to resist termites (Sudhersan and Hussain 2003). The honey collected from the flowers of *Z. spina-christi* is of excellent flavor (Sudhersan and Hussain 2003) and is normally sold at a price higher than that derived from the flowers of other trees. For Muslims, seeds are used as rosaries, while for Christians the thorns are reported to have made up the crown of thorns of Jesus. In natural religions, the tree's roots are used in superstitious practices (Bircher and Bircher 2000). The utilization of different parts, for example, fruits, leaves, roots, and bark in making useful products. Moreover, the plant is adapted to dry and hot climates which make it suitable for cultivation in an environment characterized by increasing degradation of land and water resources.

CHRIST'S THORN JUJUBE TREE OIL

Essential oils are volatile oils, oils that vaporize readily, derived from vegetation that give plants their distinctive odors. Also known as aromatic oils, these nontimber forest products must be stored with care to avoid vaporization. They vary

TABLE 36.1 Different Parts of *Ziziphus* Plants and Their Role in Human Health

S. No.	Plant part	Elemental content	Application
1.	Fruit	Ca, K, Zn, Fe, Cr, Co, Pb, Mn, Na, and Mg	Antioxidants, anticancer, refrigerant, pectoral, styptic, sedative, stomachache, tonic, loss of appetite, chronic fatigue, diarrhea, anemia, hysteria, irritability, pulmonary ailments fevers, and ulcers. Reduce eye inflammations. Purify the blood cure sunstroke and healing cuts.
2.	Leaves	Ca, K, Zn, Fe, Cr, Co, Pb, Mn, Na, and Mg	Antibacterial, antifungal, phytotoxic, cytotoxic, insecticidal activities. Liver troubles, asthma, and fever. Relieve gingivitis. Antihelminthic and antidiarrhetic properties. Skin infections
3.	Flower	Ca, K, Mg, and Na	Eye lotion
4.	Bark	Ca, K, Zn, Fe, Cr, Co, Pb, Mn, Na, and Mg	Rheumatism, stomach troubles, and coughs
5.	Root	K, Ca, Fe, Mg	Dyspepsia, chronic fevers, ulcers, nausea, vomiting, and abdominal pains in pregnancy. Headaches, tubercular gland swellings, measles, dysentery, lumbago, chest complaints, and snakebite
6.	Seeds	Ca, K, Mg, and Na	Palpitations, sedative, hypnotic, narcotic, stomachache, and insomnia, night sweats, nervous exhaustion, and excessive perspiration

in strength, but this is not to say that they ever occur as dilute substances; they usually need to be adulterated to smell best. The plants produce these oils in special secretary structures during regular growth (Thomas and Schumann, 1993).

CONCLUSION

Z. spina-christi known as Christ's thorn. Jujube is a native plant that grows in tropical and subtropical regions, especially in the Middle East. For a long time, *Z. spina-christi* has been used in alternative medicine for the treatment of fever, pain, dandruff, wounds and ulcers, inflammatory conditions, asthma, and to cure eye diseases. *Z. spina-christi* has recently been shown to have antibacterial, antifungal, antioxidant, antihyperglycemic, and antinociceptive activities.

The oil of this tree can be obtained by different means, including steam distillation, cold pressing, enfleurage, turbo distillation, solvent extraction, and carbon dioxide extraction. It is of great use for humans and their developing industry where it is used to make fragrances, scent soaps, room sprays, disinfectants, and similar products. It also was shown to have a lot of medical benefits as antibacterial, antifungal, antioxidant components, and many more applications in different felids. So, studies should focus more on analyzing and extraction of such important product.

REFERENCES

Abdelmuti, O. M. S. (1991). Biological and nutritional evaluation of Famine food of the Sudan. PhD Thesis, University of Khartoum, Sudan.

Altieri, M. A. (1999). The ecological role of biodiversity in agroecosystems. *Agriculture, Ecosystems and Environments, 74*(1–3), 19–31.

Anonymous. (1992). Nutritional study on *Ziziphus spina-christi*. Sweden: Eden Foundation. Available from http://www.eden-foundation.org/index.html

Berry-Koch, A., Moench, R., Hakewill, P., & Dualeh, M. (1990). Alleviation of nutritional deficiency diseases in refugees. *Food Nutrition Bulletin, 12*, 106–112.

Bircher, A. G., & Bircher, W. H. (2000). *Encyclopedia of fruit trees and edible flowering plants in Egypt and the subtropics*. Cairo: The American University in Cairo Press.

Borlaug, N. E. (2002). Feeding a world of 10 billion people: the miracle ahead. *In Vitro Cellular & Developmental Biology - Plant, 38*, 221–228.

Duke, J. A. (1985). *Handbook of medicinal herbs*. Boca Raton: CRC Press.

El-Kamali, H. H., & El-Khalifa, K. F. (1999). Folk medicinal plants of riverside forests of the Southern Blue Nile district, Sudan. *Fitoterapia, 70*, 493–497.

El-Siddig, K., Ebert, G., & Lüdders, P. (1999). Tamarind (*Tamarindus indica* L.): a review on a multipurpose tree with promising future in the Sudan. *Journal of Applied Botany and Food Quality, 73*, 202–205.

Eromosele, I. C., Eromosele, C. O., & Kuzhkuzha, D. M. (1991). Evaluation of mineral elements and ascorbic acid contents in fruits of some wild plants. *Plant Foods for Human Nutrition, 41*, 151–154.

Farooqi, M. I. H. (2016). Cedar or lote-tree in the light of Al-Quran-a scientific study. Available from: http://www.irfi.org/articles4/articles_5001_6000/cedar_or_lotetree_in_the_light.html

FAO. (1999). Use and potential of wild plants in farm households. FAO Information Division, Rome, Italy. Available from http://www.fao.org

Gebauer, J. (2005). Plant species diversity of home gardens in El Obeid, Central Sudan. *Journal of Agriculture and Rural Development in the Tropics and Subtropics, 106*, 97–103.

Grivetti, L. E., & Ogle, B. M. (2000). Value of traditional food in meeting macro- and micronutrients needs: the wild plant connection. *Nutrition Research Reviews, 13*, 31–46.

Hammer, K. (2001). Rhamnaceae. In P. Hanelt, & IPK (Eds.), *Mansfeld's encyclopedia of agricultural and horticultural crops* (pp. 1141–1150). (Vol. 3). Berlin: Springer.

Hammer, K., & Khoshbakht, K. (2005). Towards a 'red list' for crop plant species. *Genetic Resources and Crop Evolution, 52*, 249–265.

Harami, M. A., Abayeh, O. J., Ibok, N. N. E., & Kafu, S. E. (2006). Antifungal activity of extracts of some *Cassia, Detarium* and *Ziziphus* species against dermatopytes. *Natural Product Radiance, 5*, 361–365.

Jongbloed, M. (2003). The comprehensive guide to the wild flowers of the United Arab Emirates. Environmental Research and Wildlife Development Agency (ERWDA), Abu Dhabi.

Maxwell, S. (1991). *To cure all hunger: Food policy and food security in Sudan.* London: Short Run Press.

Miehe, S. (1986). *Acacia albida* and other multipurpose trees on the fur farmlands in the Jebbel Marra highlands, western Darfur, Sudan. *Agroforestry Systems, 4*, 89–119.

Miller, A. G., & Morris, M. (1988). *Plants of Dhofar, the southern region of Oman traditional, economic and medicinal uses.* Muscat: Adviser for Conservation of the Environment, Diwan of Royal Court Sultanate of Oman.

Nazif, N. M. (2002). Phytoconstituents of *Ziziphus spina-christi* L. fruits and their antimicrobial activity. *Food Chemistry, 76*, 77–81.

Nour, A., Ali, A. O., & Ahmed, A. H. R. (1987). A chemical study of *Ziziphus spina-christi* (Nabag) fruits grown in Sudan. *Tropical Science, 27*, 271–273.

Obeid, M., & Mahmoud, A. (1971). Ecological studies in the vegetation of the Sudan: II. The ecological relationships of the vegetation of Khartoum province. *Vegetation, 23*, 77–198.

Raman, S. (2006). *Agricultural sustainability: Principles, processes, and prospects.* New York: Food Products Press.

Said, A., Huefner, A., Tabl, E., & Fawzy, G. (2006). Two new cyclic amino acids from the seeds and antiviral activity of methanolic extract of the roots of *Zizyphus spina-christi.* Paper presented at the 54th Annual Congress on Medicinal Plant Research. *Planta Med, 72*, 222.

Sudhersan, C., & Hussain, J. (2003). In vitro propagation of a multipurpose tree, *Ziziphus spina-christi* (L.) Desf. *Turkish Journal of Botany, 27*, 167–171.

Thomas, M. G., & Schumann, D. R. (1993). *Income opportunities in special forest products: self-help suggestions for rural entrepreneurs. Agriculture Information Bulletin 666.* Washington, DC: United States Department of Agriculture.

von Maydell, H. J. (1989). Criteria for the selection of food producing trees and shrubs in semi-arid regions. In G. E. Wickens, N. Haq, & P. Day (Eds.), *New crops for food and industry* (pp. 66–75). London: Chapman and Hall.

Chapter 37

Salvadora persica Seed and Seed Oil

INTRODUCTION

Salvadora persica Linn., commonly known as miswak (tooth brush), belongs to the family Salvadoracea is a large, well-branched evergreen shrub, or small tree having soft whitish yellow wood, bark is of old stems rugose. It is an upright evergreen small tree or shrub, seldom more than 1 ft. in diameter reaching a maximum height of 3 m. Leaves are somewhat fleshy, glaucous, 3.8–6.3 by 2–3.2 cm in size. The flowers are greenish yellow in color. Drupe is 3 mm in diameter, globose, smooth and becomes red when ripe. It is widely distributed in the drier parts of India, Baluchistan, and Ceylon and in the dry regions of West Asia and Egypt (Khatak et al., 2010). Fruit spherical, fleshy, 5–10 mm in diameter, pink to scarlet when mature, single seeded; seeds turn from pink to purple-red and are semitransparent when mature (Orwa, 2009) (Fig. 37.1).

S. persica is widely distributed in the arid regions of India and often on saline soils. The fresh leaves are eaten as salad and are used in traditional medicine for cough, asthma, scurvy, rheumatism, piles, and other diseases. The use of miswak is a pre-Islamic custom, which was adhered to by the ancient Arabs to get their teeth white and shiny. The beneficial effects of miswak in respect of oral hygiene and dental health are partially due to its mechanical action and partially due to pharmacologic action. There is investigation of its different chemical constituents, which are responsible for these activities (Sadhan & Almas, 1999). Benzyl-isothiocyanate was isolated from *S. persica* root, and it contains saponins along with tannins, silica, a small amount of resin, trimethylamine, and alkaloidal constituents. β-sitosterol, *m*-anisic acid, and salvadourea were isolated from the plant (Khatak et al., 2010). Seeds have bitter and sharp taste. They are used as purgative, diuretic, and tonic seed oil is applied on the skin in rheumatism with high content of minerals in the root, 27.06%. The fruits are sweet and edible. The pulp contains glucose, fructose, and sucrose. Milk cattle when fed with fruits increase milk yield. Leaves are also eaten as vegetables and used in the preparation of sauce. Tender shoots and leaves are eaten as salad (Sujata, 2015).

Unconventional Oilseeds and Oil Sources. http://dx.doi.org/10.1016/B978-0-12-809435-8.00037-8

FIGURE 37.1 *Salvadora persica* **tree, fruits and seeds.** *(https://commons.wikimedia.org/wiki/ Category:Salvadora_persica#/media/File:Salvadora_persica-1-jodhpur-India.JPG while the red fruits and the dried ones were cited from google.com)*

S. persica L., also known as the toothbrush tree (Miswak), has been used since ancient times as a chewing stick for oral hygiene. Miswak is a natural source of many unique phytochemicals, which are described by traditional medicine as a remedy for various disease symptoms with beneficial properties. The availability and richness of biologically active compounds and minerals, related to oral and dental health, in Miswak makes it a superior tool for oral hygiene and a barrier against general pathogens that enter the human body through the mouth. The results show that Miswak contains more than one type of antimicrobial agent that inhibits the growth of both Gram-positive and -negative bacteria (Abhary & Al-Hazmi, 2016). Some of the known commercial toothpastes were produced from *S. persica* plant (Almas, Skaug, & Ahmad, 2005). Sweet variety of *S. persica* yields 35%–44% oil which has a strong odor. The odor is due the presence of benzyl isothiocyanate in the oil. There is greenish yellow fat called in India as Khakan fat. The purified fat is free from foul odor and has an agreeable taste. It is snow white. The oil is rich in lauric and myristic acids. The fat is used in soap, making up to 20% of the soap itself, and it replaces coconut oil. It is used as a resist in the dyeing industry. The oil is also used in rheumatic infection treatment. It has a high melting point and a disagreeable odor that disappears on purification. The most important aspect of the oil is the presence of a low percentage of C8 and C10 fatty acids that are of great economic significance. The oil is an alternative source of oil for soap and detergent industries (Board, 2013).

The oil content of *S. persica* seeds is very high in comparison to common oil sources found in Sudan, such as groundnut, sesame, cotton seed, and caster bean with oil contents between 20% and 45% (Mariod, Matthäus, & Hussein, 2009). The predominant fatty acids were 14:0, 16:0, and 18:1 representing 45.50, 35.12, and 10.20% for Kordofan and 45.20, 34.49, and 10.66% for Gezira samples in Sudan as in Table 37.1.

The seeds of *S. persica* yield a pale yellow solid fat. The seed forms 45%–50% of the fruit and has about 35%–40% oil content. The crude fat obtained by pressing the decorticated seeds in high pressure expellers or in ghanis is a dark colored product having a bitter taste and a bad smell. The seed fat contains 19.6% lauric acid, 54.5% myristic acid, 19.5% palmitic acid, and 5.4% oleic acid (Sujata, 2015).

The ratios of saturated/unsaturated acid are 6.1 and 5.2, respectively which is very high because of the high level of saturated fatty acid, such as C14 and C16. The fatty acid profile of *S. persica* oil is consistent with results published by other workers (Reddy, Shah, & Patolia, 2008; Tripathi, Rathore, & Kumar, 1998).

As further important criteria for the assessment of seed oils, the contents and composition of tocopherols, tocotrienols, and plastochromanol-8 (P-8) were determined; these data are presented in Table 37.2. Gamma-tocopherol was the

TABLE 37.1 Oil Content (%) and Fatty Acid Composition (%) of *S. persica* Seed Oil[a]

Sample fatty acids	Kordofan mean ± SD	Gezira mean ± SD
12:0	3.31 ± 0.01	3.32 ± 0.02
14:0	45.50 ± 0.02	45.20 ± 0.02
16:0	35.12 ± 0.21	34.49 ± 0.32
18:1	10.20 ± 0.20	10.66 ± 0.12
20:0	0.70 ± 0.11	0.70 ± 0.22
20:1	3.82 ± 0.10	4.27 ± 0.07
22:0	0.89 ± 0.04	0.89 ± 0.06
24:0	0.46 ± 0.30	0.47 ± 0.20
ΣSFA	85.98 ± 0.22	84.07 ± 0.11
ΣMUFA	14.02 ± 0.10	15.93 ± 0.11
Ratio SFA/UFA	6.13 ± 0.21	5.27 ± 0.21
Oil content (g/100 g of seed)	41.4 ± 0.60	42.8 ± 0.50

[a]*All determinations were mean of triplicates ± standard deviation (SD). SFA, saturated fatty acids; USFA, unsaturated fatty acids; MUSFA, monounsaturated fatty acids.*
Source: Data taken from Mariod, A. A., Matthäus, B., & Hussein, I. (2009). Chemical characterization of the seed and antioxidant activity of various parts of Salvadora persica. *Journal of the American Oil Chemists' Society, 86*(9), 857–865.

TABLE 37.2 Tochopherol Composition (mg/100 g) of *S. persica* Seeds From Two Different Locations in Sudan

Sample tocopherol	Kordofan mean ± SD	Gezira mean ± SD
α-T	9.76 ± 0.12[e]	9.21 ± 0.13[e]
γ-T	28.23 ± 0.15[f]	28.20 ± 0.21[f]
β-T	0.72 ± 0.22[b]	0.80 ± 0.05[b]
P-8	1.99 ± 0.11[c]	1.95 ± 0.11[c]
γ-T₃	5.4 ± 0.11[d]	5.4 ± 0.20[d]
δ-T	0.17 ± 0.03[a]	0.20 ± 0.11[a]
Total amount of tocopherol	46.27 ± 0.11[g]	45.76 ± 0.11[h]

P-8, Plastochromanol-8; T, tocopherol; T₃, tocotrienol. The superscripts letters indicate the statistical significant differences ($P < 0.05$), the values in a raw having same letter are not significantly difference. All determination was carried out in triplicates and the mean values ± SD reported.
Source: Data taken from Mariod, A. A., Matthäus, B., & Hussein, I. (2009). Chemical characterization of the seed and antioxidant activity of various parts of Salvadora persica. *Journal of the American Oil Chemists' Society, 86*(9), 857–865.

predominant tocopherol in both samples representing 61.3 and 61.7% of the total tocopherols, respectively, followed by α-tocopherol at 21.1 and 20.2%, respectively. Total sterol content was 3399.6 and 3385.3 mg/kg for Kordofan and Gezira samples, respectively. The content of total phenolic compounds was determined in *S. persica* bark (SPB), *S. persica* leaves (SPL), and *S. persica* seedcake (SPC) extracts of each sample according to the Folin–Ciocalteau method as 111.70, 132.60, and 66.10 mg GAE/g extract for the Kordofan sample. They were found to be 105.90, 129.10 and 62.90 mg GAE/g extract in the Gezira sample, respectively. The two samples were significantly ($P < 0.05$) different in total phenolic content with SPL as the highest in both samples. The methanolic extracts of SPL, SPB, and SPC in both samples were markedly effective in inhibiting the oxidation of linoleic acid and subsequent bleaching of β-carotene in comparison with the control. In *S. persica* seed oil, instead of tocopherols appropriate unsaturated derivative c-tocotrienol was found in an amount of 5.4 mg/100 g in the two samples (Table 37.2).

The analysis of the sterols provides rich information about the quality and the identity of the oil investigated, and for the detection of oil and mixtures not recognized by their fatty acids profile (Besbes et al., 2004). This fraction has been considered as the major unsaponifiable fraction in many oils (White & Armstrong, 1986). The values for sterols of both two *S. persica* seed oils are presented in Table 37.3. The most predominant sterol of the oil from two samples was β-sitosterol followed by campesterol and stigmasterol (Mariod et al., 2009).

In comparison to other oils (sesame, sunflower, and cotton seed oils) usually used in human nutrition except groundnut, the oil of *S. persica* has lower

TABLE 37.3 Distribution of Sterols (mg/kg) in *S. persica* Seed Oil

Sample sterol	Kordofan mean ± SD	Gezira mean ± SD
Cholesterol	56.1 ± 0.11	57.3 ± 0.11
Campesterol	458.0 ± 0.20	460.0 ± 0.20
Stigmasterol	435.7 ± 0.32	448.7 ± 0.32
β-Sitosterol	2156.6 ± 0.51	2122.6 ± 0.52
Δ5-Avenasterol	134.0 ± 0.41	138.0 ± 0.42
Δ7-Avenasterol	50.0 ± 0.42	50.0 ± 0.43
Δ7-Campesterol	26.8 ± 0.41	26.2 ± 0.42
24-Methylenecholesterol	10.4 ± 0.12	10.0 ± 0.11
Δ7-Stigmasterol	72.0 ± 0.11	72.5 ± 0.13
Total	*3399.6 ± 0.62*	*3385.3 ± 0.60*

All determination were carried out in triplicates ± SD.

amounts of total sterols. The oxidative stability of *S. persica* oil is expressed as the induction period determined by the Rancimat method at 120°C. The oil from Kordofan and Gezira samples showed a remarkably medium stability (3.5 and 3.1 h, respectively). These values are to some extent high in comparison to other edible oils (sesame and sunflower), but in comparison to other more stable oils, such as rapeseed or olive oil *S. persica* oil has a low oxidative stability (Mariod et al., 2009). *S. persica* oil was found suitable for use as biodiesel and it meets the major specification of biodiesel standards of USA, Germany, and European Standard Organization (Azam, Waris, & Nahar, 2005).

TOTAL AMOUNT OF PHENOLIC COMPOUNDS

From Table 37.4, it is evident that *S. persica* contains noticeable amounts of extractable compounds, and that the highest amount of total extractable compounds (TEC) was extracted from SPC followed by SPL and SPB in both samples. Table 37.4 shows the results of total phenolic compounds of the methanolic extract from different parts of *S. persica*, these results show that SPL has the highest amount of total phenolic compounds 132.60 and 129.10 mg GAE/g extract in Kordofan and Gazira samples, respectively, followed by SPB and SPC and were significantly different at ($P < 0.05$). It becomes clear that the total phenolic compounds were not related to the TEC (Mariod et al., 2009).

The total phenolics found in the TEC were high in SPB and low in SPC. The ratio of total phenolic compounds to the TEC ranged from 16.3% to 51.6% in the Kordofan sample and from 15.4% to 49.7% in the Gazira sample (Table 37.4). From these results it can be understood that in SPB, SPL, and

TABLE 37.4 Total Extractable Compounds (TEC) and Total Phenolic Compounds (TPC) in *S. persica* Extracts[a]

Sample	TEC (mg/g)	TPC (mg GAE/g extract)	TPC/TEC (%)	IC$_{50}$ (µg/mL)
Kordofan				
SPB	216.46 ± 0.12[a]	111.70 ± 0.21[c]	51.6	1.73 ± 0.13[c]
SPL	333.53 ± 0.21[b]	132.60 ± 0.50[d]	39.7	1.49 ± 0.12[b]
SPC	405.66 ± 0.31[c]	66.10 ± 0.32[a]	16.3	1.87 ± 0.11[d]
Ascorbic acid				0.02 ± 0.01[a]
Gezira				
SPB	211.80 ± 0.40[a]	105.90 ± 0.31[b]	49.7	1.70 ± 0.40[c]
SPL	330.81 ± 0.20[b]	129.10 ± 0.40[d]	39.0	1.43 ± 0.11[b]
SPC	402.90 ± 0.30[c]	62.90 ± 0.32[a]	15.4	1.82 ± 0.21[d]
Ascorbic acid				0.02 ± 0.01[a]

[a]*SPB, S.* persica *bark; SPC, S.* persica *seed cake; SPL, S.* persica *leaves. IC$_{50}$, the concentration of antioxidants required for 50% scavenging of DPPH radicles in specific period of time. The superscripts letters in columns indicate the statistical significant differences (P < 0.05), the values having same letter are not significantly difference. All determinations were carried out in triplicates and the mean values ± SD reported.*
Source: Data taken from Mariod, A. A., Matthäus, B., & Hussein, I. (2009). Chemical characterization of the seed and antioxidant activity of various parts of Salvadora persica. *Journal of the American Oil Chemists' Society, 86*(9), 857–865.

SPC of the Kordofan sample more than 49.4, 60.3, and 83.7% of the extractable compounds, respectively, were compounds other than phenolic compounds (Table 37.4).

When checking the inhibition of β-carotene cooxidation in a Linoleate model system, the methanolic extracts of SPL, SPB, and SPC were markedly effective in inhibiting the oxidation of linoleic acid and the subsequent bleaching of β-carotene in comparison with the control (Mariod et al., 2009).

REFERENCES

Abhary, M., & Al-Hazmi, A. (2016). Antibacterial activity of Miswak (*Salvadora persica* L.) extracts on oral hygiene. *Journal of Taibah University for Science, 10*(4), 513–520.

Almas, K., Skaug, N., & Ahmad, I. (2005). In vitro antimicrobial comparison of miswak extract with commercially available non-alcohol mouthrinses. *International Journal of Dental Hygiene, 3*, 18–24.

Azam, M. M., Waris, A., & Nahar, N. M. (2005). Prospects and potential of fatty acid methyl esters of some non-traditional seed oils for use as biodiesel in India. *Biomass and Bioenergy, 29*, 293–302.

Besbes, S., Blecker, C., Deroanne, C., Bahloul, N., Lognay, G., Drira, N. E., & Attia, H. (2004). Date seed oil: phenolic, tocopherol and sterol profiles. *Journal of Food Lipids, 11*(4), 251–265.

Board, N. (2013). *Modern technology of oils, fats & its derivatives.* India: National Institute of Industrial Re, Asia Pacific Business Press Inc.

Khatak, M., Khatak, S., Siddqui, A. A., Vasudeva, N., Aggarwal, A., & Aggarwal, P. (2010). Salvadora persica. *Pharmacognosy Review, 4*(8), 209–214.

Mariod, A. A., Matthäus, B., & Hussein, I. (2009). Chemical characterization of the seed and antioxidant activity of various parts of *Salvadora persica. Journal of the American Oil Chemists' Society, 86*(9), 857–865.

Orwa P, J. (2009). *Salvadora persica* Salvadoraceae L. *Agroforestry Database,* 4.0, 1–5.

Reddy, M. P., Shah, M. T., & Patolia, J. S. (2008). *Salvadora persica,* a potential species for industrial oil production in semiarid saline and alkali soils. *Industrial Crops and Products, 28*(3), 273–278.

Sadhan, A. L., & Almas, K. (1999). Miswak (chewing stick)—a cultural and scientific heritage. *Saudi Dental Journal, 11*(2), 80–87.

Sujata, M. (2015). Medicinally potent and highly salt tolerant plant of arid zone—*Salvadora persica* L. (Meswak): a review. *Journal of Plant Sciences, 3*(1–1), 45–49.

Tripathi, Y., Rathore, M., & Kumar, H. (1998). Variability in certain qualitative characters of fatty oil from seeds of *Salvadora. Advances in Plant Sciences, 11,* 253–258.

White, P. J., & Armstrong, L. S. (1986). Effect of selected oat sterols on the deterioration of heated soybean oil. *Journal of the American Oil Chemists Society, 63*(4), 525–529.

Chapter 38

Sclerocarya birrea Marula (Homeid) Seed Oil

INTRODUCTION

Sclerocarya birrea ssp. *caffra* is a Savannah tree, belonging to the family Anacardiaceae. The common English name is Marula or Cider tree, and the tree is commonly known in Sudan as Homeid, where it is widely distributed in the western and southern areas. *S. birrea* requires sandy or alluvial soils and its propagation is possible by seeds or cuttings (Voget, Steele, & Streit, 2006). It grows in a wide variety of soils but prefers well-drained soil. It exists at altitudes varying from sea level to 1800 m and an annual rainfall range of 200–1500 mm. Its major habitat limitation is probably its sensitivity to frost (Wynberg et al., 2002). The plant develops pale yellow fruits, which are plum like, 3–4 cm in diameter with a plain tough skin and a juicy mucilaginous flesh. The fruit is edible and contains a hard brown seed. The seed encloses 2–3 soft white edible kernels (nuts), which are rich in oil and protein (Mariod, Matthäus, & Eichner, 2004). The highly aromatic sweet-sour fruit can be eaten fresh, like a small mango, or used for the preparation of juices, jams, conserves, dry fruit rolls, and alcoholic beverages. The seeds named as "the kings' nut" are highly appreciated by the local population, which eats them as delicate nuts (Mariod, Matthäus, Eichner, & Hussein, 2005a) (Fig. 38.1).

FOOD USES

The in vitro protein digestibility of *S. birrea* was almost similar to that of soybean protein concentrate and less than that of lupine, where 79% of the *S. birrea* seed protein was found digestible by pancreatic enzyme, which was similar to 79% of soybean concentrate and less than 83.2% of lupine (Mariod, Ali, Elhussein, & Hussien, 2005b). With a chemical score of 33.0%, based on the essential amino acids pattern requirements for children, the limiting amino acid in *Sclerocarya* protein is lysine (Mariod et al., 2005b). Glew et al. (2004) reported a protein content of 36.4% of dry weight; however, the protein fraction contained relatively low proportions of leucine, phenylalanine, lysine, and threonine. Due to widespread occurrence, potentially high fruit production, and use of *S. birrea*,

Unconventional Oilseeds and Oil Sources. http://dx.doi.org/10.1016/B978-0-12-809435-8.00038-X

FIGURE 38.1   *Sclerocarya birrea* **tree, fruits and seeds.**

products have been produced from seed kernels by adding them to Halva confectionary (a confectionary made of sesame paste and mixed sugars) and biscuits. In the case of Halva confectionary, 10% unroasted, blanched, and roasted kernels were added, whereas in biscuits *Sclerocarya* seed oil instead of hydrogenated oils was used. The results showed a statistically significant difference ($P \leq 0.05$) in texture, flavor, and overall preference between the Halva developed products. Biscuits processed by using the seed oil instead of hydrogenated oils were found to be significantly ($P \leq 0.05$) less acceptable using sensory scores (1–9) than the conventional biscuits (Mariod et al., 2005b). *S. birrea* oil was used in blending, cosmetics, and biodiesel production with for its high stability.

S. birrea seeds obtained from Abu Gibaiha and Ghibaish provinces, Sudan, contained 53.0, 28.0, and 8.0% of oil, protein, and carbohydrate respectively (Mariod et al., 2005b). The amino acid ratios for *S. birrea* result in an amino acid score in mg/g N. All the essential amino acids, excluding tryptophan, were present in fair amounts. The total amount of the essential amino acids comprised about 38.6% of the total amino acids. The percentage of sulfur-containing amino acid (methionine and cysteine) in *S. birrea* kernel was 4.3% of the total amino acids, which was high and desirable as food or ingredient (Mariod et al., 2005b).

Sclerocarya birrea SEED OIL

The oil obtained from *S. birrea* collected in Sudan from Abu Gibaiha agricultural area was investigated. The oil content of seeds from *S. birrea* amounted to 53.5%, which is comparable with that reported by Mizrahi, Nerd, and Janick

(1996) which was 56%. The major fraction in *S. birrea* seed oil was triacyglyc-erol, representing 76.5% of the total lipid, followed by phospholipids 12.5% and diacyglyecrol 5.6% (Mariod, 2005). The *S. birrea* oil contained 67.2% oleic acid, 5.9% linoleic acid, 14.1% palmitic acid, and traces of linolenic acid (Table 38.1). The tocopherol content of the oil amounted to 13.7 mg/100 g oil (Table 38.2). γ-Tocopherol was the predominant tocopherol in the oil of *S. birrea* and the total content of sterols in the oil was 287 mg/100 g oil, whereas β-sitosterol was determined as the main compound in the oil with about 60% of the total sterols. Also, higher amounts of 5-avenasterol (4.8 mg/100 g) was found. The oxidative stability of the oil, as measured by the Rancimat test at 120°C, was 43–44 h. In other study of Nigerian, Sclerocarya reported that it contained 50.7 % stearic, 22.6% palmitic, 8.4% arachidonic acid, with an io-dine value of 102 and 3.1% unsaponifiable matter (Mariod et al., 2004; Mariod, Matthäus, Eichner, & Hussein, 2008).

TABLE 38.1 Fatty Acid Composition (%) of *S. birrea* Seed Oil

Fatty acid		*S. birrea* seed oil
Name		Mean ± SD
Lauric acid	12:0	0.31 ± 0.02
Myristic acid	14:0	0.33 ± 0.09
Palmitic acid	16:0	14.16 ± 0.21
Palmitoleic acid	16:1 n–7	0.15 ± 0.09
Margaric acid	17:0	0.11 ± 0.01
Stearic acid	18:0	8.84 ± 0.23
Oleic acid	18:1 n–9	67.25 ± 0.31
Vaccenic acid	18:1 n–11	0.84 ± 0.03
Linoleic acid	18:2 n–6	5.93 ± 0.11
Linolenic acid	18:3 n–3	0.12 ± 0.04
Eicosanoic acid	20:0	0.91 ± 0.04
Gadoleic acid	20:1 n–9	0.36 ± 0.01
Behenic acid	22:0	0.22 ± 0.01
Lignoceric acid	24:0	0.31 ± 0.02
SAFA	Σ SAFA	25.19 ± 0.63
MUFA	Σ MUFA	67.76 ± 0.44
PUFA	Σ PUFA	6.05 ± 0.15
Ratio UFA/SAFA	Ratio UFA/SAFA	29.0

The determinations were mean of three replicates ± standard deviation (SD); SAFA, saturated fatty acids; MUFA, monounsaturated fatty acids; PUFA, polyunsaturated fatty acids; UFA is the sum of MUFA and PUFA

TABLE 38.2 Distribution of Sterols and Tocopherols of *S. birrea* Seed Oil (mg/100 g)

Types of sterol	*S. birrea* seed oil mean ± SD
Cholesterol	01.0 ± 0.21
Campesterol	02.1 ± 0.12
Stigmasterol	13.41 ± 1.11
β-Sitosterol	180.1 ± 5.23
Δ5-Avenasterol	47.6 ± 3.54
Δ7-Avenasterol	04.8 ± 1.63
Δ7-Stigmasterol	04.8 ± 1.15
Total	286.6 ± 6.32
α-Tocopherol	0.4 ± 0.01
γ-Tocopherol	13.0 ± 0.14
δ-Tocopherol	0.3 ± 0.17
Total tocopherol	13.7 ± 0.14

The determinations were mean of three replicates ± standard deviation (SD).

The tocopherol content of the oil is given in Table 38.2. Oils of *S. birrea* seed oil had low amounts of α-tocopherol, γ-tocopherol, and δ-tocopherol of 0.4, 13.7, and 0.3 mg/100 g, respectively, compared to other common vegetable oils, such as sesame oil, groundnut oil, or sunflower oil, in which the amount of tocopherols was between 27.9 and 97.6 mg/100 g. The main tocopherol of *S. birrea* seed oil was γ-tocopherol at about 95% of the total tocopherols. The other tocopherols in the oil occurred at below 1 mg/100 g each. The amount of sterols in the oil was 286.6 mg/100 g and the details of types of sterols are shown in Table 38.2.

In the oil of *S. birrea* seed a remarkably high amount of Δ5-avenasterol was found to be about 16% of the total sterols. The Δ5-avenasterol is known to act as an antioxidant and as an antipolymerization agent in frying oils (White and Armstrong, 1986).

The oxidative stability of the oils is expressed as the induction period (IP) determined by the Rancimat method at 120°C (Table 38.3). Oils from *S. birrea* seed showed a remarkably high stability (43–44 h) compared to other vegetable oils those shown in Table 38.3. The oxidative stability of such commonly used oils ranges between 0.3 h for linseed oil and about 10 h for groundnut fat.

USES OF *Sclerocarya birrea* SEED OIL

S. birrea-developed products were produced by adding to Halva confectionary 10% unroasted, blanched, and roasted kernels and biscuits processed by using the seed oil instead of hydrogenated oils (Mariod et al., 2005b). The protein

TABLE 38.3 Oxidative Stability as Induction Period[a]

Oil	Temperature (°C)	Induction period (h)
S. birrea seed oil	120	43–44
Cottonseed oil	120	2–3
Palm oil	120	7–12
Corn oil	120	Approximately 5
Coconut oil	120	Approximately 33

[a]*The determinations were mean of many replicates according to various references.*

contained good level of sulfur-containing amino acids (methionine and cysteine), when compared to that of four different food materials. The in vitro protein digestibility of *S. birrea* was almost similar to that of soybean protein concentrate and lupine. With a chemical score of 33.0%, based on the essential amino acids pattern requirements for children, the limiting amino acid in *Sclerocarya* protein was lysine (Mariod et al., 2005b).

The color of the fried potatoes in *S. birrea* seed oil was evaluated as satisfactory by the taste panel during the frying experiment and no significant change was found. With respect to the softness and flavor of the potatoes, the results were only satisfactory during the first 24 h; afterward, the products being fried were not acceptable for human consumption and would be rejected by the consumer. A similar result was found for the crispness parameter. After only 12 h of frying, the crispness of the potatoes being fried was evaluated as being less than satisfactory. These results show that *S. birrea* seed oil was not suitable for deep-frying of potatoes beyond 24 h (Mariod, Matthäus, Eichner, & Hussein, 2006a).

S. birrea seed oil, long-term stable oil from Sudan, was transesterified using methanol or ethanol in the presence of sulfuric acid; the obtained biodiesel characteristics were studied in accordance with the DIN EN 14214 specifications for biodiesel. Most of the biodiesel characteristics met the DIN specifications (water content, iodine number, phosphorus content). The kinematic viscosity values of all samples were higher than those of biodiesel standard limits. Concerning the oxidative stability, *S. birrea* seed oil has an IP higher than the required limit. It was easy and simple to prepare the methyl and ethyl esters catalyzed using H_2SO_4 from *S. birrea* seed oils (Mariod, Klupsch, Hussein, & Ondruschka, 2006b). The central problem in using vegetable oil as a diesel fuel is that vegetable oil is much more viscous (thicker) than conventional diesel fuel; it is 11–17 times thicker. Vegetable oil also has very different chemical properties and combustion characteristics from those of conventional diesel fuel or biodiesel, and it does not burn the same in the engine. Generally, the higher biodiesel iodine value, the lower the temperature at which it solidifies and may cause the cold filter plugging point (CFPP). Biodiesel from *S. birrea* seed oil has a lower CFPP than diesel fuel (Mariod et al., 2006b) and that will be of great importance when biodiesel is used in a cold weather. Also, transesterification using ethanol gave lower CFPP than using methanol in all samples.

Oxidative stability of biodiesel is an important issue because fatty acid derivatives are more sensitive to oxidative degradation than mineral fuel. Therefore, in the most recent European specification for biodiesel, a minimum value of 6 h for the IP at 110°C, measured with a Rancimat instrument, is specified. *S. birrea* seed oil has an IP higher than this limit. The natural stability of biodiesel in general depends upon the fatty acid composition of the different raw materials as well as on the different contents of natural antioxidants (Mariod et al., 2006b).

CHANGES IN OXIDATIVE STABILITY

The oxidative stability of oils can be evaluated experimentally in different ways, for example, by measuring the amount of oxygen absorbed, by changes of conductivity of solutions containing secondary oxidation products or by time required to attain a predetermined peroxide value (PV). The period of time that elapses until the oxidation process accelerates can be used to measure the resistance of oil against oxidation and is expressed as IP or induction time (Wagner and Elmadfa, 1999). The initial values of IP of the *S. birrea* seed oil stability were more than 40 h.

Changes in the PV as a measure of primary oxidation in the oil during storage were studied. The initial PV of *S. birrea* seed oil was 0.2 mEq O_2/kg and at the end of the investigation the oils reached PVs of 1.0 mEq O_2/kg. The mean values of the PVs of the oils were significantly different ($P > 0.05$) at the end of the experiment. In *S. birrea* seed oil, PV remained constant throughout the first year and then there was a very small increase with progressive storage time from 0.2 to 1.0 mEq O_2/kg oil in the last 12 months. This indicates that *S. birrea* seed oil was highly stable against oxidative deterioration during long-term storage. The PV then increased slightly, which can be explained by the fact that this is a long-term oxidation study on monounsaturated oil which contains some natural antioxidants that prevent oxidation at first then by stages of time these antioxidants decreased and the PV increased.

The high stability to oxidation of *S. birrea* seed oil might be correlated with the high content of saturated fatty acids and to the high total amount of phenolic components (20.66 mg/100 g oil) as well as to the study by (Mariod, Matthäus, Eichner, & Hussein, 2015)

REFERENCES

Glew, R. S., Vander Jagt, D. J., Huang, Y. S., Chuang, L. T., Bosse, R., & Glew, R. H. (2004). Nutritional analysis of the edible pit of *Sclerocarya birrea* in the republic of Niger (daniya, Hausa). *Journal of Food Composition and Analysis, 17*, 99–111.

Mariod, A. A. (2005). Investigations on the oxidative stability of some unconventional Sudanese oils, traditionally used in human nutrition. *PhD thesis*. Münster University, Münster, Germany.

Mariod, A., Matthäus, B., & Eichner, K. (2004). Fatty acid, tocopherol and sterol composition as well as oxidative stability of three unusual Sudanese oils. *Journal of Food Lipids, 11*(3), 179–189.

Mariod, A., Matthäus, B., Eichner, K., & Hussein, I. H. (2005a). Improving the oxidative stability of sunflower oil by blending with *Sclerocarya birrea* and *Aspongopus* viduatus oils. *Journal of Food Lipids, 12*(2), 150–158.

Mariod, A., Ali, A., Elhussein, S., & Hussien, I. (2005b). Quality of proteins and products based on *Sclerocarya birrea* (Marula) seed. *Sudan Journal of Science and Technology, 6*, 184–192.

Mariod, A., Matthäus, B., Eichner, K., & Hussein, I. H. (2006a). Frying quality and oxidative stability of two unconventional oils. *Journal of the American Oil Chemists' Society, 83*, 529–538.

Mariod, A., Klupsch, S., Hussein, I. H., & Ondruschka, B. (2006b). Synthesis of alkyl esters from three unconventional Sudanese oils for their use as biodiesel. *Energy Fuels, 20*(5), 2249–2252.

Mariod, A., Matthäus, B., Eichner, K., & Hussein, I. H. (2008). Long-term storage of three unconventional oils. *Grasas y Aceites, 59*(1), 16–22.

Mariod, A., Matthäus, B., Eichner, K., & Hussein, I. H. (2015). Phenolic compounds of three unconventional Sudanese oils. *Acta Scientiarum Polonorum. Technologia Alimentaria, 14*(1), 63–69.

Mizrahi, Y., Nerd, A., & Janick, J. (1996). New crops as a possible solution for the troubled Israeli export market. *Progress in new crops: proceedings of the third national symposium, 22–25* October, 1996. Indianapolis, Indiana, USA: American Society for Horticultural Science.

Voget, S., Steele, H., & Streit, W. (2006). Characterization of a metagenome-derived halotolerant cellulase. *Journal of Biotechnology, 126*(1), 26–36.

Wagner, K. -H., & Elmadfa, I. (1999). Nutrient Antioxidants and stability of frying oils (Tocochromanols, β-Carotene, Phylloquinone, Ubiquinone 50). In D. Boskou, & I. Elmadfa (Eds.), *Frying of Food* (pp. 163–182). Lancaster-Basel: Technomics, Publishing CO. Inc.

White, P. J., & Armstrong, L. S. A. (1986). Effect of selected oat sterols on the determination of heated soybean oil. *Journal of the American Oil Chemists' Society, 63*, 525–529.

Wynberg, R., Cribbins, J., Leakey, R., Lombard, C., Mander, M., Shackleton, S. E., & Sullivan, C. A. (2002). A summary of knowledge on *Sclerocarya birrea* subsp. *caffra* with emphasis on its importance as a non-timber forest product in South and southern Africa. Part 1: taxonomy, ecology, traditional uses and role in rural livelihoods. *Southern African Forestry Journal, 196*, 67–78.

Chapter 39

Adansonia digitata Baobab Seed Oil

INTRODUCTION

Only eight species of baobabs (*Adansonia* spp.) are known *Adansonia gibbosa* is the only baobab species endemic to Australia and is restricted to the northwestern part of the country (Pavlic et al., 2008). *Adansonia digitata* has a wide natural distribution throughout tropical parts of Africa and the six other species are found on Madagascar (Pavlic et al., 2008). A study on baobabs presented the intriguing view that the distribution of these unusual trees between Africa and Australia occurred after the division of Gondwana (Baum, Small, & Wendel, 1998). The same study revealed that *A. gibbosa* in Australia is more closely related to *A. digitata* from Africa than it is to species from Madagascar. (Pavlic et al., 2008).

Baobab (*A. digitata*) is a tree plant belonging to the Malvaceae family and it is widespread throughout the hot, drier regions of tropical Africa (De Caluwé, Halamová, Van, & Damme, 2010). Baobab leaves, bark, and fruit are used as food and for medicinal purposes in many parts of Africa. The oil extracted from seed has a rich history of use by local people as a source of food, energy, medicine, and for cosmetic applications. It has been used in the production of lubricants, soaps, and personal care products, as well as in the topical treatment of various conditions, such as hair dandruff, muscle spasms, varicose veins, and wounds (Vermaak, Kamatou, Komane-Mofokeng, Viljoen, & Beckett, 2011).

The global demand for baobab raw material (e.g., seed oil, fruit pulp) by the food and beverage, nutraceutical, and cosmetic industries has increased dramatically in recent years thereby increasing the commercial value and importance of this coveted African tree. In the past few years, there has been an increased demand for nontimber forest products (NTFPs), specifically baobab seed oil for inclusion in cosmetic formulations due to its high fatty acid composition demand for seed oils as ingredients for food, cosmetics, and biofuel has greatly increased as industry seeks natural alternatives (Kamatou, Vermaak, & Viljoen, 2011).

Baobab seed cake is one of the potential low-cost and locally available protein sources in livestock diets for African Agriculture. However, *A. digitata* seed cake contains some antinutritional factors (such as oxalate, phytate, saponins, and tannins) and this lowers its use in all animal feed (Chimvuramahwe et al., 2011) (Fig. 39.1).

FIGURE 39.1 *Adansonia digatata* **tree, fruit and seeds.**

There is a renewed interest among scientists for identifying alternative natural and safe sources of food antioxidants, and their roles in the maintenance of human health. Antioxidants are inhibitors of propagation of free radical reactions in the human body. Coumarins, flavonoids, sesquiterpene lactones, and edotides are some of the antioxidants present in plant materials and responsible for the anticancer activity.

Oils extracted from plant sources have a rich history of use by local people as a source of food, energy, medicine, and for cosmetic applications. Baobab seed oil is a relatively untapped source of components that can be applied to various aspects of the food processing and dietary supplements or cosmetics.

Therefore, builds on the antioxidant activity of the oil due to high oleic acid content and it may be plausible to predict that the high levels of oleic acid in baobab seed oil will exhibit good antioxidant properties. Moreover, although there is an increase in demand for animal related proteins, baobab seed cake is consider high crude protein and essential amino acid levels and these properties may offer some antioxidant effects.

BAOBAB (*Adansonia digitata*) SEED OIL

In the past few years, there has been an increased demand for NTFPs, specifically baobab seed oil for inclusion in cosmetic formulations due to its high fatty acid composition (Kamatou et al., 2011). The seed contains relatively high amounts of protein (18.4 ± 0.5%), crude fat (12.2 ± 0.2%), and crude fiber (16.3 ± 9.3%), and low carbohydrate (45.1 ± 1.7%) (Osman, 2004). The Baobab seeds constitute relatively high protein compare with baobab cake which approximately 16.9% (Madzimure, Musimurimwa, Chivandi, Gwiriri, & Mamhare, 2011).

Baobab seed oil is an excellent source of mono- and polyunsaturated fatty acids. The oil is composed of approximately 31.7% saturated fatty acids, 37% monounsaturated fatty acids, and 31.7% polyunsaturated fatty acids. The major fatty acid is oleic acid, which comprises about 35.8%, followed by linoleic (30.7%) and palmitic (24.2%). These results were similar to those for African baobab oils (Osman, 2004).

The lipid content in baobab seed was 13% on dry weight basis. The fatty acid profile revealed the presence of oleic acid in a percentage of 42.5%. Also, baobab seeds exhibited an appreciable level of linoleic acid making up to 28% and palmitic in a percentage of 22% of total fatty acids. The high content of linoleic acid in baobab seed reflects the nutritive significance of the seed and the potential of baobab seed oil as healthy food oil (Osman, 2004). The oil from the baobab seed contains about 1–2 mg/g linoleic acid, which is an essential fatty acid required by the body for growth and development. The high content of mono and polyunsaturated fatty acids suggests that the baobab seed oil consumed may help lower cholesterol level (Ajayi, Dawodu, Oderinde, & Egunyomi, 2003). These properties make the baobab seed oil a good option for preparing healthier foods and as both additive and preservative. Baobab seed oil displayed the slowest rate of oxidation (8.2 h) compared to olive and evening primrose oil, 5.4 and 3.1 h, respectively (Arch personal care products). The seed oil composition consists β-sitosterol (≈80% of the total sterols) is one of the major sterol constituents present in baobab seed oil. Other sterols include campesterol (8.3%) and stigmasterol. β-Sitosterol is a known antioxidant able to reduce DNA damage and the level of free radicals, in addition to possibly increasing the level of typical antioxidant enzymes.

USES OF BAOBAB SEED OIL

Baobab oil is used alone or in combination with other plant parts to treat various conditions, such as fever, diarrhea, coughs, dysentery, hemoptysis, and worms (Vermaak et al., 2011). Baobab oil is extremely stable with a highly variable shelf life estimated to be between 2 and 5 years. The high saponification value of baobab oil is comparable to some of the edible oils, such as marula oil, groundnut oil, and palm oil (Vermaak et al., 2011). Another patent for an oil absorbent wipe intended for use on the skin or hair, lists baobab oil as a possible ingredient (Seth, Katagiri, & Sakurai, 2004).

Baobab oil has been added to the list of fixed oils commonly included in cosmetic products because it will not burn the skin when applied as such, and it is said to be nonirritating as well as nonsensitizing (Wren & Stucki, 2003). Baobab oil is highly penetrating, deeply nourishing and softens dry skin. It is known to restore and remoisturize the epidermis (Africa) (Vermaak et al., 2011). During the extraction of oil from oilseeds, the antioxidant compounds present in the hulls could be incorporated into the oil, as reported for peanut oil extracted from the coated seeds, which contained higher oxidative stability than the oil from dehulled seeds (Shahidi, Amarowicz, Abou-Gharbia, & Shehata, 1997).

The oil is said to alleviate pain from burns and regenerates the epithelial tissues in a short time, thereby improving skin tone and elasticity (Kamatou et al., 2011). The oil can be used as a protecting, nourishing, moisturizing, soothing, and regenerating agent. Studies have also shown that the oil contains antioxidants (Nkafamiya, Osemeahon, Dahiru, & Umaru, 2007), can protect the skin against premature aging and prevents the appearance of wrinkles. Alone or combined with other ingredients, it is also used to aid in skin healing (small cuts, chapping) or as a mask for hair care (dry, brittle hair, split ends). Linoleic acid (found in baobab seed oil) is the most frequently used fatty acid in cosmetic products as it. The oil is used in wound care therapy and bath oil preparations, as moisturizer and massage oil, and hot oil soaks are used for hair.

REFERENCES

Ajayi, I., Dawodu, F., Oderinde, R., & Egunyomi, A. (2003). Fatty acid composition and metal content of *Adansonia digitata* seeds and seed oil. *Rivista Italiana delle Sostanze Grasse, 80*(1), 41–43.

Baum, D. A., Small, R. L., & Wendel, J. F. (1998). Biogeography and floral evolution of baobabs *Adansonia*, Bombacaceae as inferred from multiple data sets. *Systematic Biology, 47*(2), 181–207.

Chimvuramahwe, J., Musara, J., Mujuru, L., Gadzirayi, C., Nyakudya, I., Jimu, L., Katsvanga, C., Mupangwa, J., & Chivheya, R. (2011). Effect of feeding graded levels of *Adansonia digitata* (baobab) seed cake on the performance of broilers 1. *Journal of Animal and Plant Sciences*(11), 1442–1449.

De Caluwé, E., Halamová, K., Van, P., & Damme (2010). *Adansonia digitata* L.—a review of traditional uses, phytochemistry and pharmacology. *Afrika Focus, 23*(1), 11–51.

Kamatou, G., Vermaak, I., & Viljoen, A. (2011). An updated review of *Adansonia digitata*: a commercially important African tree. *South African Journal of Botany, 77*(4), 908–919.

Madzimure, J., Musimurimwa, C., Chivandi, E., Gwiriri, L., & Mamhare, E. (2011). Milk yield and quality in Guernsey cows fed cottonseed cake-based diets partially substituted with baobab (*Adansonia digitata* L.) seed cake. *Tropical Animal Health and Production, 43*(1), 77–82.

Nkafamiya, I., Osemeahon, S., Dahiru, D., & Umaru, H. (2007). Studies on the chemical composition and physicochemical properties of the seeds of baobab (*Adansonia digitata*). *African Journal of Biotechnology, 6*(6), .

Osman, M. A. (2004). Chemical and nutrient analysis of baobab (*Adansonia digitata*) fruit and seed protein solubility. *Plant foods for Human Nutrition, 59*(1), 29–33.

Pavlic, D., Wingfield, M. J., Barber, P., Slippers, B., Hardy, G. E. S. J., & Burgess, T. I. (2008). Seven new species of the Botryosphaeriaceae from baobab and other native trees in Western Australia. *Mycologia, 100*(6), 851–866.

Seth, J., H., Katagiri, & Sakurai, H. (2004). *Oil absorbent wipe with rapid visual indication*. Google Patents.

Shahidi, F., Amarowicz, R., Abou-Gharbia, H., & Shehata, A. A. Y. (1997). Endogenous antioxidants and stability of sesame oil as affected by processing and storage. *Journal of the American Oil Chemists' Society*, *74*(2), 143–148.

Vermaak, I., Kamatou, G. P. P., Komane-Mofokeng, B., Viljoen, A., & Beckett, K. (2011). African seed oils of commercial importance—cosmetic applications. *South African Journal of Botany*, *77*(4), 920–933.

Wren, S., & Stucki, A. (2003). Organic essential oils, indigenous cold pressed oils, herbs and spices in Sub-Saharan Africa. *International Journal of Aromatherapy*, *13*(2), 71–81.

Chapter 40

Sterculia setigera (Karaya Gum Tree) A New Oil Source

DISTRIBUTION AND BOTANICAL DESCRIPTION

Sterculia setigera is a savanna tree. It is wide spread in the savanna area of tropical Africa; often characteristic of stony hills. *S. setigera* is found in Senegal, in the Sudan-Sahel, and in the Sudan-Guinea zone, in Togo, eastward to Sudan and Somalia, East Africa, also in Angola (Von Maydell, 1990). *S. setigera* is a multifunctional forest woody tree species in sub-Saharan Africa, especially known for its economic value, its gum has been exported since several decades. In Senegal, several studies approached about *S. setigera* were published on different aspects of the availability, the socioeconomic, the regeneration, the agroforestry, and sustainable gum harvesting. Due to its commercialization, the gum contributes also to increase women's income in the rural areas (Niang et al., 2010).

In Sudan the tree is wide spread deciduous species of Savanna woodland. Although occurring gregarious in southern parts of Blue Nile State on the stony slopes of hilly sites, it is also found sporadically in Bahar Elgazal. It is also found in Kassala, Darfur (common below Golol in Jebel Marra), Kordofan (Nuba mountains), Red Sea hills and common in southern states also in White Nile State (Jebelin), Equatoria State. It is deciduous, grows up to 12 m high and 15 m in girth. The bark is pale purplish, smooth with thin scales, which peel off expose yellowish patches, exuding gummy sap. The 3-lobed leaves, 6–20 cm broad and long, broadly ovate, and densely pubescent on both surfaces. Flowers are borne in small inflorescence in the previous year's shoots. Compound dry fruits composed of 3–5 separate boat-shaped follicles, which are arranged to form a star shape. Each follicle is 6–10 cm long and 4–5 cm across, with a pointed apex. They are velvety outside and inside with pungent bristles along the placenta line where the seeds attach. They have a thin wall, under 5 mm thick, and are greenish or brown when ripe. Each part of the fruit contains c. 12 seeds (Sacandé, Sanon, & Schmidt, 2007). The fruits are composed of 4 or 5 boat-shaped carpels, which split to reveal about 12 slate-colored seeds, 10–15 mm long (Keay, 1989). The seed is an indehiscent diaspore (11.85–12.90 mm × 7.75–8.30 mm × 7.75–8.30 mm), black in color, and ellipsoid with a smooth surface and a small hilum. The embryo is yellowish-white, axial-investing, and occupies a quarter or more of the seed volume. The seed has no perisperm. The

Unconventional Oilseeds and Oil Sources. http://dx.doi.org/10.1016/B978-0-12-809435-8.00040-8

FIGURE 40.1 *Sterculia setigera* tree, fruit and seeds.

seeds are expected to have physical dormancy, which can be overcome by scarifying the seed coat to allow water uptake and encourage germination (Fig. 40.1). The seeds of *S. setigera* are numerous, purplish-black 10–14 mm long, 7.1 mm width, 6.9 mm thickness with small yellow-brown fleshy aril at base, composed of a thin outer coat (like the seed coat of the groundnut). Weight of 100 seeds was 28.5 g.

CHEMICAL COMPOSITION AND MAIN BIOACTIVE COMPONENTS

Idu, Uzoekwe, and Onyibe (2008) reported the crude protein as 21.40%, crude fiber 11.58%, and total carbohydrate content as 21.03% of the seed of *S. setigera*. These authors recorded the levels of macro minerals (calcium, potassium, magnesium, sodium, and Iron) in the seed as 108.0, 105.0, 59.0, and 28.41 mg/100 g DM and 27.12 ppm, respectively, and the levels of the trace elements (manganese, zinc, and copper), 19.66, 18.74, and 8.69 ppm, respectively were estimated in the seed sample. Ayodele, Alao, and Olagbemiro (2000) analyzed *S. setigera* seeds for proximate composition, elemental and amino acid composition. They reported that the seeds were adequate in their protein content. Quantitative chromatographic analysis of the seed hydrolyzates revealed 18 amino acids. Comparing the amino with hens egg, they showed a higher superiority in alanine, arginine, aspartic, cystine, and histidine but a deficiency in isoleucine and leucine was reported. The seed contains a considerable amount of oil, which has not been studied and hence not utilized in Sudan.

MEDICINAL AND OTHER USES

About 75% of gum karaya produced worldwide is used in pharmaceutical applications (external use), 20% in paper and textile technological industries, and about 5% is used in foodstuffs. Atakpama et al. (2012) conducted ethnobotanical investigations in Sudanian zone of Togo to identify use values knowledge of *S. setigera* tree. They reported that plant parts values and specific uses are raised more in the Moba's ethnic group. The main quoted uses are medicinal, religious, food, and cosmetic.

The tree has different medicinal uses, for example, the water extracts of bark are used for jaundice and bilharzia, leaves and bark are used in treating cough, diarrhea, fever, leprosy, syphilis, and as diuretic. The best is used for clothing and bark fibers are used for ropes and mats. The tree, boiled leaves are used to treat malaria, and the stem bark decoction is used for the treatment of asthma, bronchitis, wound, fever, toothache, gingivitis sore, abscess, and diarrhea (Igoli, G.Ogaji, Tor-Anyiin, & Igoli, 2005). A supportive evidence of the use of the plant in folkloric medicine was provided by the study of Sunday and Sunday (2011) in southwest Nigeria. The medicinal and cosmetic use of *S. setigera* was studied in Togo focused on floristic and ethnobotany (Adjanohoun, Ahyi, & AkeAssi, 1986).

FATTY ACID COMPOSITION AND PHYSICOCHEMICAL CHARACTERISTICS OF TARTAR (*Sterculia setigera*) OIL

The saturated fatty acid content varied from 0.2% to 33.4%, while unsaturated fatty acid content varied from 0.1% to 24.2%, with no significant difference between the solvent and mechanical extraction methods. Oleic acid represented 33.4% of the total fatty acids composition, followed by palmitic 24.2% and linoleic 21.0%. Cyclopropenoid fatty acids ranged from 3.4% to 5.3% but were significantly reduced after refining. Glycerides content varied with maximum values of 78.03, 4.58, and 2.82 for tri-, di-, and monoglycerides, respectively. Sterols content varied from 0.3% to 66.0%, β-sitosterol represented 66.0% of the total sterols followed by stigmasterol (15.1%), campesterol (9.0%), and Δ-4-avenasterol (6.3%). Physicochemical investigation of the oil showed that the color, refractive index, free fatty acids, peroxide value, saponification value, relative viscosity, iodine value, and unsaponifiable matter were similar for both methods of extraction with minor exceptions (Eljack, Babiker, & El Tinay, 2004). Oil content was reported as 26.45%, crude protein as 24.2%, while the crude fiber content was 26.2% and the total carbohydrate was 11.93%. The oxidative stability at 100°C of crude Tartar seed oil was 14.5 h and it was reduced as the effect of bleaching process to be 3.12 h.

REFERENCES

Adjanohoun, E. J., Ahyi, M. R. A., AkeAssi, L. (1986). *Contribution aux études ethnobotaniques et floristiques au Togo*. Medecine Traditionnelle et Pharmacopee, Agence de Cooperation Culturelle et Technique.

Atakpama, W., Batawila, K., Dourma, M., Pereki, H., Wala, K., Dimobe, K., Akpagana, K., Gbeassor, M. (2012). Ethnobotanical knowledge of *Sterculia setigera* Del. in the Sudanian zone of Togo (West Africa) (Vol. 2012). Article ID 723157, 8 pages.

Ayodele, J. T., Alao, O. A., & Olagbemiro, T. O. (2000). The chemical composition of *Sterculia setigera*. *Nigerian Journal of Animal Science, 3*(2), 33–37.

Eljack, M., Babiker, E. E., & El Tinay, A. H. (2004). Fatty acid composition and physicochemical characteristics of Tartar (*Sterculia stigera*) oil as affected by the extraction method. *University of Khartoum Journal of Agricultural Sciences, 12*(2), 185–195.

Idu, M., Uzoekwe, S., & Onyibe, H. I. (2008). Nutritional Evaluation of *Sterculia setigera* seeds and pod. *Pakistan Journal of Biological Sciences, 11*, 139–141.

Igoli, J. O., G.Ogaji, O., Tor-Anyiin, T. A., & Igoli, N. P. (2005). Traditional medicine practice amongst the Igede people of Nigeria. Part II. *The African Journal of Traditional, Complementary and Alternative Medicines, 2*(2), 134–152.

Keay, R. W. J. (1989). *Trees of Nigeria* (1st ed.). Oxford: Clarendon Press (pp. 288–298).

Von Maydell, H. (1990). *Trees and Shrubs of the Sahel, Their Characteristics and Uses*. Weikersheim, Germany: GTZ. Verlag Josef Margraf Scientific Books.

Niang, D., Gassama, Y. K., Ndiaye, A., Sagna, M., Samba, S. A. N., & Toure, M. A. (2010). In vitro micrografting of *Sterculia setigera* Del. *African Journal of Biotechnology, 9*(50), 8613–8618.

Sacandé, M., Sanon, M., & Schmidt, L. (Eds.), (2007). *Sterculia setigera Delile*. Seed Leaflet (134).

Sunday, Y. S., & Sunday, O. (2011). Anti-nociceptive and anti inflammatory properties of the aqueous leaf extract of *Sterculia setigera* Del (Sterculiaceae). *Journal of Pharmacology and Toxicology, 6*(5), 516–524.

Chapter 41

Albizia lebbeck (L. Benth.) Lebbeck Tree Seed

BOTANICAL DESCRIPTION

Albizia lebbeck (L. Benth.), family Mimosaceae is a fast growing nitrogen fixing, heavy shade tree, recommended for reforestation and firewood plantations. It is native to tropical Africa, Asia, and northern Australia, widely planted and naturalized throughout the tropics (NAS, 1980). It is deciduous, unarmed tree to 20 m (65 ft.) tall, with a rounded, spreading crown, and pale bark. Leaves alternate, twice compound, with 2–5 pairs of pinnae, each pinna with 3–10 pairs of leaflets (even-pinnate); leaflets elliptic-oblong, 2–4 cm (1–2 in.) long, usually asymmetrical at base, dull green above, paler green below; petiole with a sessile, elliptic gland near the base above. Flowers are mimosa-like, in showy, rounded clusters near stem tips, 5–6 cm (2–2.5 in.) across, cream or yellowish-white, each flower with numerous long stamens. Fruit a flat, linear pod, to 30 cm (1 ft.) long, with many seeds; dried pods persistent after leaf-fall, often heard rattling in the wind (Fig. 41.1).

DISTRIBUTION AND HABITAT

A. lebbeck (L.) Benth. is found in dense deciduous forests in tropical and subtropical countries of Asia, such as Laos, Cambodia, Malaysia, Indonesia and Vietnam, Africa (Dy Phon, 2000), and Australia. It has been widely cultivated and is now pantropical. The species grows poorly on heavy clays, but grows well on fertile, well-drained, loamy soils, in areas that receive from 600 to 2500 mm of rain per year. However, this species is also capable of tolerating years with as little as 300 mm of rainfall. It is normally encountered below 1800 m a.s.l., and prefers mean annual temperatures from 20 to 35°C. This tree is nitrogen fixing, and tolerates acidity, alkalinity, heavy and eroded soils, waterlogged soils, and drought. Older trees can survive grass fires and intense night frost, and although these events will kill off aboveground growth of young trees, new growth normally follows (DFSC, 2000).

MEDICINAL USES

Bark containing saponin can be used in making soap, and containing tannin can be used for tanning. The tree is used in folk remedies for abdominal tumors, cough, eye ailments, flu, and lung ailments. The seed oil is used for leprosy, the

Unconventional Oilseeds and Oil Sources. http://dx.doi.org/10.1016/B978-0-12-809435-8.00041-X

FIGURE 41.1 *Albizia lebbeck.*

powdered seed to scrofulous swellings. Fernandez et al. (1996) reported the use of *A. lebbeck* seedpods for water softening by the adsorption of calcium at 25°C and they mentioned that the adsorption increased with the pH value. Noté et al. (2009) investigated and isolated two new oleanane-type saponins, named lebbeckosides A–B (1–2) from the roots of *A. lebbeck;* these compounds were evaluated for their inhibitory effect on the metabolism of high grade human brain tumor cells, the human glioblastoma U-87 MG cell lines and the glioblastoma stem-like TG1 cells were isolated from a patient tumor, and known to be particularly resistant to standard therapies. The isolated saponins showed significant cytotoxic activity against U-87 MG and TG1 cancer cells with IC_{50} values of 3.46 and 1.36 lM for 1, and 2.10 and 2.24 lM for 2, respectively.

In Southeast Asia and Australia, the stem bark is used as a folk remedy to treat abdominal tumors, boils, cough, eye disorders, and lung ailments. It is also reported to be astringent, pectoral, rejuvenating, and tonic (Hartwell, 1971). Nootropic and anxiolytic activities of a saponin fraction isolated from *A. lebbeck* leaves have been reported (Une, Sarveiya, Pal, Kasture, & Kasture, 2001). Oral administration of the saponin fraction isolated from *A. lebbeck* bark to male rats has been reported to significantly reduce fertility through reduction of sperm mobility and density (Gupta, Pradeep, & Soni, 2005). Different extracts of *A. lebbeck* showed good inhibitory effects against different pathogens. The methanolic extracts of *A. lebbeck* illustrated a high zone of inhibition against the pathogens *Bacillus subtilis* (16 mm), *Escherichia coli* (22 mm), *Klebsiella pneumonia* (11 mm), *Proteus vulgaris* (18 mm), *Pseudomonas aeruginosa* (22 mm), *Salmonella typhi* (23 mm), and *Staphylococcus aureus* (17 mm). The ethyl acetate extracts demonstrated maximum zone of inhibition against *E. coli* (26 mm), *P. aeruginosa* (22 mm), and *K. pneumonia* (16 mm) (Bobby, Wesely, & Johnson, 2012). In carrageenan-induced animal models, the petroleum ether extract of *A. lebbeck* at concentrations 100, 200, and 400 mg inhibited the edema formation in 3rd hour by 23.7% ($P < 0.05$), 46.5%, and 48.6% ($P < 0.005$) in a dose-dependent manner, respectively. This effect also extended and significantly increased up to the 5th hour ($P < 0.005$). The ethanol extract at 100, 200, and 400 mg showed 10.35, 31.62, and 59.57% of inhibition at 3rd hour and 13.23, 40.89, and 60.04% ($P < 0.005$) of inhibition at 5th hour, respectively. The chloroform extract showed a mild activity in higher concentration only. The reference drug group significantly inhibited the edema formation by 69.53 and 70.87% at 3rd and 5th hour, respectively ($P < 0.005$) (Babu, Pandikumar, & Ignacimuthu, 2009).

Albizia lebbeck SEED COMPOSITION

Seeds have yielded 5.3%–6.8% fixed oil or fat, the endosperm 11%. The oil contains 9.6% stearic, 10.9% arachidic, 39.3% oleic, and 32.9% linoleic acid. Bark contains 5%–15% tannin (leaves contain c. 4%) and saponins. The saponin from the seed yields oleanolic acid and albizziagenin (Mitchell & Rook, 1979).The seeds of *A. lebbeck* contain 33.60% crude protein and 3.13% crude fat, while the pods contain 17.86% crude protein and 2.6% crude fat. Prohibitive levels of toxic compounds were not detected in any of the plant parts analyzed (Roskoski, Gonzalez, Dias, Tejeda, & Vargas-Menay, 1980). Mariod and Matthäus (2008a) reported an oil content of 12.8% for *A. lebbeck* seeds collected from Ghibaish, North Kordofan state, Sudan; this oil content is very low compared with that of common oil seeds. Therefore, from an economical point of view, the production of oil from such seeds could not be interesting unless some genetic modifications could be applied.

MINOR COMPONENTS AND PHYSICOCHEMICAL PROPERTIES OF *Albizia lebbeck* OIL

Unsaponifiable composition of *A. lebbeck* oil was found as vitamin E, stigmastadiene, octadecane, tetradecane, nonadecane, phytol, hexadecane, and eicosane (Adewuyi & Oderinde, 2014). The content of tocopherols in freshly extracted oil of *A. lebbeck* was high (85.6 mg/100 g), among the tocopherols identified, α-tocopherol was most abundant accounting for 48.2 mg/100 g represented 56.3% of the total. The amounts of tocopherols (85.6 mg/100 g) is considered very high when compared with other common oils, such as sesame oil, groundnut oil, or sunflower oil, in which the amount of tocopherols was between 27.9 and 97.6 mg/100 g. The oil extracted from seeds showed 1.5% FFA, 1.41% unsaponifiables, with 1.4750 refractive index (30°C) and 0.9220 specific gravity (60°C) (Mariod & Matthaus 2008a).

FATTY ACID COMPOSITION OF *Albizia lebbeck* OIL

Mariod and Matthäus (2008a) studied the fatty acid composition of *A. lebbeck* oil; they found that 14 fatty acids were identified, among which linoleic acid contributed 43.8%, followed by oleic acid at 21.0% then palmitic acid at 16.4%, linolenic at 4.3%, and stearic acid at 5.2%. The remaining fatty acids contributed only few percentages to the total fatty acids present. Adewuyi and Oderinde (2014) reported fatty acid analysis of *A. lebbeck* oil as C20:1 which was represented 0.3, C22:1 was 6.30, and C18:2 was 47.2 ± 0.50 g/100 g fatty acids. The oil had 75.3 g amounts of unsaturated fatty acids.

The most dominant lipid class in the oil of *A. lebbeck* seeds was neutral lipid. The neutral lipid was 92.50 ± 0.40%. The fatty acids are distributed along the triglyceride bonds in different proportions with the phospholipids having the highest amounts of saturated fatty acids. It was also observed that the amounts of saturated fatty acid slightly increased in the glycolipids. The lipid classes accumulated C18:2 fatty acid in high amounts making these lipid classes good sources of this fatty acid. The long-chain fatty acids in *A. lebbeck* were found to be the highest in the neutral lipids except for C24:0, which was higher in the glycolipids (3.6 ± 0.20 g/100 g fatty acids) and phospholipids (3.5 ± 0.10 g/100 g fatty acids) than in the neutral lipids (1.6 ± 0.20 g/100 g fatty acids) (Adewuyi & Oderinde, 2014). The molecular species in *A. lebbeck* oil were reported based on classification of triglycerides, which was found containing most abundant fatty acids. The most abundant molecular species are LnLnLn (33.27 ± 0.5%) and LLnLn (33.86 ± 0.2%). Molecular species with equivalent carbon chain number C48 (SLnS/OOO/POP) was found the least in oil and it was 2.00 ± 0.2% (Adewuyi & Oderinde, 2014).

REFERENCE

Adewuyi, A., & Oderinde, R. A. (2014). Fatty acid composition and lipid profile of *Diospyros mespiliformis, Albizia lebbeck,* and *Caesalpinia pulcherrima* seed oils from Nigeria. *International Journal of Food Science, 2014,* 283614, Article ID 283614, 6 pages http://dx.doi.org/10.1155/2014/283614.

Babu, N. P., Pandikumar, P., & Ignacimuthu, S. (2009). Anti-inflammatory activity of *Albizia lebbeck* Benth., an ethnomedicinal plant, in acute and chronic animal models of inflammation. *Journal of Ethnopharmacology, 125,* 356–360.

Bobby, M. N., Wesely, E. G., & Johnson, M. A. (2012). In vitro anti-bacterial activity of leaves extracts of *Albizia lebbeck* Benth against some selected pathogens. *Asian Pacific Journal of Tropical Biomedicine*, *2*, S859–S862.

DFSC, (2000). Seed Leaflet No. 7, *Albizzia lebbeck* (L.) Benth, Danida Forest Seed Centre, September.

Dy Phon. (2000). *Dictionary of plants used in Cambodia*. Cambodia: Phnom Penh.

Fernandez, N., Chacin, E., Garcia, C., Alastre, N., Leal, F., & Forster, C. F. (1996). Pods from *Albizia lebbek* as a novel water softening biosorbent. *Environmental Technology*, *17*, 541–546.

Gupta, S., Pradeep, S., & Soni, P. L. (2005). Chemical modification of *Cassia occidentalis* seed gum: carbamoylethylation. *Carbohydrate Polymers*, *59*, 501–506.

Hartwell, J. L. (1971). Plants used against cancer; a survey. *LLOYDIA*, 103–160.

Mariod, A., & Matthäus, B. (2008a). Physico-chemical properties, fatty acid and tocopherol composition of oils from some Sudanese oil bearing sources. *Grasas y Aceites*, *59*(4), 321–326.

Mitchell, J. C., & Rook, A. (1979). *Botanical dermatology*. Vancouver, Canada: Green glass Ltd.

NAS. (1980). *Firewood crops: Shrub and tree species for energy production*. Washington, DC: National Academy of Sciences.

Noté, O. P., Mitaine-Offer, A. -C., Miyamoto, T., Paululat, T., Mirjolet, J. -F., Duchamp, O., & Lacaille-Dubois, M. -A. (2009). Cytotoxic acacic acid glycosides from the roots of Albizia coriaria. *Journal of natural products*, *72*(10), 1725–1730.

Roskoski, J. P., Gonzalez, G.C., Dias, M. F., Tejeda, E. P., Vargas-Menay, A. (1980). Woody tropical legumes: potential sources of forage, firewood, and soil enrichment. In: SERI, *Tree crops for energy co-production on farms* (pp. 135–155). Washington: USGPO, SERI/CP-622-1086.

Une, H. D., Sarveiya, V. P., Pal, S. C., Kasture, V. S., & Kasture, S. B. (2001). Nootropic and anxiolytic activity of saponins of *Albizzia lebbeck*. *Pharmacology Biochemistry and Behavior*, *69*, 439–444.

Part C

Unconventional Oils from Insects

Chapter 42

Tessaratoma papillosa Longan Stink Bug

DESCRIPTION

Tessaratomidae is a family of true bugs; this family resembles large stink bugs and are sometimes quite colorful. All species are exclusively plant-eaters, some of major economic importance as agricultural pests. A few species are also consumed as human food in some countries. Larger species of Tessaratomidae are known informally as giant shield bugs, giant stink bugs, or inflated stink bugs. These bugs are ovate to elongate-ovate. They range in size from 6 to 7 mm to the large ones at 43 to 45 mm (Gillott, 1995). The head of tessaratomids is generally small and triangular, with the antennae having 4–5 segments. The scutellum, the hard extension of the thorax covering the abdomen in hemipterans is triangular and does not cover the leathery middle section of the forewing but is often partially covered by the prothorax. The tarsi (the final segments of the legs) have 2–3 segments; these bugs possess a piercing-sucking mouthpart (Picker, Griffiths, & Weaving, 2004; Schuh & Slater, 1995) (Fig. 42.1).

DISTRIBUTION

They are mostly found in tropical Africa, Asia, and Oceania though a few species can be found in the Neotropics and Australia. There are about 240 species known.

INSECT COMPOSITION AND BIOACTIVE COMPONENTS

The carbohydrate content of longan stink bug was reported as 6.71%. Recently, attention has been focused on low carbohydrate–high protein (LC–HP) diets. It is clear that people who consume such diets have a reduced intake of calories, resulting in a predictable degree of weight loss (Kappagoda, Hyson, & Amsterdam, 2004). The high protein content is an indication that the insects can be of value in man and animal ration and can equally replace higher animal protein usually absent in the diet of rural dwellers in developing countries (Banjo, Lawal, & Songonuga, 2006). The protein content of longan stink bug is 50.54% which is very high when compared with other conventional sources (Raksakantong, Meeso, Kubola, & Siriamornpun, 2010). A total of 10 fatty acid

Unconventional Oilseeds and Oil Sources. http://dx.doi.org/10.1016/B978-0-12-809435-8.00042-1

FIGURE 42.1 *Tessaratoma papillosa* longan stink bug.

methyl esters were identified from the unprocessed and the traditionally processed stink bug samples, of these, 7 were unsaturated fatty acid derivatives (methyl oleate, methyl (Z)-9-hexadecenoate, methyl linoleate, methyl (Z)-9-octadecenoate, methyl octadecanoate, methyl (Z,Z)-9,12-octadecadienoate, and methyl myristoleate), while the remaining 3 were saturated fatty acid derivatives (Musundire, Osuga, Cheseto, Irungu, & Torto, 2016). The percentage of oil content of longan stink bug was 23.5%, with C20:4n–6 as the major fatty acid representing 46.7%, followed by C18:0 as 41.0% (Raksakantong et al., 2010). Lipids are essential in diets as they help in the transport of nutritionally essential fat-soluble vitamins, such as (Omotoso, 2006). The results of this study confirm the fact that insects are indeed a good source of protein and other nutrients.

DIFFERENT USES AND MEDICINAL VALUE

Insects are traditional foods in most cultures, playing an important role in human nutrition and have much nutrient to offer. Raksakantong et al. (2010) demonstrated the informative results of fatty acid composition and lipid contents of longan stink bug, especially the long-chain PUFAs. These authors have also shown that longan stink bug was a good nutritional food source, especially for fat and protein. In Thailand, the local style of longan stink bug cooking is roasting, curry, and chilli pasting (Raksakantong et al., 2010).

The edible stink bug has the potential to help lessen nutrient-deficient communities in Africa where vegetables and animal sources may be limited. Stink bug has high protein, fatty acids, and antiinflammatory chemicals, such as flavonoids content. The researchers identified 7 essential fatty acids for human nutrition and health out of the 10 they found, 4 flavonoids, and 12 amino acids, including two considered to be the most limiting in cereal-based diets (Musundire et al., 2016).

REFERENCES

Banjo, A. D., Lawal, O. A., & Songonuga, E. A. (2006). The nutritional value of fourteen species of edible insects in southwestern Nigeria. *African Journal of Biotechnology, 5*(3), 298–301.

Gillott, C. (1995). *Entomology.* Dordrecht, Netherlands: Springer, p. 604.

Kappagoda, T. C., Hyson, A. D., & Amsterdam, A. E. (2004). Low carbohydrate—high protein diets is there a place for them in clinical cardiology? *Journal of the American College of Cardiology, 43*(5), 725–730.

Musundire, R., Osuga, I. M., Cheseto, X., Irungu, J., & Torto, B. (2016). Aflatoxin contamination detected in nutrient and anti-oxidant rich edible stink bug stored in recycled grain containers. *PLoS One, 11*(1), e0145914.

Omotoso, O. T. (2006). Nutritional quality, functional properties and anti-nutrient compositions of the larva of *Cirina forda* (Westwood) (Lepidoptera: Saturniidae). *Journal of Zhejiang University, 7*(1), 51–55.

Picker, M., Griffiths, C., & Weaving, A. (2004). *Field guide to insects of South Africa.* Ithaca, New York, USA: Penguin Random House South Africa, p. 134.

Raksakantong, P., Meeso, N., Kubola, J., & Siriamornpun, S. (2010). Fatty acids and proximate composition of eight Thai edible terricolous insects. *Food Research International, 43*, 350–355.

Schuh, R. T., & Slater, J. A. (1995). *True bugs of the world (Hemiptera: Heteroptera): classification and natural history* (1st edn). Ithaca, New York, USA: Cornell University Press, p. 241.

Chapter 43

Copris nevinsoni Dung Beetle

DESCRIPTION

Dung beetles belong to family Scarabaeidae, order Coleoptera. Their sizes were varied according to the species from approximately 1.5 to 4.5 cm in length. Adult dung beetles vary significantly in size from small beetles no more than 1/8th of an inch in length (*Aphodius pseudolividus*) to large beetles measuring 1¼ in. in length (*Dichotomius carolinus*). Most dung beetles are brown to black in color. Occasionally, a bright metallic green beetle appears and can be an easily identified as *Phanaeus vindex*. Many of the male dung beetles have distinct horns, for example, *Onthophagus taurus* horns resemble bull horns, while *Onthophagus gazella* has a short spike-like horns. Horn size is generally a product of larval nutrition. Major males have large horns while minor males have short horns (Bertone et al., n.d.) (Fig. 43.1).

DISTRIBUTION

Dung beetles commonly found in tropical rain forests are of two main groups. The first group is the rollers (or paracoprids) which excavate balls from the main dung mass, roll them some distance away from the source, and then bury or conceal them within vegetation on the surface. The second group includes the tunnellers (or telecoprids), which dig tunnels directly beneath the dung mass (Boonrotpong, Sotthibandhu, & Pholpunthin, 2004).

Dung beetles are mostly found in rice fields and cattle farms, where cattle manure is found since their tunnel nests are built underneath. Dung beetle adults feed on fluid extracted from cattle dung, while the larvae live on undigested plant fibers in the dung (Bophimai & Siri, 2010). This insect is distributed in Asia, Laos, Vietnam, Cochinchina, Malaysia, Cambodia, and Thailand. These beetles are usually found in the soil of agricultural and weedy areas.

DIFFERENT USES

Although, more than 200 species of dung beetles were identified, only some are edible according to local people. Excluded head and wings, the rest portion of dung beetles can be eaten by various means, such as grilling, boiling, and streaming. Although nutritional values of some species of dung beetles were reported to contain high protein contents (Hanboonsong, 2003). Local style

Unconventional Oilseeds and Oil Sources. http://dx.doi.org/10.1016/B978-0-12-809435-8.00043-3

FIGURE 43.1 *Copris nevinsoni* **Dung beetle.**

cooking includes curry, chilli pasted, and frying. Dung beetles are one of delicious insects, although, they are unconventional food for people of different regions due to their habitant and feeding behavior.

Dung beetles (Scarabaeidae: Scarabaeinae, Coprinae) are an important group of primary decomposers in the forest ecosystem. They use dung for breeding and feeding, and are beneficial in numerous ways, playing an important part in recycling nutrients by drying out the dung, which will eventually give off minerals. They improve soil structure and water holding capacity by incorporating organic matter back into the soil (Tyndale-Biscoe, 1990). They are also important secondary seed dispersers. As much as 90% of seeds that are defecated onto the soil surface might be destroyed by rodents and other seed-eating animals if not buried by dung beetles. They play a key role in biological control by reducing the number of pests, such as dung flies in dung. Moreover, dung beetles can be used as biological indicators of changes in the ecosystem. Due to their sensitivity to changes of the physical structure of the habitat and the ease of sampling them, dung beetles have been used to determine the effects of environmental changes on the diversity and structure of the forest. It has been reported that the physical structure of the forest appears to be an important determining factor in the structure and distribution of dung beetle communities (Boonrotpong et al., 2004).

CHEMICAL COMPOSITION

Some beetles contain 19.8 g of protein, 13.6 mg of iron, and a whopping 43.5 mg of calcium per 100 g. Bophimai and Siri (2010) studied the lipid contents of six edible dung beetles and they noticed that the lipid content is similar in most species and range from 12.1% to 14.0%. The lipid content of June beetle which lives in soil is 3.34%, which considered as the lowest when compare with other insects that live in different areas. Cicada, which lives in trees, contained the highest lipid content (23.98%) (Raksakantong, Meeso, Kubola, & Siriamornpun, 2010). Bophimai and Siri (2010) reported that, the total fatty acids of the insects ranged from 3934.1 to 9832.4 mg/100 g. The concentration of total saturated fatty acids ranged from 1440.9 to 5406.3 mg/100 g. The most abundance SFA in six species of dung beetles were palmitic acid (16:0), which was similar in many edible insects, such as mole cricket, ground cricket, spur-throated grasshopper, giant water bug, true water beetle, water scavenger beetle, melon and sorghum bugs, and winged reproductive of termite (Yang, Siriamornpun, & Li, 2006; Mariod, Matthäus, & Eichner, 2004; Ekpo & Onigbinde, 2007). Dung beetle oil contains palmitic acid as the major saturated fatty acids, while oleic acid and linoleic acid were the most abundant monounsaturated and polyunsaturated fatty acids, respectively. The long-chain fatty acids (C20:4n6) were also found in these species. The concentration of MUFAs in dung beetles was ranged from 2109.5 to 4152.9 mg/100 g. Low concentrations of eicosenoic acid (20:1n9) were also detected in dung beetles, which might synthesized from 18:1n9 precursor by Δ9 desaturase (Barker, Larson, Graham, Lynn, & King, 2007). The polyunsaturated fatty acids were ranged from 273.1 to 670.4 ± 5.6 mg/100 g which accounted for up to 9.8% of total fatty acids. The content of PUFAs in all dung beetles was very low compared to the content of SFAs (36.6%–55.0%) and MUFAs (42.2%–53.6%). Among detected PUFAs, linoleic acid (C18:2n6) was the most abundance. Three PUFAs (18:2n6, 18:3n3, and 20:4n6) were detected in dung beetles (Bophimai & Siri, 2010).

REFERENCES

Barker, G. C., Larson, T. R., Graham, I. A., Lynn, J. R., & King, G. J. (2007). Novel insights into seed fatty acid synthesis and modification pathways from genetic diversity and quantitative trait loci analysis of the Brassica C Genome. *Plant Physiology, 144*(4), 1827–1842.

Bertone, M., Watson, W., Stringham, M., Green, J., Washburn, W., Poore. M., & Hucks, M. (n.d.). Dung Beetles of Central and Eastern North Carolina Cattle Pastures. Retrieved from https://www.ces.ncsu.edu/depts/ent/notes/forage/guidetoncdungbeetles.pdf. Retrieved.

Boonrotpong, S., Sotthibandhu, S., & Pholpunthin, C. (2004). Species composition of dung beetles in the primary and secondary forests at Ton Nga Chang wildlife sanctuary. *Science Asia, 30*, 59–65.

Bophimai, P., & Siri, S. (2010). Fatty acid composition of some edible dung beetles in Thailand. *International Food Research Journal, 17*, 1025–1030.

Ekpo, K. E., & Onigbinde, A. O. (2007). Characterization of lipids in winged reproductives of the termite *Macrotermis bellicosus*. *Pakistan Journal of Nutrition, 6*, 247–251.

Hanboonsong, Y. (2003). The dung beetle fauna (Coleoptera, Scarabaeidae) of Thailand. *Research Report from the National Institute for Environmental Studies, Japan, 175*, 249–258.

Mariod, A. A., Matthäus, B., & Eichner, K. (2004). Fatty acid, tocopherol and sterol composition as well as oxidative stability of three unusual Sudanese oils. *Journal of Food Lipids, 11*, 179–189.

Raksakantong, P., Meeso, N., Kubola, J., & Siriamornpun, S. (2010). Fatty acids and proximate composition of eight Thai edible terricolous insects. *Food Research International, 43*, 350–355.

Tyndale-Biscoe, M. (1990). *Common dung beetles in pastures of Southeastern Australia* (p. 72). Victoria: CSIRO publication.

Yang, L. -F., Siriamornpun, S., & Li, D. (2006). Polyunsaturated fatty acid content of edible insects in Thailand. *Journal of Food Lipids, 13*(3), 277–328.

Chapter 44

Schistocerca gregaria (Desert Locust) and *Locusta migratoria* (Migratory Locust)

INTRODUCTION

Schistocerca gregaria, the desert locust is a species of locust. Plagues of desert locusts have threatened agricultural production in Africa, the Middle East, and Asia for centuries. The desert locust is potentially the most dangerous of the locust pests because of the ability of swarms to fly rapidly across great distances. It has two to five generations per year.

The migratory locust (*Locusta migratoria*) is the most widespread locust species, and the only species in the genus *Locusta*. It occurs throughout Africa, Asia, Australia, and New Zealand. It used to be common in Europe, but has now become rare there. Due to the vast geographic area it occupies, which comprises many different ecological zones, numerous subspecies have been described. However, not all experts agree on the validity of some of these subspecies (https://en.wikipedia.org/wiki/Migratory_locust) (Fig. 44.1).

DESCRIPTION

The egg pod is 3–4 cm long and the lower end is about 10 cm below the surface of the ground. The eggs are surrounded by foam and this hardens into a membrane and plugs the hole above the egg pod. The eggs absorb moisture from the surrounding soil. The incubation period, before the eggs hatch, may be 2 weeks or much longer, depending on the temperature.

DISTRIBUTION

Locusts live in different parts of the world, on all continents except Antarctica. *S. gregaria* is probably the best-known species owing to its wide distribution (North Africa, Middle East, and Indian subcontinent) and its ability to migrate over long distances. The rain allowed swarms to develop and move north to Morocco and Algeria, threatening croplands. Swarms crossed Africa, appearing in Egypt, Jordan, and Israel, the first time in those countries for 50 years

Unconventional Oilseeds and Oil Sources. http://dx.doi.org/10.1016/B978-0-12-809435-8.00044-5

FIGURE 44.1 *Schistocerca gregaria* **swarm and adult laying.**

(Wagner, 2008). The migratory locust (*L. migratoria*), sometimes classified into up to 10 subspecies, swarms in Africa, Asia, Australia, and New Zealand, but has become rare in Europe. Locust adults and swarms regularly cross the Red Sea between Africa and the Arabian Peninsula, and are even reported to have crossed the Atlantic Ocean from Africa to the Caribbean in 10 days during the 1987–89 plague. A single swarm can cover up to 1200 km^2 and can contain between 40 and 80 million locusts per square kilometer. The locust can live between 3 and 6 months, and there is a 10- to 16-fold increase in locust numbers from one generation to the next (FAO, 2006).

FOOD USES AND OTHER USES

Edible Locust Species

The locusts are a group of grasshopper species that becomes gregarious and migratory when their populations are sufficiently dense. During the swarming phase, locusts destroy or severely damage crops. They are a major pest of historical importance, notably in Africa (North, West, Sahelian, Madagascar), Australia, and the Middle East. A locust swarm can represent a considerable amount of biomass, containing up to 10 billion insects and weighing approximately 30,000 ton (Van Huis et al., 2013). The swarming behavior makes locusts relatively easy to harvest for food. In Africa, the desert locust (*S. gregaria*), the migratory locust (*L. migratoria*), the red locust (*Nomadacris septemfasciata*), and the brown locust (*Locustana pardalina*) are commonly eaten. In Japan, China, and Korea, rice field grasshoppers (including *Oxya yezoensis*, *Oxya velox*, *Oxya sinuosa*, and *Acrida lata*) are harvested for food (Van Huis et al., 2013). In Mexico, chapulines, which are grasshoppers of the

Sphenarium genus, and notably *Sphenarium purpurascens*, a pest of alfalfa, are popular edible insects (Cohen, Sánchez, & Montiel-Ishinoet, 2009). The grasshopper *Ruspolia differens*, which is actually a katydid, is a common food source in many parts of eastern and southern Africa. Crickets are a common food in Southeast Asia, particularly in Thailand; the house cricket *Acheta domestica*, *Gryllus bimaculatus*, *Teleogryllus occipitalis*, *Teleogryllus mitratus*, the shorttail cricket *Brachytrupes portentosus*, and *Tarbinskiellus portentosus* are edible cricket species (Van Huis et al., 2013).

Locusts are consumed worldwide for human food in Africa, South America, and Asia, both in rural and urban areas. Commercial farming of locusts for food and feed is developing in Southeast Asia and rice field grasshoppers are harvested for food in Japan, China, and Korea. In Africa, the desert locust (*S. gregaria*), the migratory locust (*L. migratoria*), the red locust (*N. septemfasciata*), and the brown locust (*L. pardalina*) are commonly eaten. They are an important food source, as are other insects, adding proteins and fats to the daily diet, especially in times of food crisis. However, in many African, Middle Eastern, and Asian countries, locusts are considered a delicacy and eaten in abundance. They are also served on skewers in some Chinese food markets (Mohamed, 2015).

The grasshoppers and locusts were deep-fried, used as a cracker ingredient and fermented to make a cooking sauce. Today, the grasshopper (deep-fried) is one of the best-known and most popular edible insects in Thailand, and this species is no longer a major agricultural pest. Some farmers even grow maize crops to feed the insect, rather than harvesting the maize for sale (Hanboonsong, 2010).

Locusts may occur in swarms, which make them particularly easy to harvest. In Africa, the desert locust, the migratory locust, the red locust, and the brown locust are eaten. However, due to their status as agricultural pests, they may be sprayed with insecticides in governmental control programs or by farmers. For example, relatively high concentrations of residues of organophosphorus pesticides were detected in locusts collected for food in Kuwait (Saeed, Dagga, & Saraf, 1993).

Increasing urbanization will change insect consumption in developing regions of the world if supply to cities remains small and unreliable and urban areas westernize. For example, locust consumption in the Fertile Crescent has disappeared in areas characterized by strong westernization (Amar, 2003). People in most Western countries view entomophagy with feelings of disgust. Nomads of Arabia and of Libya greet the appearance of locust swarms with joy. They boil and eat them, dry others in the sun and pound them into flour for future consumption (Van Huis et al., 2013).

Native Americans, such as those who lived freely in, what today is called, the state of Utah, are very accustomed to eating grasshoppers, locusts, and crickets. On their first taste of shrimp, the Goshute Indians are reported to have named the creatures "sea crickets" (Lockwood, 2004). Locusts are eaten by most Africans, some Asiatics, and especially the Arabs. On their market, they appear roasted or grilled in great quantities. When salted, they keep for some time in storage. They used for supplying ships, served as dessert or with coffee. This food is in no way repugnant to look at or by association. It tastes like prawn, and is perhaps more delicately flavored, especially the females when filled with eggs (Van Huis et al., 2013).

CHEMICAL COMPOSITION AND BIOACTIVE COMPOUNDS

Proximate composition of locust meal revealed that the dry locust has a crude protein of 52.3% on dry matter basis. The ether extract, crude fiber, and ash were 12.00, 19.00, and 10.00%, respectively. Desert locust has great potential as a protein source in broiler diets without causing any physiological disorder as reflected in the hematological analysis.

Proximate analysis for the edible migratory locust revealed a range of dry matter with an average value of 96.2%. The average of crude protein was 50.42%, crude fat average was of 19.62%, while carbohydrates formed average value of 4.78%. Fiber content was 15.65%, ash value was 6.24%, and value of moisture content was 3.8%. The average range of calories in the migratory locust was 490.8 cal/100 g. Mineral content was very low except for phosphorus (29.6 ± 4.32 ppm), while other minerals have a range of 0.04–2.2 ppm. The iron content of locusts (*L. migratoria*) varies between 8 and 20 mg/100 g of dry weight, depending on their diet (Oonincx et al., 2010). The edible migratory locust contains appreciable quantities of dry matter and protein as other insects (Mohamed, 2015). The gross energy of the migratory locust (*L. migratoria*) was in the range 598–816 kJ/100 g fresh weight, depending on the insect's diet (Oonincx & van der Poel, 2011).

Fournier et al. (1995) separated six groups of phospholipids from the rectal tissues of the African locust. These phospholipids are: phosphatidylcholine, phosphatidylethanolamine, phosphatidylinositols, phosphatidylserine, sphingomyelin, and cardiolipins. Saturated and unsaturated C18 components were abundant, as is generally observed in insects. Odd-chain (15:0, 17:0, 17:1) and long-chain fatty acids were also detected. Both $C18:2n-6$ and $20:4n-6$ were metabolized into prostaglandins and hydroxyoctadecadienoic acids, respectively, by the intervention of prostaglandin-endoperoxide synthase. Incorporation of $18:2n-6$ into phospholipids and neutral lipids or active exchanges of the fatty acid between both categories of lipids reflected different phospholipase, lipase, and transferase activities. Triglycerides appeared as a major source of fatty acids for locust rectum phospholipids (Fournier et al., 1995).

REFERENCES

Amar, Z. (2003). The eating of locusts in Jewish tradition after the Talmudic period. *The Torah u-Madda Journal, 11*, 186–202.

Cohen, J. H., Sánchez, N. D. M., & Montiel-Ishinoet, F. D. (2009). Chapulines and food choices in rural Oaxaca. *Gastronomica: The Journal of Food and Culture, 9*(1), 61–65.

FAO (2006). Biological control of locusts. Food and Agriculture Organization.

Fournier, B. R., Wolff, R. L., Nogaro, M., Radallah, D., Darret, D., Larrue, J., & Girardie, A. (1995). Fatty acid composition of phospholipids and metabolism in rectal tissues of the African locust. *Comparative Biochemistry and Physiology Part B: Biochemistry and Molecular Biology, 111*(3), 361–370.

Hanboonsong, Y. (2010). Edible insects and associated food habits in Thailand. In P. B. Durst, D. V. Johnson, R. L. Leslie, & K. Shono (Eds.), *Forest insects as food: Humans bite back, Proceedings of a workshop on Asia-Pacific resources and their potential for development* (pp. 171–182). Bangkok: FAO Regional Office for Asia and the Pacific.

Lockwood, J. A. (2004). *Locust: The devastating rise and disappearance of the insect that shaped the American frontier*. New York, NY: Basic Books.

Mohamed, E. H. A. (2015). Determination of nutritive value of the edible migratory locust *Locusta migratoria*, Linnaeus, 1758 (Orthoptera: Acrididae). *International Journal of Advances in Pharmacy, Biology and Chemistry, 4*(1), 144–148.

Oonincx, D. G. A. B., & van der Poel, A. F. B. (2011). Effects of diet on the chemical composition of migratory locusts (*Locusta migratoria*). *Zoo Biology, 30*, 9–16.

Oonincx, D. G. A. B., van Itterbeeck, J., Heetkamp, M. J. W., van den Brand, H., van Loon, J., & van Huis, A. (2010). An exploration on greenhouse gas and ammonia production by insect species suitable for animal or human consumption. *PLoS One, 5*(12), e14445.

Saeed, T., Dagga, F. A., & Saraf, M. (1993). Analysis of residual pesticides present in edible locusts captured in Kuwait. *Arab Gulf Journal of Scientific Research, 11*(1), 1–5.

Van Huis, A., Van Itterbeeck, J., Klunder, H., Mertens, E., Halloran, A., Muir, G., & Vantomme, P. (2013). Edible insects: Future prospects for food and feed security. FAO Forestry Paper 171. Rome, Italy: FAO.

Wagner, A. M. (2008). Grasshoppered: America's response to the 1874 Rocky Mountain locust invasion. *Nebraska History, 89*(4), 154–167.

Chapter 45

Oecophylla smaragdina Fabricius Weaver Ant

DESCRIPTION AND DISTRIBUTION

Oecophylla smaragdina (common names include weaver ant, green ant, green tree ant, and orange gaster) is a species of arboreal ant found in Asia and Australia. They make nests in trees, made of leaves stitched together using the silk produced by their larvae. Weaver ants may be red or green (Dlussky, Wappler, & Wedmann, 2008). The weaver ants belonging to the genus Oecophylla consist of two extant species *O. smaragdina* which is distributed throughout tropical Asia, Australasia, and some Pacific islands and *Oecophylla longinoda* distributed throughout tropical Africa (Sribandit, Wiwatwitaya, Suksard, & Offenberg, 2008). *O. longinoda* is distributed in the Afro-tropics and *O. Smaragdina* from India and Sri Lanka in South Asia, through southeastern Asia to northern Australia and Melanesia (Crozier, Newey, Schlüns, & Robson, 2010). In Australia, *O. smaragdina* is found in the tropical coastal areas as far south as Broome in Western Australia and across the coastal tropics of the Northern Territory down to Yeppoon in Queensland.

The species share similar biological and ecological characteristics. They are both polydomous canopy ants that build leaf nests on their host trees. Nests are constructed by drawing together leaves and fixing them with silk produced from their larvae (Offenberg, Nielsen, Macintosh, Aksornkoae, & Havanon, 2006) (Fig. 45.1).

DIFFERENT USES

Weaver ants are one of the most valued types of insects eaten by humans (entomophagy). Weaver ants can be utilized directly as a protein and a food source since the ants (especially the ant larvae) are edible for humans and high in protein and fatty acids (Raksakantong, Meeso, Kubola, & Siriamornpun, 2010). It has furthermore been shown that the harvest of weaver ants can be maintained; while at the same time using the ants for biocontrol of pest insects in tropical plantations, since the queen larvae and pupae that are the primary target of harvest, are not vital for colony survival (Offenberg & Wiwatwitaya, 2010).

Unconventional Oilseeds and Oil Sources. http://dx.doi.org/10.1016/B978-0-12-809435-8.00045-7

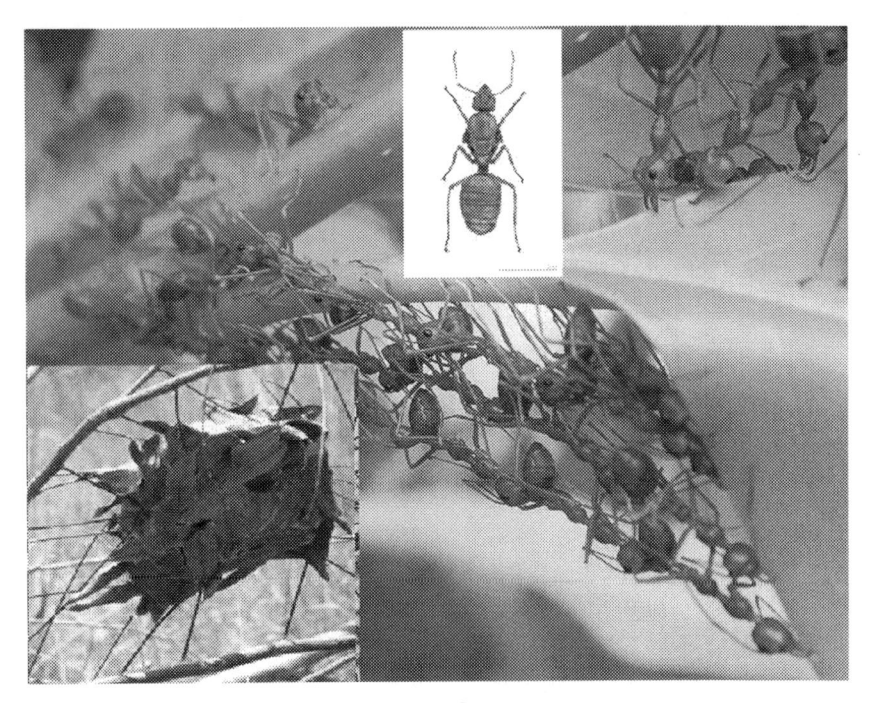

FIGURE 45.1 *Oecophylla smaragdina*, **adult and nest.**

Due to their predatory habit *Oecophylla* ants are recognized as biological control agents in tropical tree crops as they are able to protect a variety of crops against many different insect pests (Van Mele, 2008). In this way they are used indirectly as an alternative to chemical insecticides. It is less well known that the ants can be utilized directly also, as a commercial product. There exist, at least three different markets for the use of these units in Southeast Asia: (1) in Chinese and Indian traditional medicines, (2) as a valued feed for songbirds in Indonesia (Césard, 2004), and (3) as a prized human delicacy in Thailand and other Asian countries. In Chhattisgarh, India, traditional healers believe that regular intake of *O. smaragdina* will prevent rheumatism—a view shared by practitioners of traditional Chinese medicine (Oudhia, 2002). The Indian healers also prepare oils in which they dip collected ants. After 40 days, oils are used externally to cure rheumatism, gout, ringworm or other skin diseases, or else as an aphrodisiac (Oudhia, 2002).

The larvae of weaver ants are also collected commercially as an expensive feed for insect-eating birds in Indonesia and the worker ants are used in traditional medicine in, for example, India and China. According to bird lovers the larvae and pupae of *O. smaragdina* provide essential protein and vitamins to their birds and so will improve the bird's performance. For use as a bird food they are willing to pay up to US $1.4 per kg of ant brood. Lower-quality ant brood is used to feed chickens where it is believed to accelerate feather growth and flesh production (Césard, 2004). The tradition of including *Oecophylla* ants in food and/or traditional medicine has been reported from various cultures in Thailand, India, Myanmar, Borneo, Philippines, Papua New Guinea, Australia, and Congo (De Foliart, 2008). Especially in Thailand, *O. smaragdina* is considered a delicacy and has been eaten by humans for centuries. Imagos as well as brood is used in a variety of Thai dishes and are easily obtained on many local markets throughout the country during the ant harvest season. Larvae and pupae are preferred over imagos and the queen caste is preferred over the worker castes and males. The season in which *O. smaragdina* produce new queens therefore defines the ant harvest season (Sribandit et al., 2008). The ants are used as ingredients in soups, salads, and fried dishes and sometimes eaten raw together with spices as a snack. The tradition of eating ants is most prominent among the Isaan people of Northeast Thailand and the people in Northern Thailand, but has spread to other parts of the country with the migration of people from these cultures. A growing interest in the eating of ants has led to higher demand throughout Thailand with increasing prices as a result. Thus, the collection of ants is becoming more profitable and the harvest pressure on local *O. smaragdina* populations may increase accordingly, potentially leading to an unsustainable overexploitation of these ants in natural habitats (Sribandit et al., 2008). The harvest of *Oecophylla* ants in Northeast Thailand is substantial and not only for local subsistence, but an effective way of earning cash. There is an economically driven, increasing interest in harvesting ants and thus increasing pressure on natural ant populations. Ant farming may be a solution to retain sustainability and at the same time enhance profitability (Sribandit et al., 2008). Siri and Maensiri (2010) produced natural fibers of weaver ants and they demonstrated its potential application as a natural matrix to support in vitro cell adhesion. They found that the morphological structure of natural ant fibrous mat made it served as a good matrix to support cell adhesion and proliferation. Interestingly, old and freshly made fibers differed in their water wettability, which is important for its potential applications.

Van Mele and Cuc (2000) used *O. smaragdina* in citrus, as a fruit quality improver and a biological control agent; they reported that the use of ants offers benefits in terms of a better environment, fewer health risks for farmers and consumers, without affecting farmers' income.

COMPOSITION AND BIOACTIVE COMPOUNDS

Sihamala et al. (2010) determined total lipid contents, lipid classes, and fatty acid compositions of hot-air dried edible black ants (*Polyrhachis vicina*) from Wenzhou and Guizhou and edible red ant (*O. smaragdina*) from Thailand. These authors noticed that the total lipid content was 6.3, 15.2, and 9.3%, respectively, and the major lipid components were of triacylglycerol 43.4, 37.4, and 79.4%, respectively, followed by phospholipids 7.3, 6.1, and 21.5%, respectively; diacylglycerol 6.1, 10.3, and 18.1%, respectively; and cholesterol ester 4.9, 6.6, and 13.5%, respectively. While free fatty acids (1.8, 2.9, and 2.8%, respectively) and sterol (0.5, 0.8, and 0.7%, respectively) were the minor components. They reported that oleic acid was the most predominant fatty acid, accounting for 60.8, 63.0, and 52.1% in Wenzhou and Guizhou black ants and Thai red ant, respectively; followed by palmitic acid 16.5, 17.5, and 20.8%; and linoleic acid 2.4, 2.1, and 7.0%, respectively. Edible ant oil was found to be a good source of unsaturated fatty acids varying from 68.1% to 77.1%. The MUFA in edible ant oil contributed 58.7%–73.1% of the edible ants investigated; the Wenzhou black ant sample had the highest MUFA content (73.1%), while Guizhou black ant contained 72.1% (Sihamala et al., 2010). Weaver ant silks are produced by final-instar larva, not adult, for nest construction rather than for cocoon spinning. Silks are produced in the labial or larval salivary glands of the instar larva. The formation of silks in salivary glands starts and finishes in the middle and the end of the 5th instar. The secretory cells in salivary glands release a homogeneous substance, which is polymerized in the lumen to form compact birefringent tactoids. After water is absorbed, tactoids are aggregated to form spiral-shaped filament with zigzag pattern. The common architecture of fibroins contains a signal peptide and a coiled-coil region containing about 30 heptad repeats flanked by variable regions. The heptad repeat usually contains hydrophobic residues in the α and d positions and hydrophilic residues in the remaining positions (Woolfson, 2005). Weaver ant fibroins are small proteins (about 400 amino acid residues), do not contain primary sequence repeats, and has low protein identity to other fibroins. Four fibroin proteins of weaver ants (WAF1–4) were identified, which contain 391, 400, 395, and 443 amino acid residues, respectively (Sutherland et al., 2007). Their amino-acid compositions of ants (Table 45.1) are quite different from silkworm fibroin, in which it contains high level of acidic amino acids and few glycine residues, leading to a distinctive conformation and properties compared to silkworm fibroin. Differences of weaver ant fibroins to other fibroins, especially to silkworm fibroins, led to the question of their possible applications, especially for biomedical applications similar to those of silkworm fibroins (Siri & Maensiri, 2010). Ant eggs contain a glycoprotein, which was extracted with phenol-saline and purified on jacalin-Sepharose 4B. The glycoprotein contained D-mannose, o-galactose, D-glucose, and 2-acetamido-2-deoxy-D-glucose, and gave precipitin bands with the cY-o-galactosyl-specific lectins. The presence of nonreducing α-D-galactosyl end groups was corroborated by methylation analysis and enzymic degradation. The formation of the jacalin–glycoprotein complex was dependent on time, pH, and the ionic strength of the medium (Ray & Chati-Erjee, 1989).

Three lectins have been isolated from an extract of ant egg and purified by gel filtration on Sephadex G-75 of the 100% ammonium sulfate fraction from the crude extract, followed by ion-exchange chromatography on DEAE-cellulose.

TABLE 45.1 Amino Acid Composition of Natural Silk Produced by Three Ant Species

Ants	Gly	Ala	Ser	Val	Thr	Leu	Ile	Asx	Lys	Glx	Met	Phe	Arg	Tyr
Myrmecia forficata	7.7	29.1	11.5	4.5	3.4	6.5	3.3	8.2	4.8	13.3	1.4	0.5	4.4	0.5
O. smaragdina	7.3	31.8	14.2	4.3	4.2	5.6	3.8	7.2	4.0	11.7	0.4	0.6	3.3	1.1
Harpegnathos saltator	10.6	21.0	8.6	5.5	4.8	5.3	5.3	9.1	5.3	9.9	1.5	1.4	3.4	2.6

Source: Modified from Campbell, P.M., Trueman, H. E., Zhang, Q., Kojima, K., Kameda, T., & Sutherland, T. D. (2014). Cross-linking in the silks of bees, ants and hornets. *Insect Biochemistry and Molecular Biology, 48*, 40–50.

All the lectins were homogeneous as judged by polyacrylamide gel electrophoresis. The molecular weights of the lectins, AEL-I, AEL-II, and AEL-III, as estimated by gel filtration, were 195,000, 105,500, and 84,000. All the lectins were found to be composed of four subunits of unequal sizes and are glycoproteins with a neutral sugar content of 3.0%–7.5% (Hassan & Absar 1995).

Siri and Maensiri (2010) studied the properties of natural silk fibers of weaver ants and their potential application as a cell matrix. Weaver ant fibrous mat contained nonwoven mesh of fibers with an average diameter of 766 ± 326 nm. The thickness, mass, and apparent density of the fibrous mats were $39.0 \pm 9.8 \, \mu m$, $0.8 \pm 0.1 \, mg/cm^2$, and $0.22 \pm 0.03 \, g/cm^3$, respectively. Freshly made fibrous mats by weaver ants were highly hydrophilic as determined by water contact angle analysis, whereas older ones were quite hydrophobic. These authors analyzed their new product using TG–DTA; they found that the major weight loss peak from 260 up to about 330°C, similar to the decomposition peak of *Bombyx mori* fibroin. FT-IR spectrum of the ants fibrous mats showed amide I, amide II, amide III, C–H, and C–O peaks, which were attributed to random coil and β-sheet conformation in the protein structure of the weaver ant fibers. The fibrous mat was slight toxic to the fibroblast NIH 3T3 cells (37.8% cell death), probably due to some toxic particles deposited on the fibers. Nevertheless, weaver ant fibrous mat served as a good matrix for cell adhesion. The authors provided evidence for the properties and a potential application of natural weaver ant fibers as an alternative, natural, fibroin-based matrix (Siri & Maensiri, 2010). The ant silks, which have less beta sheet and less sequence outside the coiled-coil domains, could not be dissolved in LiBr and appear to be predominantly stabilized by covalent cross-linking. The isopeptide cross-linker, ε-(g-glutamyl)-lysine that is produced by transglutaminase enzymes, was demonstrated by mass spectrometry and was present at greater levels in silks of ants (Campbell et al., 2014).

REFERENCES

Campbell, P. M., Trueman, H. E., Zhang, Q., Kojima, K., Kameda, T., & Sutherland, T. D. (2014). Cross-linking in the silks of bees, ants and hornets. *Insect Biochemistry and Molecular Biology*, *48*, 40–50.

Césard, N. (2004). Harvesting and commercialization of Kroto (*Oecophylla smaragdina*) in the Malingping area, West Java, Indonesia. In K. Kusters & B. Belcher (Eds.), *Forest products, livelihoods and conservation: Case studies of nontimber product systems* (pp. 61–77). Bogor, Indonesia: Center for International Forestry Research.

Crozier, R. H., Newey, P. S., Schlüns, E. A., & Robson, S. K. A. (2010). A masterpiece of evolution—*Oecophylla* weaver ants (Hymenoptera: Formicidae). *Myrmecological News*, *13*, 57–71.

De Foliart, G. R. (2008). *The human use of insects as a food resource: A bibliographic account in progress*. Available from: www.foodinsects.com

Dlussky, G. M., Wappler, T., & Wedmann, S. (2008). New middle Eocene formicid species from Germany and the evolution of weaver ants. *Acta Palaeontologica Polonica*, *53*(4), 615–626.

Hassan, P., & Absar, N. (1995). Isolation, purification and characterization of three lectins from ant eggs (*Oecophylla smaragdina* Fabr.). *Carbohydrate Research*, *273*, 63–70.

Offenberg, J., Nielsen, M. G., Macintosh, D. J., Aksornkoae, S., & Havanon, S. (2006). Weaver ants increase premature loss of leaves used for nest construction in rhizophora trees. *Biotropica*, *38*, 782–785.

Offenberg, J., & Wiwatwitaya, D. (2010). Sustainable weaver ant (*Oecophylla smaragdina*) farming: harvest yields and effects on worker ant density. *Asian Myrmecology*, *3*, 55–62.

Oudhia, P. (2002). Traditional medicinal knowledge about red ant *Oecophylla smaragdina* (Fab.) [Hymenoptera; Formicidae] in Chhattisgarh, India. *Insect Environment*, *8*(3), 114–115.

Raksakantong, P., Meeso, N., Kubola, J., & Siriamornpun, S. (2010). Fatty acids and proximate composition of eight Thai edible terricolous insects. *Food Research International*, *43*, 350–355.

Ray, S., & Chatterjee, B. P. (1989). Purification of ant-egg glycoprotein and its interaction with jacalin. *Carbohydrate Research*, *191*, 305–314.

Sihamala, O., Bhulaidok, S., Li-rong, S., & Duo, L. (2010). Lipids and fatty acid composition of dried edible red and black ants. *Agricultural Sciences in China*, *9*(7), 1072–1077.

Siri, S., & Maensiri, S. (2010). Alternative biomaterials: natural, non-woven, fibroin-based silk nanofibers of weaver ants (*Oecophylla smaragdina*). *International Journal of Biological Macromolecules*, *46*, 529–534.

Sribandit, W., Wiwatwitaya, D., Suksard, S., & Offenberg, J. (2008). The importance of weaver ant (*Oecophylla smaragdina* Fabricius) harvest to a local community in Northeastern Thailand. *Asian Myrmecology*, *2*, 129–138.

Sutherland, T. D., Weisman, S., Trueman, H. E., Sriskantha, A., Trueman, J. W. H., & Haritos, V. S. (2007). Conservation of essential design features in coiled coil silks. *Molecular Biology and Evolution*, *24*, 2424–2432.

Van Mele, P. (2008). A historical review of research on the weaver ant *Oecophylla* in biological control. *Agricultural and Forest Entomology*, *10*, 13–22.

Van Mele, P., & Cuc, N. T. T. (2000). Evolution and status of *Oecophylla smaragdina* (Fabricius) as a pest control agent in citrus in the Mekong Delta, Vietnam. *International Journal of Pest Management*, *46*(4), 295–301.

Woolfson, D. N. (2005). In D. A. D. Parry, & J. M. Squire (Eds.), *Fibrous proteins: coiled-coils, collagen and elastomers* (pp. 79–112). San Diego, CA: Elsevier.

Chapter 46

Aspongopus viduatus
Watermelon Bug As Source
of Edible Protein and Oil

DISTRIBUTION AND DESCRIPTION

Aspongopus viduatus (Pentatomidae) is widely distributed in the Near East, Arabian Peninsula, Egypt, Israel, Iran, and Turkey. In Sudan, it is mainly distributed in Kordofan, Darfur, Kassala, Khartoum, Blue Nile, and Upper Nile states (Tarla, Yetisir, & Tarla, 2013; AbdalAlla, Mohammed, & Hammad, 2015). The melon bug is oval, rather flat, and relatively large (1.8–2 cm long and 0.8–1 cm broad) (Schmutterer, 1969). The bug belongs to the order Hemiptera, whose members are called "True bugs." They have very distinctive front wings called hemelytra, in which one half is leathery and the apical half is membranous (Anon, 1999). The adult of *A. viduatus* is black, length shorter than the width, cross eyes, which are prominent and black in color, oceli black, rostrum black, short and extending beyond the coxae of the foreleg, antennae with five black segments (Salih, 2008).

AbdalAlla et al. (2015) examined a female reproductive system of the watermelon bug at aestivation period and later in the active period. Their results showed that the females had no oocytes during aestivation; however, following the termination of aestivation period, oocytes developed. They found that the total number of eggs laid by a single female was averaged to 448 ± 0.90 eggs. Most of the eggs were laid during the first 2 weeks following the end of the aestivation period. The main oviposition period was 64 ± 17.7 days, preoviposition period and postoviposition periods were 5.17 ± 1.7 and 10.7 ± 1.7, respectively. The authors reported an incubation period of 8.5 ± 0.75 days and hatchability percentage as 94.4 ± 3.01%. The mean developmental periods of the nymphal instars were 7.9 ± 0.75, 10.9 ± 1.3, 7.14 ± 1.6, 9.8 ± 1.1, and 13.3 ± 1.9 for the 1st, 2nd, 3rd, 4th, and 5th nymphal instars, respectively (AbdalAlla et al., 2015).

The watermelon bugs are found in most African countries, where they cause damage to watermelon and other cucurbit shoots. The adult bugs can usually be found by lifting the young melon plants from the ground and inspecting the undersides of the leaves. The nymphs pierce the leaves, stems, and young fruits and suck the sap, resulting in wilting, fruit drop, and the death of the plant

Unconventional Oilseeds and Oil Sources. http://dx.doi.org/10.1016/B978-0-12-809435-8.00046-9

(Mariod, Matthäus, & Eichner, 2004). Melon bugs are widely distributed in Kordofan and Darfur states of Sudan (locally known as Um-buga), where field watermelons are one of the most important crops for the traditional rain-fed agriculture (Mariod, Bushra, Abdel-Wahab, & Ain, 2011a). Tons of melon bug adults can be collected there in infested fields. El-Obeid Agricultural Research Station (North Kordofan state of Sudan) designed a hand picking program for melon bug adults in plots of about 5000 ha in four different areas of the state, for two seasons. A total of 15 tons of melon bug adults were collected in the first season and 226 tons in the second one (Bashir, Ali, & Ali, 2002).

Tarla et al. (2013) investigated and determined the biological stages of the melon bug; they conducted their studies in both laboratory and natural conditions in 2007 and 2008. Furthermore, damage of the pest on four different cucurbit species [watermelon (*Citrullus lanatus*), cucumber (*Cucumis sativus* L.), squirting cucumber (*Ecballium elaterium*), and bottle gourd (*Lagenaria siceraria*)] was investigated. In these studies, it was observed that the pest was able to complete its life cycle when only squirting cucumber and watermelon were used as food sources.

DIFFERENT USES AND MEDICINAL VALUE

Melon bugs are edible and in the last nymph stage, which is a relatively soft stage, the bugs are cooked and eaten. In Namibia, they collect the adults and use them as a relish or as a spice in a ground form for cooking meals (http://www.natmus.cul.na). In the western Kordofan state of Sudan, the bug is known locally as Um-buga and is used in nutrition by collecting the oil from them after a hot water extraction. The oil is used in cooking (during the famine and shortage of food) and in some medicinal applications, for example, to cure skin lesion (Mariod et al., 2004). The oil has traditional medicinal uses in human and animal skin diseases and is also used for meat dressing before drying in western regions of Sudan. The oil has long stability against rancidity because of its high oleic acid content (Mustafa, Mariod, & Matthäus, 2008).

Mustafa et al. (2008) determined the antibacterial activities of melon bug crude oil, silicic acid column–purified oil, and phenolic compounds–free oil, by agar diffusion assay against seven bacterial isolates; four of them are food-related bacterial species: *Staphylococcus aureus*, *Salmonella enterica* serovar Paratyphi, *Escherichia coli*, and *Bacillus cereus*, and the three other isolates are: *Bacillus subtilis*, *Enterococcus faecalis*, and *Pseudomonas aeruginosa*. The crude oil and the phenolic compounds–free oil showed high antibacterial activities against some test species, while the silicic acid column–purified oil showed no antibacterial activity. The study highlights the possibility of using this oil in food preservation. The antimicrobial activity of the oil was lost when the oil was applied to the silicic column; the authors believe that this may indicate that the effective materials are pigments, tocopherols, or phospholipids, which were lost as a result of this binding to the column (Mustafa et al., 2008). Mariod,

Matthäus, Eichner, and Hussein (2005) investigated the improvement of the oxidative stability of sunflower kernel oil (SKO) by blending with melon bug oil (MBO). They blended sunflower oil with MBO with respect to the fatty acid composition and the oxidative stability. These authors found that, by increasing the proportion of MBO in SKO, the linoleic acid content decreased from 46.3% to 30.1%, while the oleic acid content increased from 41.3% to 43.9%. As a result of blending the oxidative stability was improved from 5% to 68% of MBO compared to the SKO as a control, with increasing parts of MBO. Storage of the blends at 70°C showed that the increase of the peroxide value as a measure of oxidation was remarkably lower for the mixtures of MBO with SKO than for pure SKO. These authors concluded the success of blending of sunflower oil with MBO and they obtained high stable cooking oil (Mariod et al., 2005). Crude oils have to be refined before the consumption, in order to remove undesirable accompanying substances; usually, deodorization is the last step in the edible oil refining (Martins, Ito, Batistella, & Maciel, 2006). In a laboratory for all-glass deodorizer experiments, degummed, refined, and bleached oils from MBO; water extracted (MBOH$_2$O); and melon bug oil solvent extracted (MBO-SOL) were deodorized for different periods of times (0.5, 1, and 2 h) at different temperatures (190, 210, and 250°C). Quality changes (free fatty acids, peroxide values, tocopherols, sterols, and phosphatides) were determined, and also stability against oxidation (Rancimat test) beside the fatty acid composition. It is clear that there was no change in the phosphatides according to deodorization temperature and time; peroxides and free fatty acids content were significantly ($P < 0.05$) reduced in all samples with the increase of deodorization temperature and nearly completely removed at 250°C. Tocopherols were decreased as a result of elevated temperature, and high decrease happened in high temperatures of 210 and 250°C in all samples except MBOH$_2$O, where tocopherols were completely removed during deodorization temperatures. The fatty acid composition did not undergo any changes during deodorization. The oil oxidative stability was increased as affected by elevated deodorization temperature in all studied samples (Mariod, Matthäus, Eichner, & Hussein, 2012).

Manzano-Agugliaro, Sánchez-Muros, Barroso, and Bañón (2012) studied the fat content of insects for its utilization in the production of biodiesel. The study has shown the great fat potential of insects, highlighting a large number of species with an ether extract higher than 25%, including a large number in excess of 30% and some even reaching levels close to or above 77%. They carried out a review of the main criteria to be considered for the selection of insect species for biodiesel production. These authors observed that the fat content varies widely between orders, species, and their stages of development— larva, prepupa, pupa, nymph, or adult—with the larval stage being that at which the most fat is accumulated. Furthermore, they observed variations in the fat content within the same species due to factors, such as origin or type of diet. They concluded that insects, through the development of their life cycle, can be fed with agricultural, industrial, or urban by-products in order to accumulate a large amount of fat with potentially excellent quality (fatty acids C16–18), for conversion into energy through biodiesel production (Manzano-Agugliaro et al., 2012). Insects and insect-derived products have been widely used in folk healing in many parts of the world since ancient times. Maggots and honey have been used to heal chronic and postsurgical wounds and have been shown to be comparable to conventional dressings in numerous settings. Honey has also been applied to treat burns. Honey has been combined with beeswax in the care of several dermatologic disorders. Royal jelly has been used to treat postmenopausal symptoms. Bee and ant venom have reduced the number of swollen joints in patients with rheumatoid arthritis. Propolis, a hive sealant made by bees, has been utilized to cure aphthous stomatitis. Cantharidin, a derivative of the bodies of blister beetles, has been applied to treat warts and molluscum contagiosum. Combining insects with conventional treatments may provide further benefit (Cherniack, 2010). In Botana area of central Sudan, nomads use tar obtained from highly heated melon bugs for their camels against dermatological infections (Mariod et al., 2005).

Li et al. (2011) evaluated the larvae of black soldier fly (*Hermetia illucens*) (BSFL), for biodiesel production. The extracted crude fat was then converted into biodiesel by acid-catalyzed (1% H_2SO_4) esterification and alkaline-catalyzed (0.8% NaOH) transesterification, resulting in 35.5, 57.8, and 91.4 g of biodiesel produced from 1000 BSFL growing on 1 kg of cattle manure, pig manure, and chicken manure, respectively. The major ester components of the resulting biodiesel were lauric acid methyl ester (35.5%), oleic acid methyl ester (23.6%), and palmitic acid methyl ester (14.8%). Fuel properties of the BSFL fat–based biodiesel, such as density (885 kg/m^3), viscosity (5.8 mm^2/s), ester content (97.2%), flash point (123°C), and cetane number (53) were comparable to those of rapeseed oil–based biodiesel. Their results demonstrated that the organic waste–grown BSFL could be a feasible nonfood feedstock for biodiesel production (Li et al., 2011).

Biodiesel is a renewable, biodegradable, and a nontoxic fuel that can be derived from vegetable oils by transesterification. Mariod, Klupsch, Hussein, and Ondruschka (2006) transesterified MBO, a long-term stable oil using methanol or ethanol in the presence of sulfuric acid; the obtained biodiesel characteristics were studied in accordance with the DIN EN 14214 specifications for biodiesel. Most of the biodiesel characteristics met the DIN specifications (water content, iodine number, phosphorus content). The kinematic viscosity value of MBO was higher than those of biodiesel standard limits. It was possible to prepare the methyl and ethyl esters catalyzed by H_2SO_4 from MBO. Under the described transesterification conditions, MBO seems to be suitable for biodiesel (Mariod et al., 2006).

Mariod, Matthäus, Eichner, and Hussein (2008) stored MBO in the dark at $30 \pm 2°C$ for 24 months. They withdrew oil aliquots every 2–4 month for analyses of changes in four quality indexes, namely fatty acid composition, tocopherol content, peroxide value, and oxidative stability index by Rancimat. They noticed that after 24 months of storage the fatty acid composition of the MBO showed no change while tocopherol content was decreased. MBO showed only

FIGURE 46.1 *Aspongubus viduatus* (melon bug), adult insects, oil and meal.

slight changes in its oxidative stability as indicated by the peroxide value and induction period during the 24 months of storage (Mariod et al., 2008) (Fig. 46.1).

INSECT COMPOSITION AND BIOACTIVE COMPONENTS

Insects are a highly nutritious and healthy food source with high fat, protein, vitamin, fiber, and mineral content (Van Huis et al., 2013). The approximate analysis of *A. viduatus* adults showed 8.3% moisture, 27.0% crude protein, 54.2% fat, and 3.5% ash on a dry-matter basis. The total of crude protein and fat constituents was high and represented 81.1% of the total proximate composition, which makes this insect a good source of protein and oil, which can be used as a supplement to fat and protein foods (Mariod et al., 2011a).

The nutritive value of any food depends on protein content, protein quality, and protein digestibility, which refers to the digestibility of the amino acids present in the food. Amino acids are the building blocks required for the biosynthesis of all proteins through human metabolism to ensure proper growth, development, and maintenance. Essential amino acids are indispensable because the body cannot synthesize them and so must obtain them through food. Eight amino acids are classified as essential: phenylalanine, valine, threonine, tryptophan, isoleucine, methionine, leucine, and lysine (Van Huis et al., 2013). In some insect species, the amino acids are very well represented, so in order to make recommendations regarding the use of edible insects as food enrichments in diets, it is important to look into their amino acid composition (Van Huis et al., 2013). The melon bug cake, after extraction of oil, was analyzed for amino acid using HPLC; the results showed 16 amino acids, including all of the essential ones. The total amino acids were found to be 360.5 mg/g crude protein. The percentage of sulfur-containing amino acid (methionine and cysteine) in *A. viduatus* was 57.1 mg/g crude protein, which can be considered a reasonable amount as cysteine plays an important function in the cell processes in the oxidation and reduction system while methionine plays a role as a methyl donor in metabolism (Hoshi & Heinemann, 2011). The total amount of the essential amino acids found in *A. viduatus* was 208.5 mg/g crude protein, which was higher than the values recommended by FAO/WHO (1991) (113 g protein for adults). According to the amino acid scores, the first limiting amino acid of the melon bug protein was lysine and the second was leucine followed by phenylalanine and tyrosine (Mariod et al., 2011b).

The oil content of melon bug was 45%, which is very high when compared to vegetable oils; it is similar to sesame (45%–55%). The major fatty acids of the crude oil, as determined by gas chromatography, were palmitic, stearic, oleic, and linoleic acids. The oil of melon bugs showed a low amount of saturated fatty acids (37.4%) and a high content of unsaturated fatty acids (61.5%), mainly oleic acid (46.63%).

Mariod, Matthäus, and Abdelwahab (2011b) analyzed the development of total fatty acids of MBO during different maturity stages and as reported for the first time in this study (Table 46.1), palmitic acid (16:0) was found to be $33.6 \pm 0.3\%$ at the first maturity stage in January (MBO Jan) and dropped during the rest of the stages to reach 30.9% by the end of the maturity process (MBO May), while oleic and linoleic acids were increased from 45.3 and 3.6% to 46.6 and 3.9%, respectively. The fatty acid composition of the oil from melon bug exhibited greater variation during insect maturity stages (MBO Jan to MBO May). Although variations were clear between total saturated and total unsaturated acids, the differences were quite pronounced in saturated, which decreased from 39.0% to 37.3%, while total unsaturated fatty acids increased from 59.3% to 61.9%. The tocopherol content of oils is important to protect their lipids against autoxidation and, thereby, to increase their storage life and their value as wholesome foods (Mariod et al., 2005). Total tocopherol content varied significantly ($P < 0.05$) between various insect maturity stages. In the first maturity stage the total tocopherol content was 0.9 mg/100 g oil, then it decreased to reach 0.3 mg/100 g in May. A significant difference was found within the tocopherol content during maturity stages (Mariod et al., 2011b). The decrease of tocopherol during harvesting can be as a result of consumption of tocopherol to protect lipids as the major function of tocopherols is to protect stored lipids from oxidation (Mariod, Matthaus, Idris, & Abdelwahab, 2010).

MBO had low amounts of tocopherols (0.3 mg/100 g) compared with other common oils, such as sesame oil, groundnut oil, or sunflower oil, in which the amount of tocopherols was between 27.9 and 97.6 mg/100 g (Mariod et al., 2005).

TABLE 46.1 Change in Fatty Acid Composition (%) and Total Tocopherol Content (mg/100 g) of *Aspongubus viduatus* (Melon Bug) Oil During Different Developmental Stages[a]

Fatty Acid	MBO$_{Jan}$	MBO$_{Feb}$	MBO$_{March}$	MBO$_{April}$	MBO$_{May}$
12:0	0.1 ± 0.1	0.2 ± 0.1	0.1 ± 0.1	0.2 ± 0.1	0.0 ± 0.01
14:0	0.3 ± 0.1	0.4 ± 0.1	0.3 ± 0.1	0.4 ± 0.1	0.3 ± 0.01
16:0	33.6 ± 0.3	33.0 ± 0.3	32.1 ± 0.3	32.0 ± 0.3	30.9 ± 0.3
16:1n–7	9.5 ± 0.2	9.9 ± 0.2	10.2 ± 0.2	10.9 ± 0.2	10.7 ± 0.2
18:0	4.6 ± 0.1	5.1 ± 0.1	5.3 ± 0.1	5.5 ± 0.1	5.9 ± 0.1
18:1n–9	45.3 ± 0.3	46.2 ± 0.3	47.0 ± 0.3	48.2 ± 0.3	46.6 ± 0.3
18:1n–11	0.8 ± 0.1	0.2 ± 0.1	0.4 ± 0.1	0.0 ± 0.0	0.5 ± 0.1
18:2n–6	3.6 ± 0.1	3.7 ± 0.1	3.5 ± 0.1	3.2 ± 0.1	3.9 ± 0.1
20:0	0.3 ± 0.1	0.3 ± 0.1	0.3 ± 0.1	0.4 ± 0.1	0.2 ± 0.1
20:1n–9	0.1 ± 0.1	0.1 ± 0.1	0.1 ± 0.1	0.1 ± 0.1	0.2 ± 0.1
24:0	0.1 ± 0.1	0.2 ± 0.1	0.2 ± 0.1	0.2 ± 0.1	0.0 ± 0.0
Others	1.7 ± 0.1	0.7 ± 0.1	0.6 ± 0.1	0.8 ± 0.1	0.8 ± 0.1
SFA	39.0	39.2	38.2	38.3	37.3
USFA	59.3	60.1	61.2	62.4	61.9
Ratio USFA/SFA	1.5	1.5	1.6	1.6	1.6
Total tocopherols	0.9 ± 0.2	0.9 ± 0.2	0.7 ± 0.1	0.5 ± 0.1	0.3 ± 0.1

MBO, Melon bug oil; PUFA polyunsaturated fatty acids; SFA, saturated fatty acids.
[a]*All determinations were carried out in triplicate and mean value ± standard deviation (SD) reported.*
Source: Mariod, A. A., Bushra, M., Abdel-Wahab, S. I., & Ain, N. M. (2011a). Proximate, amino acid, fatty acid and mineral composition of two Sudanese edible insects. *International Journal of Tropical Insect, 31*(3), 145–153.

The amount of sterols in the MBO was 17.5 mg/100 g. The main sterol of the oil was β-sitosterol (10.6 mg/100 g), representing more than 58%, followed by campesterol and cholesterol as 1.8 and 1.4 mg/100 g, respectively. Compared with other oils usually used in human nutrition, MBO had the lowest content of sterols.

The oxidative stability of MBO expressed as the induction period determined by the Rancimat method at 120°C is remarkably high (38.0 h) in comparison to other edible oils. This value is very high compared with those of other edible oils. The oxidative stability of such commonly used oils ranges between 0.3 h for linseed oil and about 10 h for groundnut oil. The high oxidative stability of MBO is perhaps a result of the low amounts of polyunsaturated fatty acids, such as linoleic and linolenic acids (Mariod et al., 2004). Rumpold and Schlüter (2013) reported that insects are considered food with high contents of a variety of micronutrients, such as the minerals copper, iron, magnesium, manganese, phosphorous, selenium, and zinc. Mineral analyses by scanning electron microscopy (SEM) seem to be a very quick method, but it is a quantitative rather than a qualitative method, and many authors use this method to analyze different samples, for example, medicinal plants (Abbdewahab, Ain, Abdul, Taha, & Ibrahim, 2009). The mineral concentration of major elements such as Mg, Ca, K, P, and Na in the defatted sample of *A. viduatus* was reported as 301.10, 1021.21, 200.08, 1234.33, and 401.10 mg/100 g) significantly. Calcium, sodium, phosphorus, and potassium contents were higher in the melon bug than that of snails (Ozogul, Ozogul, & Olgunoglu, 2005), which were 726.3, 90.5, 104.5, and 82.2 mg/100 g, respectively. Mariod et al. (2011a) reported that the defatted, dried, and ground melon bug contained fewer minerals than the undefatted one; and magnesium, aluminum, sodium, and ferrous contents were not detected in the defatted insect.

REFERENCES

Abbdewahab, S. I., Ain, N. M., Abdul, A. B., Taha, M. M. E., & Ibrahim, T. T. A. (2009). Energy-dispersive X-ray microanalysis of elements' content and antimicrobial properties of *Pereskia bleo* and *Goniothalamus umbrosus*. *African Journal of Biotechnology*, 8, 2375–2378.

AbdalAlla, M. I., Mohammed, E. E., & Hammad, A. M. A. (2015). Biology and fecundity of the melon bug *Aspongopus viduatus* (Fabricius), in the laboratory. *International Journal of Science, Environment and Technology*, 4(2), 414–423.

Anon (1999). Biology of Insects in Biology 4FF3 Entomology.

Bashir, Y. G. A., Ali, M. K., & Ali K. M. (2002). IPM options for the control of field watermelon pests in Western Sudan, presented at the 67th meeting of the pests and diseases committee. Medani, Sudan: Agricultural Research Corporation, ARC.

Cherniack, E. P. (2010). Bugs as drugs, Part 1: insects: the "new" alternative medicine for the 21st century? *Alternative Medicine Review*, 15(2), 124–135.

FAO/WHO (1991). Protein quality evaluation (Report of Joint FAO/WHO Expert Consultation. FAO Food and Nutrition Paper 51). Rome: FAO/WHO.

Hoshi, T., & Heinemann, S. H. (2011). Regulation of cell function by methionine oxidation and reduction. *The Journal of Physiology*, 531(1), 1–11.

Li, Q., Zheng, L., Cai, H., Garza, E., Yu, Z., & Zhou, S. (2011). From organic waste to biodiesel: black soldier fly, *Hermetia illucens*, makes it feasible. *Fuel*, 90, 1545–1548.

Manzano-Agugliaro, F., Sánchez-Muros, M. J., Barroso, F. G., & Bañón, L. P. C. (2012). Insects for biodiesel production. *Renewable and Sustainable Energy Reviews*, 16(6), 3744–3753.

Mariod, A. A., Bushra, M., Abdel-Wahab, S. I., & Ain, N. M. (2011a). Proximate, amino acid, fatty acid and mineral composition of two Sudanese edible insects. *International Journal of Tropical Insect*, 31(3), 145–153.

Mariod, A. A., Klupsch, S., Hussein, I. H., & Ondruschka, B. (2006). Synthesis of alkyl esters from three unconventional Sudanese oils for the use as biodiesel. *Journal of Energy and Fuels*, 20(5), 2249–2252.

Mariod, A., Matthäus, B., & Abdelwahab, S. I. (2011b). Fatty acids, tocopherols of *Aspongupus viduatus* (melon bug) oil during different maturity stages. *International Journal of Natural Product and Pharmaceutical Sciences*, 2(1), 20–27.

Mariod, A. A., Matthäus, B., & Eichner, K. (2004). Fatty acid, tocopherol and sterol composition as well as oxidative stability of three unusual Sudanese oils. *Journal of Food Lipids, 11*, 179–189.

Mariod, A. A., Matthäus, B., Eichner, K., & Hussein, I. H. (2005). Improving the oxidative stability of sunflower oil by blending with *Sclerocarya birrea* and *Aspongopus viduatus* oils. *Journal of Food Lipids, 12*, 150–158.

Mariod, A. A., Matthäus, B., Eichner, K., & Hussein, I. H. (2008). Long-term storage of three unconventional oils. *Grasas y Aceites, 59*(1), 16–22.

Mariod, A. A., Matthäus, B., Eichner, K., & Hussein, I. H. (2012). Effects of deodorization on the quality and stability of three unconventional Sudanese oils. *GIDA, 37*(4), 189–196.

Mariod, A. A., Matthaus, B., Idris, Y. M., & Abdelwahab, S. I. (2010). Fatty acids, tocopherols, phenolics and the antimicrobial effect of *Sclerocarya birrea* kernels with different harvesting dates. *Journal of the American Oil Chemists' Society, 87*, 377–384.

Martins, P. F., Ito, V. M., Batistella, C. B., & Maciel, M. R. W. (2006). Free fatty acid separation from vegetable oil deodorizer distillate using molecular distillation process. *Separation and Purification Technology, 48*(1), 78–84.

Mustafa, N. E. M., Mariod, A. A., & Matthäus, B. (2008). Antibacterial activity of *Aspongopus viduatus* (melon bug) oil. *Journal of Food Safety, 28*, 577–586.

Ozogul, Y., Ozogul, F., & Olgunoglu, A. I. (2005). Fatty acid profile and mineral content of the wild snail (*Helix pomatia*) from the region of the south of the Turkey. *European Food Research and Technology, 221*, 547–549.

Rumpold, B. A., & Schlüter, O. K. (2013). Nutritional composition and safety aspects of edible insects. *Molecular Nutrition and Food Research, 57*(5), 802–823.

Salih, M. H. M. M. (2008). Studies on the biology, morphology and host range of the melon bug *Coridius viduatus* (Fabricius) (Hemiptera, Dinidoridae) in the northern state. MSc thesis. Khartoum, Sudan: Department of Crop Protection, Faculty of Agriculture, University of Khartoum.

Schmutterer, H. (1969). *Pests of crops in North-East and Central Africa*. Stuttgart Portland, USA: Gastar Fisher Verlage.

Tarla, S., Yetisir, H., & Tarla, G. (2013). Black watermelon bug, *Coridius viduatus* (F.) (Heteroptera: Dinidoridae) in Hatay Region of Turkey. *Journal of Basic and Applied Sciences, 9*, 31–35.

Van Huis, A., Van Itterbeeck, J., Klunder, H., Mertens, E., Halloran, A., Muir, G., & Vantomme, P. (2013). Edible insects: Future prospects for food and feed security. FAO Forestry Paper 171. Rome, Italy: FAO.

Chapter 47

Agonoscelis pubescens Sorghum Bug as a Source of Edible Protein and Oil

DISTRIBUTION AND DESCRIPTION

Agonoscelis pubescens belongs to the order Hemiptera (family Pentatomidae); the English common names are Sudan millet bug and cluster bug; the insect occurs in the African countries south of the Sahara (Sudan, Uganda, Kenya, Tanzania, Somalia, Malawi, Zimbabwe, South Africa, Nigeria, Ghana, Madagascar, Cape Verde Island) (Schmutterer, 1969). The bug is commonly known in Sudan as Dura andat, where it is one of the main pests of sorghum (Dura) in both rain-fed and irrigated areas. The insect is now considered a major pest of sorghum and sesame in the Sudan, especially at the central rain belt, which is the major area for sorghum and sesame production. This area includes Gedarif, Damazine, Renk, Nuba mountains, and parts of Southern Darfur State (Mustafa, 2004).

Adult *A. pubescens* is shield-shaped, about 11–13 mm long and 6–7 mm wide. It is characterized by the presence of the silvery pubescence covering the dorsal and ventral surfaces of the body. The head is triangular with prominent eyes, antennae are five-segmented covered with fine hairs. The pronotum is convex, much wider than long. Eggs are barrel-shaped whitish to creamy in color (Schmutterer, 1969). The insect is of a medium size and the color is dark brown with silvery pubescens. Insects of the order Hemiptera are distinguished from those of other orders by having: two pairs of wings, almost always present with the front wings as hemelytra, which are thickened at the base, and having a membranous apical portion, and overlapping when not in use. Mouth parts: typically of the piercing and sucking type without palpi and arising from the front part of the underside of the head (Mustafa, 2004). The bugs have piercing–sucking mouthparts at all stages. They injure host plants by sucking the plant juices resulting in a characteristic musty flavor, which is detectable when a large number of the bugs is present or when the bugs are crushed (Mariod, Matthäus, Eichner, & Hussein, 2005). Schmutterer (1969) reported that sorghum bug eggs were laid in clusters of about 30 on the leaves, inflorescences, and other parts of the plants and 1 female may produce more than 100 eggs. In September the young adults migrate from the weeds to the sorghum crop (Fig. 47.1).

Unconventional Oilseeds and Oil Sources. http://dx.doi.org/10.1016/B978-0-12-809435-8.00047-0

FIGURE 47.1 *Agonoscelis pubescens oil* **adult and cluster of adults.**

DIFFERENT USES AND MEDICINAL VALUE

In western Sudan, sorghum bug adults are collected and eaten after frying, while in some areas of Sudan the collected bugs were extracted and the obtained oil was used for cooking and some medicinal uses (Mariod, Matthäus, Eichner, & Hussein, 2008). The behavior of crude sorghum bug oil (SBO) during deep-frying of par-fried potatoes was studied with regard to chemical, physical, and sensory parameters, such as content of FFA, tocopherols, polar compounds, oligomer TG, volatile compounds, oxidative stability, and total oxidation value. Potatoes prepared in SBO that had been used for 6–12 h of deep-frying at 175°C were not suitable for human consumption, considering the sensory evaluation. Looking at the chemical and physical parameters, SBO exceeded the limits, after no later than 18 h of use, for the amount of polar compounds, oligomer TG, and FFA recommended by the German Society of Fat Sciences (DGF) as criteria for the rejection of used frying oils. Only the amount of FFA was exceeded; this was because the amount of FFA at the beginning of the experiment was higher than for refined oils. The results showed that SBO was suitable for deep-frying of potatoes and it can be used in industrial scale as frying oil (Mariod, Matthäus, Eichner, & Hussein, 2006b).

SBO was transesterified using methanol and ethanol in the presence of sulfuric acid. The resultant fatty acid esters were compared with the DIN 51606 specifications for biodiesel. The SBO biodiesel characteristics met the DIN specifications (water content, iodine number, phosphorus). However, the kinematic viscosity value of the sample was much higher than that for biodiesel standards. These can be reduced by blending with other low-viscosity biodiesel. Mariod, Klupsch, Hussein, and Ondruschka (2006a) concluded the possibility of preparing biodiesel by using methyl and ethyl esters catalyzed by H_2SO_4 from SBO, under the described transesterification conditions. The concomitant determination of oxidized triglyceride monomers and polymerization compounds provided a complete picture of the oxidation process during long storage of sunflower oils (Martín-Polvillo, Márquez Ruiz, & Dobarganes, 2004). Mariod et al. (2008) stored SBO in the dark at $30 \pm 2°C$ for 24 months. They withdrew oil aliquots every 2–4 months for analyses of changes in four quality indexes, namely fatty acid composition, tocopherol content, peroxide value (PV), and oxidative stability index by Rancimat. They noticed that after 24 months of storage, the fatty acid composition of the sorghum oil showed no change while tocopherol content was decreased. The stored oil showed a periodical increase in the PV and had less stability as measured by the Rancimat in comparison to other oils (Mariod et al., 2008).

Mariod, Matthäus, Eichner, and Hussein (2006c) in a refining experiment, on a laboratory scale, crude-processed SBO by alkali refining. Their results showed that crude SBO had a sharp decrease in PV and showed a removal of more than 95% of their FFA content after the refining process, which indicates clear oil keepability. Chemical refining of the SBO does not result in any significant changes in the fatty acid composition. In SBO studied, the tocopherol content decreased during each step of processing, but the maximum reduction was observed after deodorization. The oxidative stabilities of the various processed oil varied considerably, whereby the stability of SBO samples decreased during the different processing steps up to bleaching, then it improved on deodorization. The highest reduction happened in the bleaching step where the tocopherol loss reached its maximum, and this loss resulted in a concurrent reduction in oxidative stability. The total sterols were decreased gradually in the processed oils compared with the crude oils, and the most affected steps were bleaching and deodorization where a high reduction in total sterols happened. The color index, FFA, tocopherol, and phytosterol contents of SBO could be used, together with other parameters, to monitor the oil quality during refining. Minor component contents progressively decreased during refining, especially during deodorization of SBO.

Minor components (tocopherols, sterols, and phenolic compounds) in edible oils affect both their autooxidative and photooxidative stabilities. Thus, stripping of minor components may influence the stability of oils being studied. PVs of non-stripped borage and evening primrose oils were significantly higher ($P \leq 0.05$) than corresponding values for stripped oils. Each stripping process or treatment has a specific function and removes certain components, which can act as prooxidants or antioxidants (Khan & Shahidi, 2002). Mariod, Matthäus, and Hussein (2011) investigated stripped and nonstripped oils from SBO, traditionally used for nutritional applications in Sudan, for their fatty acid and tocopherol composition, and their oxidative stability. They used three stripping methods, phenolic compounds extraction, silicic acid column, and aluminum oxide column. They found that, the stripping methods did not affect the fatty acid composition. Nonstripped SBO contained oleic, palmitic, stearic, and linoleic acids, which were not significantly ($P \leq 0.05$) different than stripped SBO. The stripping methods showed clear effect on the tocopherol composition of the studied oils, as the total amount of tocopherol in nonstripped oil decreased by extraction of phenolic compounds, mean that part of the tocopherols was extracted with the phenolic compounds. No traces of tocopherols were found in oils stripped using silicic and aluminum columns and the tocopherols were eliminated during the stripping processes. The stability of SBO was 5.1 h; this stability decreased by 23.5%, after extraction of phenolic compounds. This stability decreased by 90.2%, when stripped using the aluminum column and decreased by 86.3% when stripped by the silicic column. It is possible to assume that the tocopherols and phenolic compounds play a more active role in the oxidative stability of the oils than the fatty acid composition and phytosterols.

Mariod and Fadul (2015) used three methods for extraction of gelatin from sorghum bug; the yield of extracted gelatin ranged 1.5%–3.0%. They characterized the obtained gelatin by FTIR and the spectra of insect's gelatin seem to be similar when compared with commercial gelatin. These authors made ice cream using 0.5% insect's gelatin and compared with that made using 0.5% commercial gelatin as stabilizing agent. The properties of the obtained ice cream produced using insects gelatin were significantly different when compared with that made using commercial gelatin. In another study, Mariod et al. (2011a) reported that the extraction of gelatin from sorghum bug adults using hot water gave a high yield followed by mild acid. These authors concluded sorghum bug gelatin contained 40 kDa as the main components.

INSECT COMPOSITION AND BIOACTIVE COMPONENTS

The approximate analyses of *A. pubescens* adults showed 7.6% moisture, 28.2% crude protein, 57%–60% fat, and 2.5% ash on a dry matter basis, respectively. The bug protein contained 16 amino acids, including all of the essential ones. Compared with the amino acid profile recommended by FAO/WHO, the protein was of medium quality. The total amino acids of *A. pubescens* cake were 268.8 mg/g crude protein, which was less than the 864.2 mg/g crude protein in chicken egg that is considered as a main protein source in the human diet. The percentage of sulfur-containing amino acid (methionine and cysteine) in *A. pubescens* was 7.2% mg/g crude protein. The total amount of the EAAs found in *A. pubescens* was 119.3 mg/g crude protein, which was higher than the values recommended by FAO/WHO (1991) (113 g protein for adults). According to the amino acid scores, the first limiting amino acid of the sorghum bug protein was lysine and the second was methionine and cysteine followed by leucine. The mineral analysis indicated high P and K content (Mariod, Bushra, Abdel-Wahab, & Ain, 2011).

The oil content of sorghum bug was 60% on a dry basis, which is very high when compared with that of common oil seeds, such as groundnut, soybean, sunflower, or rapeseed with oil content between 20% and 45%. The recovered oil showed yellow color with distinctive flavor. Little information is available about this bug oil, although they are widely used in Sudan and some other African countries for generations not only as food, but also as a remedy for disease. The fatty acid composition is important for the use of the oil in different applications. The major fatty acids of the SBO were palmitic (12.1%), oleic (40.9%), and linoleic (34.5%) acids (Mariod, Matthäus, Eichner, & Hussein, 2007). Tocopherols react with lipid radicals to convert them into more stable products. The SBO contains 34.0 mg/100 g tocopherols with γ-tocopherol as the major analog (32.16 mg/100 g), followed by α-tocopherol (Table 47.1) (Mariod, Matthäus, & Eichner, 2004). Total lipid in the

TABLE 47.1 Lipid Classes (%), Tocopherols (mg/100 g), and Sterols (mg/100 g) of Sorghum Bug Oil (SBO)[a]

Parameters	SBO
Triacylglycerol (%)	71.6
Diacylglycerol (%)	13.6
Free fatty acids (%)	2.3
Polar lipids (%)	11.7
Steryl esters (%)	0.8
α-Tocopherol	0.88
β-Tocopherol	0.00
γ-Tocopherol	32.16
P8	0.21
δ-Tocopherol	0.78
Total	34.03
Cholesterol	2.2
Campesterol	11.6
Stigmasterol	25.4
β-Sitosterol	268.8
Δ5-Avenasterol	16.3
Δ7-Avenasterol	1.6
Δ7-Stigmasterol	2.8
Others[b]	121.2
Total sterols	449.9

[a]Data are means of triplicate results.
[b]Others include 24-methylcholesterol; campestanol; chlerosterol; sitostanol; 5,24-stigmastadienol.
Source: Mariod, A. A., Matthäus, B., & Eichner, K. (2004). Fatty acid, tocopherol and sterol composition as well as oxidative stability of three unusual Sudanese oils. Journal of Food Lipids, 11, 179–189.

female insect was reported to be higher than that of male and decreased according to resting time (Bilal, 2003).

The major fraction in SBO was triacylglycerol representing 71.6% of the total lipid, and higher percentages of polar lipids representing 11.7% (Table 47.1). The amount of sterols in the SBO was 449.9 mg/100 g (Table 47.1), which was higher compared with other oils usually used in human food. The main sterol of the oil was β-sitosterol, representing 59.7% followed by others that include 24-methylcholesterol; campestanol; chlerosterol; sitostanol; and 5,24-stigmastadienol (Mariod et al., 2007). The oxidative stability of the oils is expressed as the induction period determined by the Rancimat method at 120°C. The induction period of oil from SBO was 5.1 h. This value is moderate, compared with those of other edible oils. The oxidative stability of such commonly used oils ranges between 0.3 h for linseed oil and about 10 h for groundnut oil (Mariod et al., 2004).

Supercritical fluid extraction (SFE) has been applied to the extraction of oil from a large number of plant sources. The main advantages of using carbon dioxide fluid are: a reduced potential for the oxidation of extracted solutes, higher selectivity, increased sample throughput, shorter extraction time, and a low critical temperature (Louli, Folas, Voutsas, & Magoulas, 2004). Mariod, Abdel-Wahab, Gedi, and Solati (2010) extracted SBO by SFE using CO_2 and they compared it with oil extracted by Soxhlet extraction using hexane. They used response surface methodology (RSM) to determine the effects of pressure and temperature on the SBO yield. These authors reported that high extraction yield was obtained at 300 bar and 60°C followed by 400 bar and 70°C, while the low yield was obtained at 159 bar and 60°C. No differences were found in the fatty acid compositions of the various extracts, while the α-tocopherol extracted by the Soxhlet method was less than that extracted by the SFE process using CO_2. It can be observed that the conventional solvent extraction method exhibited notable DPPH radical-scavenging activity, with an efficacy slightly lower (IC_{50} 7.45 ± 0.3) than that of the SFE extracts. These authors concluded that the solvent-extracted oils exhibited notable DPPH radical-scavenging activity, with an efficacy (IC_{50} 7.45 ± 0.3) slightly lower than that of the SC-CO_2 extracted oils.

REFERENCES

Bilal, A. F. (2003). Some aspects of the biology, physiology and ecology of *Agonoscelis pubescens* (Thnb.) Hemiptera: Pentatomidae and the environmental implications of its chemical control. PhD thesis. Khartoum, Sudan: Graduate College, University of Khartoum.

FAO/WHO (1991). Protein quality evaluation (Report of Joint FAO/WHO Expert Consultation. FAO Food and Nutrition Paper 51). Rome: FAO/WHO.

Khan, M. A., & Shahidi, F. (2002). Photooxidative stability of stripped and non-stripped borage and evening primrose oils and their emulsions in water. *Food Chemistry, 79,* 47–53.

Louli, V., Folas, G., Voutsas, E., & Magoulas, K. (2004). Extraction of parsley seed oil by supercritical CO_2. *Journal of Supercritical Fluids, 30,* 163–174.

Mariod, A. A., Abdel-Wahab, S. I., Gedi, M. A., & Solati, Z. (2010). Supercritical carbon dioxide extraction of sorghum bug (*Agonoscelis pubescens*) oil using response surface methodology. *Journal of the American Oil Chemists' Society*, 87, 849–856.

Mariod, A. A., Abdel-Wahab, S. I., Ibrahim, M. Y., Mohan, S., Abd Elgadir, M., & Ain, N. M. (2011a). Preparation and characterization of gelatin from two Sudanese edible insects. *Journal Food Science and Engineering*, 1, 45–55.

Mariod, A. A., Bushra, M., Abdel-Wahab, S. I., & Ain, N. M. (2011b). Proximate, amino acid, fatty acid and mineral composition of two Sudanese edible insects. *International Journal of Tropical Insect*, 31(3), 145–153.

Mariod, A. A., & Fadul, H. (2015). Extraction and characterization of gelatin from two edible Sudanese insects and its applications in ice cream making. *Food Science and Technology International*, 21(5), 380–391.

Mariod, A. A., Klupsch, S., Hussein, I. H, & Ondruschka, B. (2006a). Synthesis of alkyl esters from three unconventional Sudanese oils for the use as biodiesel. *Journal of Energy & Fuels*, 20(5), 2249–2252.

Mariod, A. A., Matthäus, B., & Eichner, K. (2004). Fatty acid, tocopherol and sterol composition as well as oxidative stability of three unusual Sudanese oils. *Journal of Food Lipids*, 11, 179–189.

Mariod, A. A., Matthäus, B., Eichner, K., & Hussein, I. H. (2005). Improving the oxidative stability of sunflower oil by blending with *Sclerocarya birrea* and *Aspongopus viduatus* oils. *Journal of Food Lipids*, 12, 150–158.

Mariod, A. A., Matthäus, B., Eichner, K., & Hussein, I. H. (2006b). Frying quality and oxidative stability of two unconventional oils. *Journal of the American Oil Chemists' Society*, 83, 529–538.

Mariod, A. A., Matthäus, B., Eichner, K., & Hussein, I. H. (2006c). Effects of processing steps on the quality and stability of three unconventional Sudanese oils. *European Journal of Lipid Science and Technology*, 108(4), 298–308.

Mariod, A. A., Matthäus, B., Eichner, K., & Hussein, I. H. (2007). Fatty acids composition, oxidative stability and transesterification of lipids recovered from melon and sorghum bugs. *Sudan Journal of Science and Technology*, 8, 16–20.

Mariod, A. A., Matthäus, B., Eichner, K., & Hussein, I. H. (2008). Long-term storage of three unconventional oils. *Grasas y Aceites*, 59(1), 16–22.

Mariod, A., Matthäus, B., & Hussein, I. H. (2011c). Effect of stripping methods on oxidative stability of three unconventional oils. *Journal of the American Oil Chemists' Society*, 88, 603–609.

Martín-Polvillo, M., Márquez Ruiz, G., & Dobarganes, M. C. (2004). Oxidative stability of sunflower oils differing in unsaturation degree during long-term storage at room temperature. *Journal of the American Oil Chemists' Society*, 81(6), 577–583.

Mustafa, G. O. (2004). Ecological and behavioral studies on *Agonoscelis pubescens* (Thunberg) (Hemiptera: Pentatomidae) during the resting period. MSc thesis. Khartoum, Sudan: Faculty of Agriculture, University of Khartoum.

Schmutterer, H. (1969). *Pests of crops in North-East and Central Africa*. Stuttgart Portland, USA: Gastar Fisher Verlage.

Chapter 48

Acheta domesticus
House Cricket

DESCRIPTION

The house cricket is typically gray or brownish in color and grows up to 16–21 mm in length. Males and females look similar, but females have an ovipositor emerging from the rear, around 12-mm long. The ovipositor is brownblack, and is surrounded by two appendages. On females, the cerci are also more prominent, with wings that cover the abdomen. It has three dark transverse bands on the top of the head and between the eyes. All house crickets have long hind wings when they become adult, but they sometimes shed them later (Walker, 2007) (Fig. 48.1).

DISTRIBUTION

The house cricket is probably native to Southwestern Asia, but has been widely distributed by man. In the United States, it occurs wherever it is sold, but it survives in feral populations only in the Eastern United States and Southern California (Walker, 2007). The insect is distributed worldwide in Europe, Northern Africa, Western Asia, Indian Subcontinent, Canada, USA, Mexico, and Australasia.

COMPOSITION

The range of protein content of adult house cricket is 64.4%–70.8%, which is very high when compared with other edible insects like watermelon bugs, 27% and sorghum bug, 28% (Mariod, Bushra, Abdel-Wahab, & Ain, 2011). The lipid content of adult house cricket is 18.6%–22.8%, while the fiber is ranged from 16.4% to 19.1%. The crude protein and fat constituents are high (82%–92%) on a dry matter basis, which makes the house cricket a good source of protein and oil, which can be a supplement to fat and protein foods.

The major fatty acids of adult cricket are linoleic (30%–40%), oleic (23%–27%), palmitic (24%–30%), and stearic acids (7%–11%). There are smaller amounts of palmitoleic (3%–4%), myristic (~1%), and linolenic acids (<1%). The fatty acid composition of the cricket lipids reflects but is not identical to the fatty acids of the dietary lipids: linoleic (53%), oleic (24%), palmitic (15%), stearic (3%), myristic (2%), and linolenic acid (2%) (Hutchins & Martin, 1968). The composition of

Unconventional Oilseeds and Oil Sources. http://dx.doi.org/10.1016/B978-0-12-809435-8.00048-2

FIGURE 48.1 *Acheta domesticus* **House cricket female and male.**

the fatty acids associated with the tri-, di-, and monoglycerides and the free fatty acid fraction are all about the same. The fatty acids associated with the simple esters are high in stearic acid (Hutchins & Martin, 1968). Calcium content of adult house cricket is 132–210, potassium is 1126.6, magnesium is 109.42, phosphorus is 957.8, sodium is 435.0, iron is 11.23, zinc is 21.79, manganese is 3.73, copper is 2.01, and selenium is 0.06 mg/100 g. The vitamins analysis of the house cricket showed that the adult insect contains high amount of vitamin A (24.3 µg), vitamin E (63–81 IU/kg), vitamin C (9.74 mg), and vitamin B complex (85 mg). Vitamin B12 occurs only in food of animal origin and is well represented in house crickets, *Acheta domesticus* (5.4 µg/100 g in adults and 8.7 µg/100 g in nymphs) (Bukkens, 2005). These amounts are very high when compared with other insects. The total essential amino acids of the adult house cricket were 396.8 mg/g protein, while the nonessential amino acids were 412.8 mg/g. The percentage of sulfur-containing amino acid (Met + Cys, isoleucine, leucine, lysine) in house cricket was 228.9 mg/g crude protein (Rumpold & Schluter, 2013).

FOOD USES AND DIFFERENT USES

House crickets are farmed in Thailand for human consumption, where they have proven to be more popular than the native cricket species due to their superior taste and texture. They are most commonly eaten as a deep-fried snack and are also sold as a protein powder or protein extract (Gahukar, 2011). House crickets have the capability of recycling poultry manure into a protein-rich feedstuff for poultry on an economically competitive basis. An experiment was conducted by Nakagaki, Sunde, and Octolirrt (1987), which showed that the corn–cricket mixture contained enough of the essential amino acids to permit maximum gains' but addition of both arginine and methionine improved the feed utilization. These data demonstrated that the dried house crickets are a good source of high-quality protein for chickens, and in fact, may be better than that indicated by amino acid analysis (Nakagaki et al., 1987). In Thailand, 400 families in two villages produce 10 metric tons of crickets in the peak production period, both for the domestic market and export (Raubenheimer & Rothman, 2013).

REFERENCES

Bukkens, S. G. F. (2005). Insects in the human diet. In M. G. Paoletti (Ed.), *Ecological implications of minilivestock: Potential of insects, rodents, frogs and snails* (pp. 545–577). Enfield, NH: Science Publ..

Gahukar, R. T. (2011). Entomophagy and human food security. *International Journal of Tropical Insect Science, 31*, 129–144.

Hutchins, R. F. N., & Martin, M. M. (1968). The lipids of the common house cricket, *Acheta domesticus* L. I. Lipid classes and fatty acid distribution. *Lipids, 3*(3), 247–249.

Mariod, A. A., Bushra, M., Abdel-Wahab, S. I., & Ain, N. M. (2011). Proximate, amino acid, fatty acid and mineral composition of two Sudanese edible insects. *International Journal of Tropical Insect, 31*(3), 145–153.

Nakagaki, B. J., Sunde, M. L., & Octolirrt, G. R. (1987). Protein quality of the house cricket, *Acheta domesticus*, when fed to broiler chicks. *Poultry Science, 66*, 1367–1371.

Raubenheimer, D., & Rothman, J. M. (2013). Nutritional ecology of entomophagy in humans and other primates. *Annual Review of Entomology, 58*, 141–160.

Rumpold, B. A., & Schluter, O. K. (2013). Nutritional composition and safety aspects of edible insects. *Molecular Nutrition & Food Research, 57*(5), 802–823.

Walker, T. J. (2007). House cricket, *Acheta domesticus*. *Featured Creatures*. University of Florida/IFAS.

Chapter 49

Alphitobius diaperinus the Lesser Mealworm and the Litter Beetle

Alphitobius diaperinus is a species of beetle in the family Tenebrionidae, the Darkling beetles. It is known commonly as the lesser mealworm and the litter beetle. It has a cosmopolitan distribution, occurring nearly worldwide. It is known widely as a pest insect of stored food grain products, such as flour. It has been associated with wheat, barley, rice, oat, soybean, pea, and peanuts. It is also reported in connection with linseed, cottonseed, oilseeds, tobacco, etc. It is the most encountered beetles in the litter of poultry facilities. It is also called darkening or night due to its highest activity at dusk.

DESCRIPTION

The adult lesser mealworm beetle is roughly 6-mm long and widely oval in shape. It is shiny black or brown with reddish brown elytra, the color variable among individuals and changing with age. Much of the body surface is dotted with puncture-like impressions. The antennae are paler at the tips and are covered in tiny, yellowish hairs. The elytra have shallow longitudinal grooves (Dunford & Kaufman, 2006). The eggs are narrow, whitish or tan, and about 1.5-mm long. The larva somewhat resembles other mealworms, such as the common (*Tenebrio molitor*), but it is smaller, measuring up to 11-mm long as its final subadult stage. It is tapering and segmented, with three pairs of legs toward the front end. It is whitish when newly emerged from the egg and it darkens to a yellow-brown. It becomes pale when preparing to molt between instar stages. There are 6–11 instars (Dunford & Kaufman, 2006) (Fig. 49.1).

DISTRIBUTION

This species has long been known throughout the world as a common pest, so its origins are uncertain, but it may have originated in Sub-Saharan Africa. It moved into Europe long ago, and was likely introduced to North America from there (Dunford & Kaufman, 2006). As a pest, the beetle is more damaging to the poultry industry. This is the most common beetle found in poultry litter. The larvae damage poultry housing structures when they search for suitable pupation spots, chewing through wood, fiberglass, and polystyrene insulation.

Unconventional Oilseeds and Oil Sources. http://dx.doi.org/10.1016/B978-0-12-809435-8.00049-4

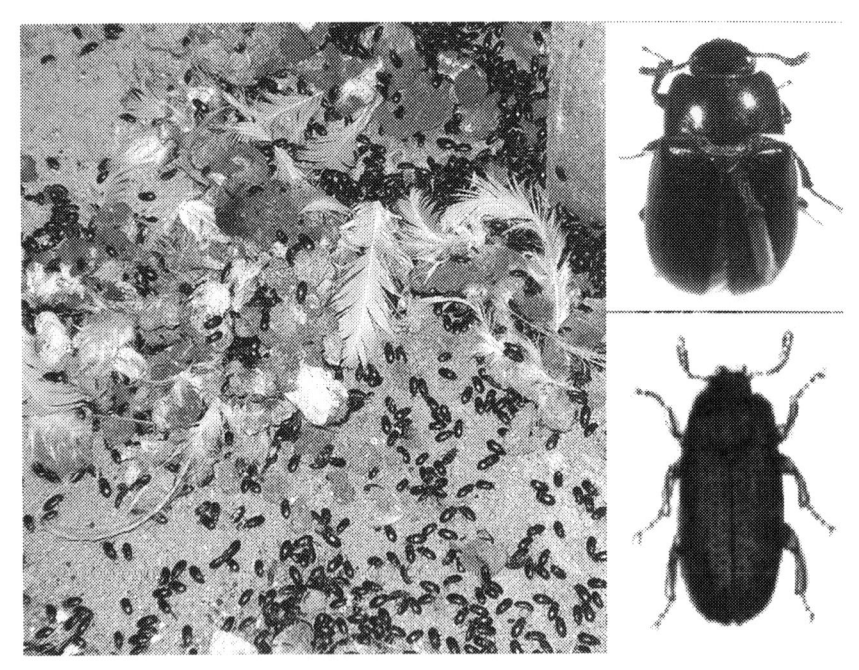

FIGURE 49.1 *Alphitobius diaperinus* in poultry facility and adults.

CHEMICAL COMPOSITION AND BIOACTIVE COMPOUNDS

Hassemer et al. (2015) identified 23 compounds from *A. diaperinus* abdominal gland extracts, of which 6 were quinones: the 2-methyl-1,4-benzoquinone and the 2-ethyl-1,4-benzoquinone were the major components, and 1,4-benzoquinone and 3 hydroquinones were registered for the first time for this species. The GC-EAD analysis using the crude extracts from abdominal glands showed that male and female antennae responded to the three major benzoquinones. For the olfactometer bioassays, both genders were repelled either by the abdominal gland extracts or by synthetic solutions containing the three benzoquinones. The results suggest that the 1,4-benzoquinones play a role as a repellent to *A. diaperinus* (Hassemer et al., 2015). The crude protein content of lesser mealworms was in general higher than that in soybean meal and close to that in poultry meat meal and fish meal. Fat content was 22.0% of DM. Crude ash content was 4.1 %. The amino acid score of Phe and Met contents was 5.2, which was higher than that of poultry meat meal (4.3), and similar to that of fish meal (5.4). The total indispensable amino acids were 42.7, which was higher than that of poultry meat meal and fish meal (Bosch, Zhang, Oonincx, & Hendriks, 2014).

The larvae of *A. diaperinus* contained all the essential amino acids (EAA) in quantities that are necessary for humans. Also, the sum of the amount of total EAA for *A. diaperinus* was comparable to that of soybean protein, but slightly lower than that of casein. The sum of total amount of amino acids per gram crude protein of *A. diaperinus* was 927 mg/g. The calculated EAA index of *A. Diaperinus* was somewhat higher than that of soybeans, but lower than that of casein, also indicating that the quality of the insect protein for this insect was comparable to conventional food protein sources. Supernatants from *A. diaperinus* had similar protein concentration, but it showed significant differences in gel strength (Yi et al., 2013). The fatty acid profile of *A. diaperinus* insect species as analyzed by GC-FID is in accordance with the profiles found in previous studies on lipids where high amounts of unsaturated fatty acids (USFA). The most abundant USFA in this insect lipid extract were C18:1 *cis*9 and C18:2 *cis*9, 12; the most abundant SFA was C16:0. These three FA accounted from 84.7 to 89.8 g/100 g of the FA in the lipid extracts. In *A. diaperinus* the average amount of USFA ranged was 64 g/100 g. Other minor FA present in the insect lipid extract included trans, ω-3 and CLA FA (Tzompa-Sosa, Yi., Van Valenberg, Van Boekel, & Lakemond, 2014).

Lipid classes of *A. diaperinus* are separated in silica gel TLC plates according to their degree of polarity. As expected, triacylglycerols were found to be the largest and strongest spot at the upper part of all the TLC plates, indicating that this is the most important lipid in the extraction. Phospholipids were present in organic solvent extracts, but not in the aqueous extraction. Monoacylglycerols, diacylglycerols, and FFA were extracted with Soxhlet and Folch methods. TLC results showed that FFA, monoacylglycerols and diacylglycerols were present in insect lipid extract (Tzompa-Sosa et al., 2014).

USES

Species of this genus are used in museums to clean the tissues off of carcasses during the preparation of zoological specimens (Dunford & Kaufman, 2006). The larvae, like certain other mealworms, are fed to captive reptiles. As pet food they have been called buffalo worms and lesser mealworms. They have been reported as a good first food for Central American wood turtle hatchlings, because they are more active than common mealworms and their movement stimulates feeding behavior (https://en.wikipedia.org/wiki/Alphitobius_diaperinus).

REFERENCES

Bosch, G., Zhang, S., Oonincx, D. G. A. B., & Hendriks, W. H. (2014). Protein quality of insects as potential ingredients in dog and cat foods. *Journal of Nutritional Science, 3*, e29.

Dunford, J. C., & Kaufman P. E. (2006). Lesser mealworm, *Alphitobius diaperinus*. Entomology and Nematology. Gainesville, FL, USA: University of Florida, IFAS.

Hassemer, M. J., Sant'Ana, J., de Oliveira, M. W., Borges, M., Laumann, R. A., Caumo, M., & Blassioli-Moraes, M. C. (2015). Chemical composition of *Alphitobius diaperinus* (Coleoptera: Tenebrionidae) abdominal glands and the influence of 1,4-benzoquinones on its behavior. *Journal of Economy Entomology, 108*(4), 2107–2116.

Tzompa-Sosa, D. A., Yi, L., Van Valenberg, H. J. F., Van Boekel, M. A. J. S., & Lakemond, C. M. M. (2014). Insect lipid profile: aqueous versus organic solvent-based extraction methods. *Food Research International, 62,* 1087–1094.

Yi, L., Lakemond, C. M. M., Sagis, L. M. C., Eisner-Schadler, V., van Huis, A., & van Boekel, M. A. J. S. (2013). Extraction and characterization of protein fractions from five insect species. *Food Chemistry, 141,* 3341–3348.

Chapter 50

Tenebrio molitor Mealworm

DESCRIPTION

Common English names are: mealworm, yellow mealworm, dried mealworms, and mealworm meal.

Mealworms are the larvae of two species of darkling beetles of the Tenebrionidae family: the yellow mealworm beetle (YMB) (*Tenebrio molitor* Linnaeus), and the smaller and less common dark or mini mealworm beetle (*Tenebrio obscurus* Fabricius). Mealworm beetles are indigenous to Europe and are now distributed worldwide. Mealworms go through four life stages: egg, larva, pupa, and adult. Larvae typically measure about 2.5 cm or more, whereas adults are generally between 1.25 and 1.8 cm in length (Bryning et al., 2005). Small (about 3-mm long), whitish larvae hatch from the mealworm's eggs. After a few days, they turn yellowish and produce a hard, chitinous exoskeleton. An adult larva weighs about 0.2 g and is 25–35 mm long. And it is the adult larvae that are used as human food in some parts of the world. The pupa is a free-living creature, 12–18 mm long and of creamy white color. Mealworms start to lay eggs 4–17 days after copulation (Aguilar-Miranda et al., 2002) (Fig. 50.1).

FOOD USES AND OTHER USES

T. molitor is a pest of grain, flour, and food stores, but often not of much importance since populations are quite small (Ramos-Elorduy et al., 2002). Mealworms are easy to breed and feed, and have a valuable protein profile. For these reasons, they are produced industrially as feed for pets and zoo animals, including birds, reptiles, small mammals, batrachians, and fish. They are usually fed live, but they are also sold canned, dried, or in powder form (Veldkamp et al., 2012). In feeding experiments, larvae have been dried at 50°C for 24 h (Klasing et al., 2000), dried at 100°C for 200 min, dried in the sun for 2 days or boiled in water for 3 min and then oven-dried at 60–100°C (Aguilar-Miranda et al., 2002).

The EU of mealworm production is higher than for milk or chicken and similar to pork and beef. However, mealworms, when considered as a human protein source, produce much less GHGs and require much less land than chickens, pigs, and cattle. With land availability being the most stringent limitation in sustainably feeding the world's population, mealworm should be considered

Unconventional Oilseeds and Oil Sources. http://dx.doi.org/10.1016/B978-0-12-809435-8.00050-0

FIGURE 50.1 *Tenebrio molitor* mealworm

as a more sustainable alternative to milk, chicken, pork, and beef (Oonincx and de Boer, 2012).

The mealworm species can be grown successfully on diets composed of organic by-products. Diet affects mealworm growth, development, and feed conversion efficiency, where diets high in yeast-derived protein appear favorable with respect to reduced larval development time, reduced mortality, and increased weight gain. Dietary protein content had a minor effect on mealworm protein content, whereas larval fat content and fatty acid composition varied over a wider range (Van Broekhoven et al., 2015).

Yi et al. (2013) investigated the potential of *T. molitor* as an alternative meat (protein) source for food applications by studying the effect of pH and NaCl on the extraction yield of water-soluble proteins, while preventing browning due to polyphenol oxidation. Minimum protein solubility (29.6%) was at pH 4–6 and maximum (68.6%) at pH 11. Protein extraction at 0.1 M NaCl, pH 10 gave a recovery of 100%. He observed that increase in browning at pH 8–11 corresponded to a lower monomeric phenol content. The author noticed that sodium bisulfite (studied from 0.5% to 4%) could prevent browning, whereas ascorbic acid (studied in the range 0.01%–0.04%) could not prevent as strong as sodium bisulfite. The author obtained a yield of 22% of total protein after acid precipitation (pH 4) of an isolate with a protein content of 74%. This author observed that proteins from *T. molitor* behave more or less the same as proteins from meat and fish with respect to aqueous extraction.

Three edible mealworm species (*T. molitor*, *Zophobas atratus*, and *Alphitobius diaperinus*) were subjected to processing and in vitro digestion and were analyzed for IgE cross-reactivity. Immunoblot and MALDI-MS/MS analyses revealed that IgE from crustaceans or house dust mite (HDM) allergic patients showed cross-reactivity to mealworm tropomyosin or α-amylase, hexamerin 1B precursor and muscle myosin, respectively. Heat processing as well as in vitro digestion did diminish, but did not eliminate HDM or tropomyosin

IgE cross-reactivity. Results show that individuals allergic to HDM or crustaceans might be at risk when consuming mealworms, even after heat processing (Van Broekhoven, Bastiaan-Net, de Jong, & Wichers, 2016).

Mealworms are edible for humans. Baked or fried mealworms are marketed as a healthful snack food. They may be easily reared on fresh oats, wheat bran or grain, with sliced potato, carrots, or apple as a moisture source. Mealworms have been incorporated into tequila-flavored novelty candies.

Mealworms are typically used as a pet food for captive reptiles, fish, and birds. They are also provided to wild birds in bird feeders, particularly during the nesting season. Mealworms are useful for their high protein content. They are also used as fishing bait. They are commercially available in bulk and are sold in containers of bran or oatmeal. In 2015, it was discovered that mealworms are capable of degrading polystyrene into usable organic matter at a rate of about 0.35–0.40 mg/day (Finke & Winn, 2004). The yellow mealworm (*T. molitor*) is very efficient at bioconverting organic waste. For this reason, this species is receiving increasing attention, as they could collectively convert 1.3 billion tons of biowaste per year (Veldkamp et al., 2012).

Zhao, Vázquez-Gutiérrez, Johansson, Landberg, and Langton (2016) established a protocol for extraction of yellow mealworm larvae proteins; they evaluated conditions and characterized the resulting protein extract. These authors found that, freeze-dried yellow mealworm larvae contained around 33% fat, 51% crude protein, and 43% true protein on a dry matter basis. The true protein content of the protein extract was about 75%, with an extraction rate of 70% under optimized extraction conditions. They reported that protein extract was a good source of essential amino acids. The lowest protein solubility in distilled water solution was found between pH 4 and 5, and increased with either increasing or decreasing pH. They observed lower solubility in 0.5 M NaCl solution compared with distilled water. The rheological tests indicated that temperature, sample concentration, addition of salt and enzyme, incubation time, and pH alterations influenced the elastic modulus of yellow mealworm protein extract (YMPE). The authors demonstrated that the functional properties of YMPE could be modified for different food applications (Zhao et al., 2016).

Buqadilla is an innovative snack under development for the Dutch market. It is a spicy Mexican leguminoceous food product made of chickpeas and lesser mealworms (40%). It was well received in several restaurants and canteens, where the product was tested, for its taste and smooth structure. The sustainable, healthy, and exotic snack is an example of an accessible and culturally acceptable way for Western consumers to experience and appreciate edible insects as food (Van Huis, Van Gurp, & Dicke, 2012). Crikizz is another example of European products made with insects; Crikizz are spicy, popped snacks based on mealworms and cassava. The mealworm composition varies from 10% to 20% in accordance with the product line. According to focus groups, the taste is very pleasing and differs from other snacks, while the texture is as crunchy as other snacks. The prototype was made without preservatives or taste enhancers, and the high fat composition of mealworms removes the need for added fat (Van Huis et al., 2013).

Zheng et al. (2013) extracted larval grease from a YMB (*T. molitor* L.), a postharvest scavenger; these authors investigated the extracted grease for finding its potential as a substitute of oilseeds. They obtained 34.2 g biodiesel from 234.8 g dried YMB larval biomass. The main fatty acids of the obtained biodiesel were linolenic acid (19.7%), palmitic acid (17.6%), linoleic acid (16.3%), and stearic acid (11.4%). Most of the properties of the obtained biodiesel fed on decayed vegetables met the standard EN 14214, including ester content (96.8%), density (860 kg/m^3), flash point (127°C), cetane number (58), water content (300 mg/kg), and methanol content (0.2%). From comprehensive analysis on the effect to society, economy, and environment, the authors concluded that YMB can recycle organic wastes into clean energy with low cost. It is indicated that the YMB grease is a proper feedstock for biodiesel. The properties of the YMB biodiesel fed on decayed vegetables, which has a desirable methyl ester profile met the standard EN 14214, including ester content (96.8%), density (860 kg/m^3), flash point (127°C), cetane number (58), water content (300 mg/kg), and methanol content (0.2%).

COMPOSITION AND BIOACTIVE COMPONENTS

The nutritional value of live mealworm is composed of 20% protein, 13% fat, 2% fiber, and 62% moisture, while the dried mealworm's nutritional value is composed of 53% protein, 28% fat, 6% fiber, and 5% moisture. The mealworm grease has an acid value of 7.6 (mg KOH/g), iodine value of 96 (g I/100 g), and saponification value of 162 (mg KOH/g) with low peroxide value of 0.27 (mEq./kg) (Zheng et al., 2013). Fresh larvae contained 56% of water, 18% of total protein, 22% of total fat, and 1.55% of ash. High contents of minerals were found in the larvae: magnesium (87.5 mg/100 g), zinc (4.2 mg/100 g), iron (3.8 mg/100 g), copper (0.78 mg/100 g), and manganese (0.44 mg/100 g). The proportion of $n-6/n-3$ fatty acids was very advantageous and amounted to 6.76. Larvae powder contained twice higher content of protein, fat, ash, and minerals. Larva of mealworm is a valuable source of nutrients in amounts more profitable for human organism than traditional meat food. Powdered larva is a high-grade product to be applied as a supplement to traditional meals (Siemianowska et al., 2013). Lysine was the most abundant indispensable amino acid in the insect larval meals (*T. molitor* meal), whereas glutamic acid was the most abundant dispensable one. As in *T. molitor* meal, glutamic acid was the most abundant of the dispensable amino acids. The insect larval meal was a good source of methionine and threonine. The *T. molitor* meal showed higher lysine, methionine, and threonine contents (De Marcoa et al., 2015).

The gross energy (which is normally higher than metabolizable energy) of the mealworm was 206 kcal/100 g fresh weight (recalculated from dry matter), depending on the insect's diet. The range of protein content of mealworm was

14–25 g/100 g fresh weight). Vitamin B12 occurs only in food of animal origin and is well represented in mealworm larvae as 0.47 µg/100 g (Finke, 2002). The removal of fat (and ash) in the production of insect products, such as insect meal, reduces the "stickiness" of the protein concentrate and prevents fatty acids (mostly unsaturated) from exposure to undesirable oxidation processes. The extracted fat can then be used for other purposes. Traditionally, the fat (e.g., oil) of some insect species is used extensively for frying meat and other food products (Van Huis et al., 2013). Glutamate dehydrogenase was purified 4300-fold from the yellow mealworm fat body by gel filtration. The purified enzyme had a maximal specific activity of 2.1 µkat/mg protein when assayed in the direction of glutamate synthesis with NADH as a coenzyme (Teller, 1988).

REFERENCES

Aguilar-Miranda, E. D., López, M. G., Escamilla-Santana, C., & Barba de la Rosa, A. P. (2002). Characteristics of maize flour tortilla supplemented with ground *Tenebrio molitor* larvae. *Journal of Agricultural and Food Chemistry, 50*(1), 192–195.

Bryning, G. P., Chambers, J., & Wakefield, M. E. (2005). Identification of a sex pheromone from male yellow mealworm beetles, *Tenebrio molitor. Journal of Chemical Ecology, 31*(11), 2721–2730.

De Marcoa, M., Martínez, S., Hernandez, F., Madrid, J., Gai, F., Rotolo, L., Belforti, M., Bergeroa, D., Katz, H., Dabboud, S., Kovitvadhi, A., Zoccarato, I., Gascoc, L., & Schiavone, A. (2015). Nutritional value of two insect larval meals (*Tenebrio molitor* and *Hermetia illucens*) for broiler chickens: apparent nutrient digestibility, apparent ileal amino acid digestibility and apparent metabolizable energy. *Animal Feed Science and Technology, 209,* 211–218.

Finke, M. D. (2002). Complete nutrient composition of commercially raised invertebrates used as food for insectivores. *Zoo Biology, 21*(3), 269–285.

Finke, M., & Winn, D. (2004). Insects and related arthropods: a nutritional primer for rehabilitators. *Journal of Wildlife Rehabilitation, 27,* 14–17.

Klasing, K. C., Thacker, P., Lopez, M. A., & Calvert, C. C. (2000). Increasing the calcium content of mealworms (*Tenebrio molitor*) to improve their nutritional value for bone mineralization of growing chicks. *Journal of Zoo and Wildlife Medicine, 31*(4), 512–517.

Oonincx, D. G. A. B., & De Boer, I. J. M. (2012). Environmental impact of the production of mealworms as a protein source for humans—a life cycle assessment. *PLoS One, 7,* e51145.

Ramos-Elorduy, J., Avila Gonzalez, E., Rocha Hernandez, A., & Pino, J. M. (2002). Use of *Tenebrio molitor* (*Coleoptera: Tenebrionidae*) to recycle organic wastes and as feed for broiler chickens. *Journal of Economic Entomology, 95*(1), 214–220.

Siemianowska, E., Kosewska, A., Aljewicz, M., Skibniewska, K. A., Polak-Juszczak, L., Jarocki, A., & Jędras, M. (2013). Larvae of mealworm (*Tenebrio molitor* L.) as European novel food. *Agricultural Sciences, 4*(6), 287–291.

Teller, J. K. (1988). Purification and some properties of glutamate dehydrogenase from the mealworm fat body. *Insect Biochemistry, 18*(1), 101–106.

van Broekhoven, S., Oonincx, D. G. A. B., van Huis, A., & van Loon, J. J. A. (2015). Growth performance and feed conversion efficiency of three edible 4 mealworm species (Coleoptera: Tenebrionidae) on diets composed of organic by-products. *Journal of Insect Physiology, 73,* 1–10.

Van Broekhoven, S., Bastiaan-Net, S., de Jong, N. W., & Wichers, H. J. (2016). Influence of processing and in vitro digestion on the allergic cross-reactivity of three mealworm species. *Food Chemistry, 196,* 1075–1083.

Van Huis, A., Van Gurp, H., & Dicke, M. (2012). Het insectenkookboek [The insect cookbook]. Amsterdam: Atlas Uitgeverij Atlas.

Van Huis, A., Van Itterbeeck, J., Klunder, H., Mertens, E., Halloran, A., Muir, G., & Vantomme, P. (2013). Edible insects: Future prospects for food and feed security. FAO Forestry Paper 171. Rome, Italy: FAO.

Veldkamp, T., van Duinkerken, G., van Huis, A., Lakemond, C. M. M., Ottevanger, E., Bosch, G., van Boekel, M. A. J. S. (2012). Insects as a sustainable feed ingredient in pig and poultry diets—a feasibility study. Rapport 638—Wageningen Livestock Research.

Yi, L., Lakemond, C. M. M., Sagis, L. M. C., Eisner-Schadler, V., van Huis, A., & van Boekel, M. A. J. S. (2013). Extraction and characterization of protein fractions from five insect species. *Food Chemistry, 141*, 3341–3348.

Zhao, X., Vázquez-Gutiérrez, J. L., Johansson, D. P., Landberg, R., & Langton, M. (2016). Yellow mealworm protein for food purposes—extraction and functional properties. *PLoS One, 11*(2), e0147791.

Zheng, L., Hou, Y., Li, W., Yang, S., Li, Q., & Yu, Z. (2013). Exploring the potential of grease from yellow mealworm beetle (*Tenebrio molitor*) as a novel biodiesel feedstock. *Applied Energy, 101*, 618–621.

Chapter 51

Blaptica dubia Cockroaches

DESCRIPTION

Blaptica dubia is a medium-sized species of cockroach, which grows to around 4.0–4.5 cm. They are sexually dimorphic; adult males have full wings covering their body, while females have only tiny wing stubs, their tegmina (forewings) being around a fourth of their body length. Adults are dark brown to black with somewhat lighter orange spot/stripe pattern, sometimes visible only in bright light. Coloration does differ slightly with environment and diet from one colony to another. *B. dubia* are ovoviviparous, giving birth to live young, and can give birth to 20–40 nymphs per month under favorable conditions (Hao, Arthur, & Ping, 2013) (Fig. 51.1).

DISTRIBUTION

The Dubia cockroach is found in Central and South America, beginning in Costa Rica. It is common from French Guyana and Brazil to Argentina. Documented specimens have been found in Brazil, Argentina, and Uruguay. The species is found in central Asia, the Caucasus Mountains, north-eastern Africa, and its distribution includes the following countries: Afghanistan, Azerbaijan, Egypt, India, Iran, Iraq, Israel, Jordan, Kashmir, Libya, Palestine, Pakistan, Saudi Arabia, Sudan, the United Arab Emirates, and the United States.

FOOD USES AND OTHER USES

B. dubia has become a popular feeder insect, particularly among tarantula, amphibian, and reptile enthusiasts. Keeping or breeding the insect is made easier by their inability to jump or climb smooth surfaces, relatively slow movement, and rarity of flying. They are also quiet, unlike crickets, and tropical environmental requirements reduce the likelihood of establishment of escapees in colder, dryer climates. *B. dubia* can cause allergic reactions in humans, although they produce relatively little odor compared to many cockroaches (Friederich & Volland, 2004; Alamer, 2013).

Dubia roaches are a better feeder species for pets than most other insects. They are popular for feeding reptiles and amphibians because they contain a high amount of protein (a high-quality herb food source). They are soft-bodied and have more meat for example, than crickets making them an excellent food

FIGURE 51.1 *Blaptica dubia* cockroaches adults and nymphs.

item for example, for arachnids and small lizards (Alamer, 2013). A review discovered different cockroaches gave a high-protein, low-fat nutrition composition similar to crickets, more than that provided by mealworms or super worm larvae. The gut contents of the cockroach, depending on its diet, may provide essential nutrients unavailable from a cockroach with an empty gut. Vitamin and mineral content in studied cockroaches was well represented except for low calcium:phosphorus ratios typical in cockroaches, and relatively low vitamin A and E in captive cockroaches. Supplementation of these nutrients in feeder cockroaches may be advisable (Oonincx & Dierenfeld, 2012). *B. dubia* is a cockroach that is used as feed for most reptiles and amphibians because of its nutritive content. Its alternative protein sources can replace the current high-priced imported animal feed, such as fishmeal, soybeans, and meat. It has potential as animal diet and minimizes high livestock farming cost of the farmers (Yee, Kumara, Abdul Latif, & Rao, 2015).

COMPOSITION AND BIOACTIVE COMPONENTS

The total lipid content of *B. dubia* was 7.6%, while the crude protein was 19.3%. The measured protein content of *B. dubia* is comparable with that of beef (18.4%), chicken (22.0%), and fish (18.3%), and it was higher than that of lamb (15.4%), pork (14.6%), eggs (13%), and milk (3.5%), but lower in comparison to soy (36.5%). High amounts of unsaturated fatty acids (76.31%) were found relative to saturated fatty acids (23.30%) (Rumpold & Schluter, 2013). The most abundant unsaturated fatty acids in *B. dubia* lipid extracts was C18:1 *cis*9 and C18:2 *cis*9,12; the most abundant saturated fatty acids was C16:0. These three fatty acids accounted for 51.4, 17.4, and 18.1 g/100 g, respectively; ω-3 fatty acids, which are related to health benefits, were most abundant in lipids from aqueous extraction. The method of extraction affected the quantity and types of lipids and the fatty acids profile of the insect lipid extracts. Higher *omega-6/omega-3* fatty acids ratios were extracted by Folch method while aqueous extraction had the lowest ratio, hence having the highest omega-3 fatty acids content (Tzompa-Sosa, Yi, Van Valenberg, Van Boekel, & Lakemond, 2014). A study was conducted to determine the nutritional composition of *B. dubia* in different growth stages and a preliminary digestibility test on animal model using *B. dubia* as feed. The proximate analysis of adults and nymph of *B. dubia* contained 59.06%–62.70% moisture, 2.47%–4.17% ash, 47.50%–54.32% protein, 3.83%–5.58% chitin, and 35.49%–44.22% fat on dry weight. A preliminary protein digestibility trial using three domesticated cats was conducted to assess apparent digestibility of crude protein in prepared *B. dubia* feed pellets. High apparent crude protein digestibility (81.57 ± 3.30%) was found. Thus, the *B. dubia* tested feed pellets could be an alternative protein source as the animal feed especially in the pet industry (Yee et al., 2015). The composition of cuticular free fatty acids in the three species of cockroaches was also studied. The analysis indicated that all of the fatty acids were mostly straight-chain, long-chain, saturated, or unsaturated. Cuticular lipids of three species of cockroaches contained 19 free fatty acids, ranging from C14:0 to C24:0. The predominant fatty acids found in the three studied species of cockroaches were oleic, linoleic, palmitic, and stearic acid. Only in adults of *Blaptica orientalis*, myristoleic acid, γ-linolenic acid, arachidic acid, dihomolinoleic acid, and behenic acid were identified. Lignoceric acid was detected only in nymphs of *B. orientalis*. Heneicosylic acid and docosahexaenoic acid were identified in adults of *B. dubia* (Gutierrez et al., 2015). The total amount of amino acids per gram crude protein of *B. dubia* was 776 mg/g. The sum of essential amino acids for *B. dubia* was 361 mg/g, which was lower than casein and soybean protein, but essential amino acids were available in quantities that are necessary for human requirement. The fact that the sum of the total amount of amino acids did not add up to 1000 mg/g crude protein is likely mainly explained by the presence of nonprotein nitrogen in the form of chitin. The extraction method of protein is easy and feasible to apply, but the yield of extracted supernatant fractions is relatively low. The purity of measured protein content expressed as percentage of dry matter ranged from

50% to 61% of supernatant fractions, from 65% to 75% of pellet fractions, and from 58% to 69% of residue fractions depending on the insect species. The soluble protein fractions of cockroach had poor foaming capacity at pH 3, 5, 7, and 10, but could form gels at a concentration of 30% w/v. The gelation temperature ranged from about 51 to 63°C for cockroach species at pH 7. In addition, all insect gels had a comparable maximum linear strain at this concentration, with a value of around 50% (Yi, 2015). Yi et al. (2013) extracted and fractionated proteins from *B. dubia* and they evaluated the protein purity and yield of the obtained fractions. Around 20% of total protein was found back in the supernatant, the rest of the protein was divided about equally over the residue and the pellet fraction for all five insect species after aqueous extraction. They reported that the extraction method is easy and feasible to apply, but the yield of extracted supernatant fractions is relatively low. The purity of the measured protein content expressed as percentage of dry matter ranged from 50% to 61% of supernatant fractions, from 65% to 75% of pellet fractions, and from 58% to 69% of residue fractions depending on the insect species (Yi et al., 2013).

REFERENCES

Alamer, A. H. (2013). Endocrine control of fat body composition and effects of the insect growth regulators methoprene and pyriproxyfen on the development and reproduction of the Argentinean cockroach, *Blaptica dubia* Serville (Blattaria: Blaberidae). Thesis. Germany: University of Bayreuth.

Friederich, U., & Volland, W. (2004). *Breeding food animals: Live food for vivarium animals*. Malabar, FL: Krieger Pub., p. 66.

Gutierrez, A. C., Gołbiowski, M., Pennisi, M., Peterson, G., García, J. J., Manfrino, R. J., & Lastra, C. C. L. (2015). Cuticle fatty acid composition and differential susceptibility of three species of cockroaches to the entomopathogenic fungi *Metarhizium anisopliae* (Ascomycota, Hypocreales). *Journal of Economic Entomology, 108*(2), 752–760.

Hao, W., Arthur, G. A., & Ping, H. X. (2013). Instar determination of *Blaptica dubia* (Blattodea: Blaberidae) using Gaussian mixture models. *Annals of the Entomological Society of America, 106*(3), 323–328.

Oonincx, D. G. A. B., & Dierenfeld, E. S. (2012). An investigation into the chemical composition of alternative invertebrate prey. *Zoo Biology, 31*(1), 40–54.

Rumpold, B. A., & Schluter, O. K. (2013). Nutritional composition and safety aspects of edible insects. *Molecular Nutrition & Food Research, 57*(5), 802–823.

Tzompa-Sosa, D. A., Yi, L., Van Valenberg, H. J. F., Van Boekel, M. A. J. S., & Lakemond, C. M. M. (2014). Insect lipid profile: aqueous versus organic solvent-based extraction methods. *Food Research International, 62*, 1087–1094.

Yee, L. P., Kumara, T. K., Abdul Latif, N. S., & Rao, P. V. (2015). Dubia cockroach (*Blaptica dubia*) as an alternative protein source for animal feed. *Journal of Biochemical Biopharmaceutical and Biomedical Sciences, 1*, 31–39.

Yi, L. (2015). A study on the potential of insect protein and lipid as a food source. PhD thesis. Wageningen, The Netherlands; Wageningen University.

Yi, L., Lakemond, C. M. M., Sagis, L. M. C., Eisner-Schadler, V., van Huis, A., & van Boekel, M. A. J. S. (2013). Extraction and characterization of protein fractions from five insect species. *Food Chemistry, 141*, 3341–3348.

Chapter 52

Macrotermes bellicosus Termite

DESCRIPTION

Termites (*Macrotermes bellicosus*) are usually small, measuring between 4 and 15 mm in length. The largest of all extant termites are the queens of the species *M. bellicosus*, measuring up to over 10 cm in length. Like other insects, termites have a small tongue-shaped labrum and a clypeus; the clips are divided into postclypeus and anteclypeus. The mouthparts contain a maxillae, a labium, and a set of mandibles. The maxillae and labium have polyps that help termites' sensational food and handling (Bignell, Roisin, & Lo, 2010). On alates, the wings are located at the mesothorax and metathorax. The mesothorax and metathorax have well-developed exoskeletal plates; the prothorax has smaller plates. Termites have a 10-segmented abdomen with two plates, the tergites and the sternites (Bignell et al., 2010).

There are two worker castes in *M. bellicosus*, the major worker and the minor worker. In both cases, the workers begin their lives by taking care of the queen and later on leave the nest to begin foraging. The point at which the worker leaves the nest to begin gathering food differs slightly between the two castes. Major workers will leave at any point between 13 and 25 days after moulting and the minor worker exits between 9 and 32 days (Hinze & Leuthold, 1999) (Fig. 52.1).

DISTRIBUTION

Termites are found on all continents except Antarctica. The diversity of termite species is low in North America and Europe, but is high in South America, where over 400 species are known. Of the 3000 termite species currently classified, 1000 are found in Africa, where mounds are extremely abundant in certain regions. In Asia, there are 435 species of termites, which are mainly distributed in China. Within China, termite species are restricted to mild tropical and subtropical habitats south of the Yangtze River. In Australia, all ecological groups of termites are endemic to the country, with over 360 classified species (Meyer, 1999).

M. bellicosus constructs mounds differently based on the surrounding habitat. In the forest, the mounds are constructed with thick walls and are dome shaped, whereas the mounds that are constructed in the savanna have thin walls and deviate from the simple dome construction to more complicated structures. Heating experiments demonstrated that this difference in structure is due to

Unconventional Oilseeds and Oil Sources. http://dx.doi.org/10.1016/B978-0-12-809435-8.00052-4

FIGURE 52.1 *Macrotermes bellicosus* mound with a minor soldier and major workers present.

different thermal properties of each mound as a response to the unique habitats. Mounds in the cooler forest habitat will retain their temperature for longer periods of time while mounds in the warmer savanna will shed heat faster. *M. bellicosus* individuals will burrow themselves in the subsoil and collect clay in their mouths. The clay is moistened by their saliva. Mound building is usually most labor-intensive in the wet months (Mackean & Mackean, 2016)

FOOD USES AND DIFFERENT USES

Entomophagy, the practice of using insects as a part of the human diet, has played an important role in the history of human nutrition in Africa, Asia, and Latin America. Another important use of insects by humans is medicinal use, featuring a practice known as entomotherapy (Srivastava, Babu, & Pandey, 2009; Costa-Neto, 2002). Forty-three termite species are used as food by humans or are fed to livestock. These insects are particularly important in less-developed countries where malnutrition is common, as the protein from termites can help improve the human diet. Termites are consumed in many regions globally, but this practice has only become popular in developed nations in recent years (Figueirêdo, Vasconcellos, Policarpo, & Alves, 2015). Termites are consumed by people in many different cultures around the world. In Africa, the alates are an important factor in the diets of native populations. Tribes have different ways of collecting or cultivating insects; sometimes tribes will collect soldiers from several species. Though harder to acquire, queens are regarded as a delicacy. Termite alates are high in nutrition with adequate levels of fat and protein. They are regarded as pleasant in taste, having a nut-like flavor after they are cooked (Nyakupfuka, 2013).

Alates are collected when the rainy season begins. During a nuptial flight, they are typically seen around lights to which they are attracted, and so nets are set up on lamps and captured alates are later collected. The wings are removed through a technique that is similar to winnowing. The best result comes when they are lightly roasted on a hot plate or fried until crisp. Oil is not required as their bodies usually contain sufficient amounts of oil. Termites are typically eaten when livestock is lean and tribal crops have not yet developed or produced any food, or if food stocks from a previous growing season are limited (Nyakupfuka, 2013). The nutritionally weak grain, frequently consumed in many African countries, is low in proteins and fats and lacks several essential amino acids, such as lysine. For this reason, fortifying the grain with highly nutritious flying termites (*Macrotermes* species), easily gathered at the start of the rainy season, makes sense. The fermented mixture can be consumed as porridge at breakfast, lunch, or dinner, depending on local preferences. Both raw materials are easily obtained locally (Van Huis et al., 2013).

In addition to Africa, termites are consumed in local or tribal areas in Asia and North and South America. In Australia, Indigenous Australians are aware that termites are edible but do not consume them even in times of scarcity; there are few explanations as to why. Termite mounds are the main sources of soil consumption (geophagy) in many countries including Kenya, Tanzania, Zambia, Zimbabwe, and South Africa (Nchito et al., 2004). Researchers have suggested that termites are suitable candidates for human consumption and space agriculture, as they are high in protein and can be used to convert inedible waste of consumable products for humans (Geissler, 2011).

At least 45 species of termites, belonging to 4 families, are used in the world, with 43 species used in human diets and/or in livestock feeding. Nine termite species are used as a therapeutic resource. There is an overlapping use of seven species. The use of termites was registered in 29 countries over 3 continents. Africa is the continent with the highest number of records, followed by America and Asia (Figueirêdo et al., 2015). In addition to their ecological importance, termites are a source of medicinal and food resources to various human populations in various locations of the world, showing their potential for being used as an alternative protein source in human or livestock diets, as well as a source for new medicines (Figueirêdo et al., 2015).

Winged termites are often fried in their own fat. Fried termites contain 32%–38% protein. In Uganda, termites are steamed in banana leaves. Termites are boiled or roasted after swarming and then sun-dried or smoke-dried, or both, depending on the weather. Sometimes termites are crushed with a pestle and mortar and eaten with honey (Van Huis et al., 2013).

For medicinal purposes, the use of 10 species of termites was recorded. These species are used as an alternative treatment for physiological and spiritual problems. Termites are widely used in different countries and nations as a therapeutic resource for the treatment of asthma, hoarseness and sinusitis, child malnutrition, suture wounds, sickness in pregnant women, wounds, spiritual protection against witches and wizards, rituals protection and promotion (in jobs, trade), appealing to gods and witches, safe delivery of baby, soothsaying asthma, bronchitis, influenza, whooping cough, flu, asthma, and sore throat, antibiotic activity, antimicrobial, chickens' asthma, enhancement of lactation, and antifungal and antibacterial properties (Figueirêdo et al., 2015).

Evidence of antimicrobial activity of products isolated from termites has been reported, such as peptides like espinigerine and termicine, isolated from *Pseudocanthotermes spiniger*, which showed antifungal and antibacterial activities (Ramos-Elorduy, 2005). Their use as protein and lipid source reveals that these insects may play important roles in human and other animals' diets, reinforcing a tendency indicated in previous studies.

CHEMICAL COMPOSITION AND BIOACTIVE COMPONENTS

Ntukuyoh, Udiong, Ikpe, and Akpakpan (2012) analyzed queen, soldiers, and workers of termites (*M. bellicosus*) for proximate composition, vitamin, mineral elements, and antinutrient content. Proximate composition showed that crude protein content of the soldiers (54%–56%) was higher than those of workers (25%–29%) and queen (31%–32%). The high mineral element was sodium in queen, while the least mineral was manganese in the soldiers. The termites were rich in vitamins A (6.3–7.7 mg/kg) and C (1.22 ± 0.10 mg/kg). Worker termites had the highest vitamin C content, while queen had the highest vitamin A content. Antinutrient compositions in *M. bellicosus* were considerably low. The highest oxalate content values were found in the soldiers caste that is, 233.20 ± 3.0 mg/kg. The queen of all termites analyzed has the least oxalate content value of 44.0 ± 2.0 mg/kg. Hydrocyanide acid (HCN) concentration in the termites ranges from 0.403 ± 0.003 to 2.488 ± 0.02 mg/kg. The highest phytic acid value among termites, 1.037 ± 0.007 mg/kg, was obtained from the soldier caste (Ntukuyoh et al., 2012).

Banjo, Lawal, and Songonuga (2006) analyzed two termites species *M. bellicosus* and *Macrotermes notalensis* (Table 52.1). They reported nitrogen-free extracts above 50%, fat content of 28.2 and 22.5%, and crude protein 20.4 and 22.1%, respectively.

Ekpo and Onigbinde (2007) extracted oil of *M. bellicosus* and they characterized its physical and chemical properties. They reported that the lipid content

TABLE 52.1 Proximate Analysis (%) of Two Nigerians Termites Species

Parameters	M. bellicosus	M. notalensis
Crude protein	20.4	22.1
Fat	28.2	22.5
Ash	2.90	1.90
Crude fiber	2.70	2.20
Dry matter	90.6	89.5
Nitrogen-free extract	2.82	2.98
Moisture	43.3	42.8
Vitamin A (µg/100 g)	2.89	2.56
Vitamin B2 (mg/100 g)	1.98	1.54
Vitamin C (mg/100 g)	3.41	3.01
Ca (mg/100 g)	21	18
P (mg/100 g)	136	114
Fe (mg/100 g)	27	29
Mg (mg/100 g)	0.15	0.26

of the insect was 31.46 (wet weight). This lipid value is found to be lower when compared to lipids derived from *Aspongopus viduatus* (45%) and *Agonoscelis pubescens* (60%) (Mariod, Matthäus, & Eichner, 2004). The oil was a clear, odorless liquid, of a light yellow color and it was fluid at room temperature (26°C). Lipid analysis revealed that the insect oil comprised 69.87% neutral lipid, 19.14% phospholipid, and 10.81% glycolipids. Further analysis revealed a refractive index of 1.20 ± 0.01, specific gravity of 0.90, solidification value of 10–14°C, total lipid phosphorus of 47.18 (mg/g lipid), acid value of 3.60, iodine value of 108, saponification value of 193.40, unsaponifiable matter of 12.04, free cholesterol of 8.73 (mg/100 g lipid), and total cholesterol of 47.18. The unsaturated fatty acids accounted for 51.02% of the total fatty acids, whereas the saturated fatty acids constituted 48.98% of the fatty acids. These values when compared with that observed in oils, which have been considered to be of high quality, suggest that *M. bellicosus* oil has potentials that could be exploited by the nutritional and pharmaceutical companies (Ekpo & Onigbinde, 2007).

REFERENCES

Banjo, A. D., Lawal, O. A., & Songonuga, E. A. (2006). The nutritional value of fourteen species of edible insects in southwestern Nigeria. *African Journal of Biotechnology*, 5(3), 298–301.

Bignell, D. E., Roisin, Y., & Lo, N. (2010). *Biology of termites: a modern synthesis* (1st ed.). Dordrecht: Springer.

Costa-Neto, E. M. (2002). The use of insects in folk medicine in the State of Bahia, Northeastern Brazil, with Notes on insects reported Elsewhere in Brazilian folk medicine. *Human Ecology, 30,* 245–263.

Ekpo, K. E., & Onigbinde, A. O. (2007). Characterization of lipids in winged reproductives of the termite *Macrotermis bellicosus. Pakistan Journal of Nutrition, 6,* 247–251.

Figueirêdo, R. E. C. R., Vasconcellos, A., Policarpo, I. S., & Alves, R. R. N. (2015). Edible and medicinal termites: a global overview. *Journal of Ethnobiology and Ethnomedicine, 11*(1), 1–17.

Geissler, P. W. (2011). The significance of earth-eating: social and cultural aspects of geophagy among Luo children. *Africa, 70*(4), 653–682.

Hinze, B., & Leuthold, R. H. (1999). Age related polyethism and activity rhythms in the nest of the termite *Macrotermes bellicosus* (Isoptera, Termitidae). *Insectes Sociaux, 46,* 392–397, 4.

Mackean, D. G., & Mackean, I. (2016). The termite (*Macrotermes bellicosus*).

Mariod, A. A., Matthäus, B., & Eichner, K. (2004). Fatty acid, tocopherol and sterol composition as well as oxidative stability of three unusual Sudanese oils. *Journal of Food Lipids, 11,* 179–189.

Meyer, V. W. (1999). Distribution and density of termite mounds in the northern Kruger National Park, with specific reference to those constructed by *Macrotermes Holmgren* (Isoptera: Termitidae). *African Entomology, 7*(1), 123–130.

Nchito, M., Geissler, P. W., Mubila, L., Friis, H., & Olsen, A. (2004). Effects of iron and multimicronutrient supplementation on geophagy: a two-by-two factorial study among Zambian school children in Lusaka. *Transactions of the Royal Society of Tropical Medicine and Hygiene, 98,* 218–227.

Ntukuyoh, A. I., Udiong, D. S., Ikpe, E., & Akpakpan, A. E. (2012). Evaluation of nutritional value of termites (*Macrotermes bellicosus*): soldiers, workers, and queen in the Niger Delta Region of Nigeria. *International Journal of Food Nutrition and Safety, 1*(2), 60–65.

Nyakupfuka, A. (2013). *Global delicacies: discover missing links from Ancient Hawaiian teachings to clean the plaque of your soul and reach your higher self.* Bloomington, Indiana: Balboa Press, pp. 40–41.

Ramos-Elorduy, J. (2005). Insects: a hopeful food source. In M. G. Paoletti (Ed.), *Ecological implications of minilivestock* (pp. 263–291). Enfield, NH, USA: Science Pub.

Srivastava, S. K., Babu, N., & Pandey, H. (2009). Traditional insect bioprospecting—as human food and medicine. *Indian Journal of Traditional Knowledge, 8,* 485–494.

Van Huis, A., Van Itterbeeck, J., Klunder, H., Mertens, E., Halloran, A., Muir, G., & Vantomme, P. (2013). Edible insects: Future prospects for food and feed security. FAO Forestry Paper 171. Rome, Italy: FAO.

Chapter 53

Principles of Oil Extraction, Processing, and Oil Composition

The extraction of oil from an oil-bearing commodity involves some steps, depending on the particular commodity to be processed, the scale of operation, and the technical options available to the processor. Many steps of oil processing are involved, for example, seed pretreatment which deals with the stages prior to the extraction stage, such as cleaning, crushing, and scorching; extraction, this step involves the separation of the raw material into oil and residue (cake); and finally refining step. Postextraction treatments involve the packaging of the oil and cake for marketing. Five fundamental specialized alternatives exist for oil extraction: hot water floatation, ghanis, manual presses, powered expellers, and solvent extraction (Bernardini, 1985). One of the more established techniques for oil extraction is utilizing mechanical intends to squeeze oil out of the seeds of oil products, which might possibly be filtered a short time later. While the utilization of heat mixed with pressure can possibly separate moderately more oil; it likewise can possibly degrade the oil's quality and modify other attractive attributes, similar to taste, odor, color, and texture (http://www.ehow.com).

The first stage of the process sometimes requires the separation of the oil-bearing part of the plant, be it a nut, fruit, or seed. The process is generally referred to as decortication or dehulling. Seed dehulling is generally performed before oil extraction to diminish the amount of material to be prepared, in this way expanding the throughput of the downstream processing equipment. Dehulling likewise lessens maintenance cost connected with the war of the covering bars and the shaft of the screw press. Moreover, dehulling expands the protein substance of the meal and lessens the amount of wax that gets extracted with the oil (Williams & Hron, 1996). Seeds with thin testa similar to rapeseed or sesame seed can be processed without decortication. In most cases the oil-bearing material is then broken into smaller pieces by pounding in a pestle and mortar or by manual or motorized grinding. When motorized oil expellers are used the decorticated or undecorticated material can often be fed directly to the expeller. The next step involves heating the oil-bearing material, sometimes with the addition of a little water to assist in the rupture of oil-bearing cells and in the liberation of oil. On a small scale, this is carried out in a pan over a fire, but mechanized heaters or kettles are available and used in commercial plants.

Unconventional Oilseeds and Oil Sources. http://dx.doi.org/10.1016/B978-0-12-809435-8.00053-6

OIL EXTRACTION

Oil extraction is defined as an isolation of oil from animal by-products, fleshy fruits, such as the olive and palm, and oilseeds, such as cottonseed, sesame seed, soybeans, and peanuts. Oil is extracted by three general methods: rendering, used with animal products and oleaginous fruits; mechanical pressing, for oil-bearing seeds and nuts; and extracting with volatile solvents, employed in large-scale operations for a more complete extraction than is possible with pressing (http://global.britannica.com). In brief, five basic technical options will be reviewed: hot water floatation, ghanis, manual presses, powered expellers, and solvent extraction.

TRADITIONAL METHODS

1. Hot water floatation is probably the simplest method and is still used in many rural areas. The ground material is placed in boiling water and simmered for several hours. On removing from the fire and cooling, the oil floats to the surface and is skimmed off. In general, the oil is then heated in a shallow pan to drive off the last traces of water. This improves the keeping quality of the oil as water has a catalytic role in the development of rancidity in oils. The extraction efficiency is generally low and problems often occur with the formation of oil–water emulsions, which make the final separation difficult. In some cases, salt is used to break such emulsions.

    In many countries a large rotating pestle and mortar system is used for oil extraction. These systems are powered by animals or motors, although sometimes human power is used. The mortar is firmly fixed in the ground and as the pestle rotates, oil is released by friction and pressure and runs out of a small aperture at the base of the mortar. A typical one bullock ghani can process 40 kg of the material/day. In the case of power-ghanis either the pestle or mortar is fixed, the other rotating. Power ghanis usually are operated in pairs and have a typical capacity of 100 kg/day. The extraction efficiency is generally greater than animal units (FAO, 1992). Kamal-Eldin, Yousif, Iskander, and Appelqvist (1992) described the traditional camel-powered ghani method for grinding sesame seeds (12 kg, oil content 53.1% MFB) with 0.5 L of water. Oil release was observed after 30 min when the temperature of the mass was 41°C. After 40 min, 2.0 L of oil previously extracted (temperature 46°C) were added to assist extraction. Extraction was complete in 55–60 min, giving approximately 5 L of oil (temperature 50°C). The crushing system that presses oilseeds, which developed in the form of a mortar-and-pestle arrangement powered by animals is commonly called ghani in India (Achaya, 1993) and Assara in Sudan. The device is widely used in the Sudan to crush sesame seeds, though its antiquity needs to be documented. The camel tradition mill was composed of wooden bowl (Assara) approximately 75 cm in diameter and 100 cm in length, woody stick (Walad), woody connector (Hawam), and a camel; milling was started with putting enough cleaned sesame seeds (approximately 50 kg) in the woody bowl, which was then pressed by the woody stick (Walad) driven by a blindfolded camel plodding round and round, and oil is squeezed out of sesame seeds. Pressed seeds squeeze out oil into two layers: the clean, light-colored upper one, which is known as Normal sesame oil and the lower layer that produced cloudy, dark color and strong flavor known as Walad sesame oil (Umarabi, 2011; Mariod, Idris, & Mohamed, 2014).

2. Other traditional methods are still used to extract oil from oil-bearing materials. Such systems include the use of heavy stones, wedges, levers, and twisted ropes to apply pressure to the material so as to squeeze out the oil. They are inefficient, of low capacity, and labor intensive.

REFINING

The last stage in the processing of an oil is refining, which includes some or all of the following treatments: filtering, neutralization, winterizing, bleaching, deodorization, and degumming and filtering. In many cases, refining is not a necessary stage in traditional processing systems as local palates are accustomed to the flavor of unrefined oils and in many parts of the world, these flavors are in fact preferred to the blandness of a fully refined oil. Many commercial plants, in fact bleach crude oil as routine and then add a controlled amount of color in order to produce a standard final product (FAO, 1992).

Vegetable food oils are of great importance in the diet; they are a concentrated form of energy vital to people all over the world. In developing countries, poor people rarely have access to or can afford cooking oils from larger refineries and still rely on minor oil crops to meet their needs. In some cases, viable and important local industry and trading are involved. Plant seeds are important sources of oils of nutritional, industrial, and pharmaceutical importance. No oil from any single source has been found to be suitable for all purposes because oils from different sources generally differ in their fatty acid composition. Plant seed oils are also important sources of lipid soluble antioxidants, such as phenols where they have many health benefits. This necessitates the search for new sources of novel oils (Ramadan & Mörsel, 2002).

Fats and oils are composed of triglycerides, which are esters composed of three fatty acid units linked to glycerol. An increase in the percentage of shorter-chain fatty acids and/or unsaturated fatty acids lowers the melting point of a fat or oil. The hydrolysis of fats and oils in the presence of a base makes soap and is known as saponification. Double bonds present in unsaturated triglycerides can be hydrogenated to convert oils (liquid) into margarine (solid). The oxidation of fatty acids can form compounds with disagreeable odors. This oxidation can be minimized by the addition of antioxidants (Ball, Hill, & Scott, 2011).

Vegetable oils contain a wide range of minor components besides triacylglycerols, the major component of most oils. Sterols, tocopherols, carotenoids, phenolic compounds, chlorophyll, free fatty acids, and other minor derivatives, as well as trace metal ions are the most recognized of these compounds. Minor lipid components are of interest because they affect the technical properties of the oil, possibly provide health benefits, and/or might be used for the authentication of different oils and their mixtures. The tocopherols are interesting, as they are primarily responsible for the stability of the polyunsaturated fatty acids of vegetable oils. Tocopherols and carotenoids are important for human nutrition as they provide vitamins E and A activities. Phytosterols are currently receiving much interest as cholesterol-lowering agents derived from, inter alia, vegetable oil refining. The phenolic compounds in certain oils, for example, olive and sesame oils, are interesting from oil stability and health points of view. Gas chromatography or gas chromatography–mass spectrometry of the unsaponifiable fractions of vegetable oils can provide a fingerprint for the identity of the oils (Kamal-Eldin, 2005).

REFERENCES

Achaya, K. T. (1993). *Ghani: The traditional oil mill of India.* Kemblesville, PA: Olearius Editions.

Bernardini, E. (1985). *Oilseeds Oils and Fat* (Vol. II, Oil and Fat Processing, 2nd Ed.). Rome: Interstampa.

Ball, D. W., Hill, J. W., & Scott, R. J. (2011). *The basics of general, organic, and biological chemistry.* Irvington, NY, USA: Flat World Knowledge.

FAO (1992). FAO agricultural services bulletin No. 94. Minor oil crops. Rome, Italy: Food and Agriculture Organization of The United Nations.

Kamal-Eldin, A. (2005). Minor components of fats and oils. In *Bailey's industrial oil and fat products.* Oak Brook, Illinois IL, USA: John Wiley & Sons, Inc.

Kamal-Eldin, A., Yousif, G., Iskander, G. M., & Appelqvist, L. A. (1992). Seed lipids of *Sesamum indicum* L. and related wild species in Sudan I: fatty acids and triglycerols. *Fat Science Technology, 94*(7), 254–259.

Mariod, A. A., Idris, Y. M. A., & Mohamed, N. (2014). Composition and stability of traditional processed sesame oil. *Innovative Romanian Food Biotechnology, 15*(11), 51–57.

Ramadan, M. F., & Mörsel, J. T. (2002). Neutral lipid classes of black cumin (*Nigella sativa* L.) seed oils. *European Food Research and Technology, 214*, 202–206.

Umarabi, M. (2011). Limaza adat assarat aljamel lilwageha min jaded? [Translated from Arabic]. Akhir Lahza daily newspaper, Khartoum, Sudan.

Williams, M. A., & Hron, R. J. (1996). Obtaining oils and fats from source materials. In Y. H. Hui (Ed.), *Bailey's industrial oil and fat products, edible oil and fat products: processing technology* (pp. 61–156). (Vol. 4). New York, NY: Wiley.

Chapter 54

Unconventional Oils From Annuals, Herbs, and Vegetables

INTRODUCTION

The simultaneous increase in human population has brought about the demand and additionally the cost of edible oils, prompting the search for alternative unconventional sources of oils, especially in the developing countries. There are several un- or underexplored plant seeds rich in oil reasonable for consumable or industrial purposes. Huge numbers of them are rich in polyunsaturated essential fatty acids, which build up their utility as "healthy oils." Some agrowaste items, for example, rice bran have picked up significance as a potential source of edible oil. Hereditary modification has made ready for expanding the oil yields and enhancing the fatty acid profiles of traditional as well as unconventional oilseeds. Some of these unconventional oils may have magnificent potential for medicinal and therapeutic uses, regardless of the fact that their low oil contents do not promote commercial production as edible oils (Sabikhi & Kumar, 2012).

Many unconventional oil sources, mainly plants and animals that produce oil-bearing materials have been identified, but the potential of some of them as oil sources has not been assessed. Among these sources are some annuals, herbs, vegetables, trees, and insects (Mariod, 2005). The purpose of this book is to present three parts on unconventional oil sources for food and nonfood uses as well as for use as fixed or essential oils. These sources were studied intensively concerning source description, food and medicinal uses, oil physicochemical properties, fatty acid composition, minor components for example, tocopherols and sterols, as well as phenolic compounds and oxidative stability of some of them. The food and industrial applications like deep-fat frying and interesterification into biodiesel were also studied. As oil processing steps are very important from consumer behavior point of view and how these steps will affect the quality and stability of the processed oils, some of studied oil sources were subjected to lab-scale processing steps and the effect on quality and stability was investigated (Mariod, Matthäus, Eichner, & Hussein, 2006). In compiling these chapters, it became clear that the information available on these unconventional oil sources varied to a considerable extent, depending on the source. In many instances, there were little or no scientific and technical data on

Unconventional Oilseeds and Oil Sources. http://dx.doi.org/10.1016/B978-0-12-809435-8.00054-8

aspects of general information concerning the physicochemical properties and fatty acid composition of these oils, besides production, processing, and utilization. Major gaps of knowledge were recorded concerning studying fatty acids, minor components, stability, and food and industrial applications.

There are considerable gaps in information available on the oils from the sources researched, so hopefully this book will encourage workers to provide further information and build up knowledge of what appears to have been somewhat ignored information for many unconventional oil sources, which are of vital dietary and economic importance to large numbers of poor people in Sudan, and some it is suggested could prove of much wider application.

REFERENCES

Mariod, A. A. (2005). Investigations on the oxidative stability of some unconventional Sudanese oils, traditionally used in human nutrition. PhD thesis. Münster, Germany: University of Münster.

Mariod, A. A., Matthäus, B., Eichner, K., & Hussein, I. H. (2006). Effects of processing steps on the quality and stability of three unconventional Sudanese oils. *European Journal of Lipid Science and Technology, 108*(4), 298–308.

Sabikhi, L., & Kumar, M. H. S. (2012). Fatty acid profile of unconventional oilseeds. *Advances in Food and Nutrition Research, 67,* 141–184.

Index

Printed in the United States
By Bookmasters